C0-ATB-721

In Search of Sustainability

Benjamin Cashore, George Hoberg,
Michael Howlett, Jeremy Rayner,
and Jeremy Wilson

In Search of Sustainability: British Columbia Forest Policy in the 1990s

UBCPress · Vancouver · Toronto

Printed in Canada on acid-free paper

Canadian Cataloguing in Publication Data

In search of sustainability
Includes bibliographical references and index.
ISBN 0-7748-0830-6

1. Sustainable forestry—Government policy—British Columbia. 2. Forest policy—British Columbia. 3. Forest management—British Columbia. I. Cashore, Benjamin William, 1964-

SD146.B7I5 2000 333.75'09711 C00-911156-5

This book has been published with the help of a grant from the Humanities and Social Sciences Federation of Canada, using funds provided by the Social Sciences and Humanities Research Council of Canada.

UBC Press acknowledges the financial support of the Government of Canada through the Book Publishing Industry Development Program (BPIDP) for our publishing activities.

Canadä

We also gratefully acknowledge the support of the Canada Council for the Arts for our publishing program, as well as the support of the British Columbia Arts Council.

UBC Press
University of British Columbia
2029 West Mall
Vancouver, BC V6T 1Z2
(604) 822-5959
Fax: (604) 822-6083
E-mail: info@ubcpress.ubc.ca
www.ubcpress.ubc.ca

Contents

Acronyms

AAC	allowable annual cut
ACF	advocacy coalition framework
BBF	billion board feet
BCEN	BC Environmental Network
CAAC	Coast Appraisal Advisory Committee
CFFG	Coastal Fisheries/Forestry Guidelines
COFI	Council of Forest Industries
CORE	Commission on Resources and Environment
CSA	Canadian Standards Association
CVP	comparative value pricing system
FML	Forest Management Licence
FPC	Forest Practices Code
FRBC	Forest Renewal BC
FRC	Forest Resources Commission
FSC	Forest Stewardship Council
FSSC	Forest Sector Strategy Committee
GIS	Geographical Information Systems
IAAC	Interior Appraisal Advisory Committee
IMA	Interim Measures Agreement
ISO	International Standards Organization
IWA-Canada	International Woodworkers of America-Canada, until the mid-1990s; then Industrial, Wood and Allied Workers of Canada
LRMP	Land and Resource Management Plan
LTHL	long-term harvest level
LUCO	Land Use Coordination Office
MAI	mean annual increment
MB	MacMillan Bloedel
MIWS	Management of Identified Wildlife Species
MOELP	Ministry of Environment, Lands and Parks

MOF	Ministry of Forests
MOU	Memorandum of Understanding
NAFA	National Aboriginal Forestry Association
NDP	New Democratic Party
OTT	Old Temporary Tenures
PAS	Protected Areas Strategy
PSYU	Public Sustained Yield Units
SBFEP	Small Business Forest Enterprise Program
SFU	Simon Fraser University
SLA	Canada-US Softwood Lumber Agreement
SLDF	Sierra Legal Defence Fund
SMZ	Special Management Zone
TFL	Tree Farm Licence
TLA	Truck Loggers Association
TSA	Timber Supply Area
TSR	Timber Supply Review
UBC	University of British Columbia
UVic	University of Victoria
WAC	Wilderness Advisory Committee
WCWC	Western Canada Wilderness Committee
WWF	World Wildlife Fund Canada

Preface

The origins of this book lie in a disagreement among its authors about the capacity of ideas to change public policy. As we explored this disagreement, it became clear that we had much more in common than we had originally supposed. Close to the top of our list of shared assumptions was a feeling that the academic study of forest policy in BC had, for far too long, been the almost exclusive preserve of resource economists. Policy analysis informed by political science might add a fresh perspective to an increasingly sterile debate about forest policy options, a perspective that puts the politics of forest policy front and centre. *In Search of Sustainability* is the outcome of the challenge we set ourselves: to review the barrage of forest policy initiatives undertaken since the election of the NDP government in 1991; to determine where those initiatives have led to genuine policy change; and to pull together the factors that have promoted or inhibited change.

Our goal has been to describe and explain the evolution of forest policy during the 1990s, *not* to promote a particular vision of forest policy in the province or recommend specific changes to programs. But we do hope our work is of specific interest to those who are shaping forest policy directly, both in the government and in the broader forest policy community. Whoever takes on the task of crafting a provincial forest policy after the next election will occupy a decision space bounded by the opportunities and constraints that we have identified here.

Throughout this process, the authors have accumulated a number of debts, both individually and collectively. For their comments on earlier versions of the analysis or their contribution to our understanding of particular policy developments, we would like to thank Mike Apsey, Graeme Auld, Steven Bernstein, Clark Binkley, David Boyd, Bob Cairns, John Cashore, Linda Coady, Jim Cooperman, Brian Gilfillan, Mark Haddock, Johanna den Hertog, Will Horter, Jake Kerr, Hartley Lewis, Evert Lindquist, Jean-Pierre Martel, Keith Moore, Gordon Morrison, Deanna Newsom,

Larry Pedersen, Kim Pollock, Brian Scarfe, Detmar Schwitchenberg, Grace Skogstad, Tom Tevlin, Jeffrey Waatainen, George Weyerhaeuser, Jr., Michael Whybrow, Bill Wilson, Garry Wouters, Karen Wristen, and Daowei Zhang. For their help with research, we would like to thank Jessica Alford, Paola Baca, Tim Dueck, Max Guirguis, Robyn Jarvis, and especially Scott Morishita. For their institutional support or research funding, we are very grateful to the Social Sciences and Humanities Research Council of Canada; Humanities and Social Sciences Federation of Canada; Bruce Clayman of Simon Fraser University; Resources for the Future; the Alabama Agricultural Experiment Station; and Auburn University's Center for Forest Sustainability.

Finally, we are each, in our own particular way, profoundly grateful for the support and patience of our families.

In Search of Sustainability

1
Policy Cycles and Policy Regimes: A Framework for Studying Policy Change

The forests of British Columbia have become a battleground over sustainable resource development, focused primarily on the conflict between the economic development of the timber resource and the protection of a variety of environmental amenities provided by the forests. Although the dispute is decades old, dating back to the surge of environmental concern in the late 1960s, it has intensified in recent years and has become the subject of increasing international attention. Beginning in 1991 with the rule of the New Democratic Party, the 1990s witnessed significant policy initiatives in virtually all major aspects of forest policy. Among the most prominent initiatives:

- Land use policies, using innovative consensus-based processes, promised to double the province's protected areas to 12 percent of the land base.
- The new Forest Practices Code intensified the regulation of logging practices.
- A review of timber supply was designed to update inventories and to establish harvest rates on a more sustainable footing.
- The Forest Renewal Plan was introduced to compensate for job losses resulting from environmental restrictions, with investment in more intensive silviculture and ecological restoration.

In Search of Sustainability brings together a group of political scientists to examine this recent transformation of BC forest policy. It provides a much needed assessment of the provincial government's efforts to reform forest policy, and it uses the case of change in BC forest policy to advance broader understanding of theories of public policy. The analysis focuses on the questions of how much change and why. The core of the book is the analysis of specific components of BC forest policy. For each area, the authors will apply an analytical framework developed by the group, combining policy cycle and policy regime frameworks. Indicators of policy change will be developed for each area, the magnitude of policy change will be assessed, and the extent of change will be explained within the regime framework. One of the major challenges is to examine whether the policy changes of the 1990s reflect a fundamental change in direction for provincial forest policy or whether they are better conceived of as relatively minor departures from the status quo.

This introductory chapter begins with an overview of the policy cycle model. It proceeds to an overview of policy theories and the development of our distinctive approach. It then provides an overview of the forest policy regime in British Columbia and the pressures for change in the regime that emerged in the late 1980s and early 1990s. The chapter concludes with an outline of the remainder of the book.

The Policy Cycle Model

One of the most influential models of the policy process is the so-called policy cycle framework, developed originally by American scholars in the 1960s and 1970s[1] and more recently applied to Canada.[2] The model consists of five stages:

(1) *agenda setting,* how problems come to the attention of government (what Charles O. Jones calls "problems to government")
(2) *policy formulation,* the development of policy alternatives
(3) *decision making,* the adoption of a particular course of action or inaction
(4) *implementation,* putting the policy into effect (what Jones calls "government to problems")
(5) *evaluation,* assessing the consequences of the policy through monitoring and analysis.[3]

Policy evaluation frequently results in serious criticisms of existing policy and leads to pressures for revision, thus creating feedback that can put the policy issue back on the government agenda and restart the policy cycle.

Although the policy cycle model has been very influential in policy research and education, scholars have become frustrated with its limited explanatory potential.[4] A major reason for this disenchantment has been a shift in policy studies from an emphasis on describing and analyzing the policy-making process to an emphasis on explaining policy outcomes; that is, why governments adopt the policies they do. Nonetheless, the policy cycle model still has significant utility in illuminating policy development, and for that reason, we rely on it as an organizing framework in the individual analyses. Moreover, little work has been done to integrate the policy cycle model with more ambitious explanatory models, and we hope this book will be a contribution in that direction.

Perhaps the most important reason for continued reliance on the policy cycle model is that we cannot understand the substance and consequences of a policy without following the cycle through to the implementation and the evaluation stages. In our application of the model, we will focus on particular aspects of each stage in the cycle.

In agenda setting, our main focus will be how issues get on the formal governmental agenda and how relevant actors construct those issues. In his influential study of agenda setting, John Kingdon emphasizes the interaction of three relatively distinct streams: the problem stream, which consists of changing indicators and focusing events; the politics stream, which consists of public opinion, interest group pressures, and elections; and the policy stream, which consists of solutions promoted by policy specialists.[5] Although issues can reach the agenda through significant changes in any one of the streams, Kingdon stresses the importance of the confluence of all three streams to create a "window of opportunity" that will be seized by policy entrepreneurs promoting particular solutions.

One of the most important aspects of the agenda-setting stage is the social construction of the issue that emerges: who or what is thought to be responsible for the creating the policy problem and what types of solutions may be effective or appropriate in addressing the problem. This process of problem definition influences the agenda status as well as subsequent stages of the cycle, including formulation, decision making, and implementation. It is frequently one of the most important subjects of dispute among political adversaries in the policy process.[6]

The policy formulation stage analyzes how the alternatives to be seriously considered by decision makers get narrowed down and refined.[7] The serious consideration of alternatives is constrained by a variety of factors, among them:

- Technical feasibility: Some options that may be attractive for some reasons (for example, political desirability) may simply not be technically feasible.
- Fiscal restraint: Governments tend to prefer alternatives that minimize budgetary implications.
- Political feasibility: Some technically or economically desirable options may not be politically feasible because they are opposed by an actor with a de facto veto or because they violate widely held values.

Kingdon considers policy formulation to be the arena of professionals, expert bureaucrats, and policy entrepreneurs;[8] Howlett and Ramesh view the process as potentially more open to a variety of political forces.[9]

Much of the study of decision making compares the competing models of rational-comprehensive decision making, in which actors follow a careful, deliberate consideration of all available alternatives and their consequences, with the model of incrementalism championed by Charles Lindblom in his aptly named seminal article, "The Science of Muddling Through."[10] In Lindblom's model, decision makers don't bother to clarify objectives; they narrow the range of alternatives to those closest to the status quo, and they settle on policies for which they can get adequate agreement rather than those that maximize their objectives. In recent years, scholarship on decision making has advanced beyond these dichotomous models to emphasize a continuum of approaches, the location along which is a function of the context confronted by the decision maker.[11] In this book, we will examine the decision process to determine the major factors influencing the outcome, and we will attempt (within the regime framework) to explain how that outcome is linked with the context from which it emerged.

Another fundamental objective of our analysis of the decision-making stage is to characterize the content of policy in terms of the following grounds:

- the broad *goals* of policy, such as promoting a healthy forest industry or protecting biodiversity
- the more specific *objectives* embodied in policies, such as doubling park land or creating a specific number of jobs
- the *instruments* adopted, some of which are substantive because they are designed to directly promote the policy objectives (for example, coercive regulations, subsidies), others of which are more procedural because they are more oriented toward the manipulation of process and actors (for example, persuasion, intervenor funding)[12]
- the specific *instrument settings* (for example, a forty-hectare-maximum clearcut size).

This description of policy content allows some judgment about the magnitude of policy change in each area of forest policy. In addition to determining the substantive content of policy, decision makers also establish the organizational arrangements through which the policy is to be implemented. The work might involve assigning new tasks to existing organizations, changing the relationships among existing organizations, or creating new organizational entities.

The study of implementation focuses largely on the determinants of the success or failure of the policy in meeting its goals.[13] One of the most influential discussions of these factors can be found in Mazmanian and Sabatier.[14] They list six conditions for effective implementation: clear and consistent objectives, a sound causal theory, adequate administrative authority, committed and skilful implementing officials, support from key interest groups and sovereigns, and facilitative socioeconomic conditions. The study of implementation is important for several reasons. Without effective implementation, policy change is more symbolic than real. The implementation stage determines the consequences of the policy decision for the behaviour of concern: How far has the policy gone in achieving its objectives?

Moreover, the meaning of policy is frequently clarified or simply changed during implementation. Thus, although policy change cannot be assessed without examining its implementation stage, the very significance of the stage has the effect of complicating the distinction between the decision-making and the implementation stages. It is frequently very difficult to establish exactly where decision making ends and implementation begins, particularly in political systems like Canada where legislation is so generally worded. Thus, when assessing the magnitude and significance of policy change, we need to examine both the policy content emerging from the decision stage (and as redefined in the implementation stage) and the policy consequences emerging from the implementation stage.

The final stage of the policy cycle is evaluation, including monitoring. This stage is fundamental to our understanding of the policy because it concerns how governments (and other actors) learn about the consequences of policies as they are implemented. This learning process is fundamentally linked to the problem definition and issue construction that are so essential to the agenda-setting and subsequent stages of the policy cycle.

The policy cycle model is useful for analyzing the process by which policies are made and implemented as long as its limitations are understood. A cycle for a particular policy decision needs to be kept in historical context. Most new policies are revisions of previous ones, so it is perhaps more appropriate to refer to policy succession; that is, the historical evolution of policy through multiple cycles.[15] This complication need not preclude the focus on one or more specific cycles, however. In this book, we identify for each area of policy one of more crucial decisions (or nondecisions) that occurred in the 1990s as the focus of our analysis. In addition, once policy issues enter a cycle (for example, get to the agenda stage), they do not inevitably complete the policy cycle. Because of resistance to change, some policy cycles may be interrupted. This interruption usually occurs before the decision stage, as in the tenure case presented here, but it can occur afterward, as in the case of the Timber and Jobs Accord. Finally, the cycle model is not oriented toward explaining why governments adopt the policies that they do. For this task, we have developed a policy regime framework.

Approaches to Studying Policy Change

The development of empirical theories of public policy, understood as approaches to explaining why governments adopt the policies they do, has been relatively limited. Approaches to the study of public policy have been influenced by broader trends in the discipline of political science. Before the 1980s, scholars concentrated on societal forces underlying policy development, be they interest groups or classes. The 1980s witnessed a substantial shift in focus, with the institutions of the state becoming the focus of explanatory theories, both in the historical institutionalism influenced by sociology and in the rational choice institutionalism inspired by economics. In the 1990s, scholars seemed to have shifted their focus again, this time to the significance of ideas in explaining policy outcomes.

Because the discipline has lurched from category to category, the advancement of our understanding has not been what it should have been. However, an intriguing convergence does seem to be emerging among theories. Although a number of characterizations of the theoretical landscape are possible, we have identified six streams of research influential in recent years.

First, perhaps the framework most influential in Canadian and British policy studies is that of policy networks and communities, both defined loosely as a structural pattern of interaction between interest groups and state institutions.[16] Early versions of this approach were mainly focused on explaining the organizational arrangement of actors, not the policies that result from the interaction of those actors. These versions also had a strong static tendency and thus have not been oriented toward explaining institutional or policy change. Nor did this approach treat ideas seriously. Each of these problems has been addressed in recent works within this tradition.[17] As a result, a promising convergence of approaches within the tradition seems to be emerging.

Second, one of the most significant developments in social science in the last several decades has been the emergence of a focus on political institutions. Historical institutionalism emerged as a corrective to the relative neglect of institutions in the behavioural revolution after the Second World War, which focused a great deal of attention on societal groups but far less on institutional arrangements. This theoretical stream makes three major contributions by recognizing that government actors frequently have considerable autonomy from societal interests; that the organization and structure of political institutions can have important influences on the strategies, resources, and even preferences of societal actors; and that the development of public policy is not just about the resolution of societal conflicts but also the intellectual process of social learning.[18]

The third important stream of theoretical development has been the rational choice approach adopted by a number of American policy scholars.

Beginning with the premise of rational actors pursuing their self-interests within constraints, scholars in this tradition have had a number of important insights into the dynamics of institutional structure, the behaviour of actors, and policy outcomes.[19] With one exception, this approach has largely ignored the role of ideas.[20]

Fourth, a discourse approach, influenced by postmodern thinking, has become far more prominent in recent years.[21] Its distinctive contribution is an emphasis on the politics of discourse and social construction, including how particular social constructions of policies can promote certain interests and disadvantage others. Its drawback, however, is that it tends to downplay the importance of actors and institutions.

The fifth major stream in policy studies is the concept of punctuated equilibrium developed by Baumgartner and Jones.[22] It builds on the tradition of issue network and subsystem models developed in the United States but advances them with a very sophisticated analysis of the political strategies of actors. The approach emphasizes how lengthy periods of equilibrium in public policy can be followed by dramatic changes, usually caused by one or both of the following: a shift in the institutional venue in which policy is made or a shift in the policy "image" that frames how the public and elites think about policy. It thus combines the insights of institutionalism with discourse analysis.

The sixth, and perhaps the most comprehensive, approach to the study of policy change is the advocacy coalition framework (ACF) developed by Paul Sabatier and colleagues.[23] Within the ACF, policy-oriented learning is defined as "relatively enduring alterations of thought or behavioral intentions which result from experience and are concerned with the attainment or revision of policy objectives."[24] The ACF is based on several important, and widely shared, premises: understanding policy change requires a decade or more; the appropriate unit of analysis is the policy subsystem; and policies can usefully be interpreted as sets of value priorities and causal assumptions or belief systems. The most distinctive aspect of the ACF is its emphasis on the importance of characterizing the subsystem as consisting of distinctive advocacy coalitions "composed of people from various governmental and private organizations who share a set of normative and causal beliefs and who often act in concert."[25] Advocacy coalitions adopt strategies in the attempt to influence government, and conflict between advocacy coalitions is typically mediated by "policy brokers." The subsystem is influenced by exogenous factors, such as constitutional rules and socioeconomic conditions, which act as constraints on and resources for advocacy coalitions. Although strongly influenced by the ACF approach, the policy regime framework set out below departs from it in several significant ways.[26]

A promising convergence seems to be emerging from these different theoretical streams. This trend seems to recognize the importance of subsystems acting within a broader environment and the need for a multicausal approach emphasizing the interaction of actors, institutions, and ideas. Within this larger pattern of convergence, there is still a great deal of terminological and conceptual diversity along with a marked difference of emphasis over which of the many variables are the most important.

The Policy Regime Approach

The theoretical approach employed in this book is deliberately multicausal in orientation. Depicted in Figure 1.1, the policy regime framework consists of regime components, background conditions, and policy outcomes that are produced through the interactions of regime components and background conditions. The three components of policy regimes are actors, institutions, and ideas. Actors are individuals and organizations, both public and private, that play an important role in the formulation and implementation of public policies. Actors have interests they attempt to pursue through the political process, resources they bring to bear in their efforts to influence public policy, and strategies they use to plan how to employ those resources in the pursuit of those interests. Unlike the advocacy coalition approach, the regime framework draws an explicit distinction between government and private actors because of the importance of authority as a resource.[27]

Figure 1.1

The policy regime framework

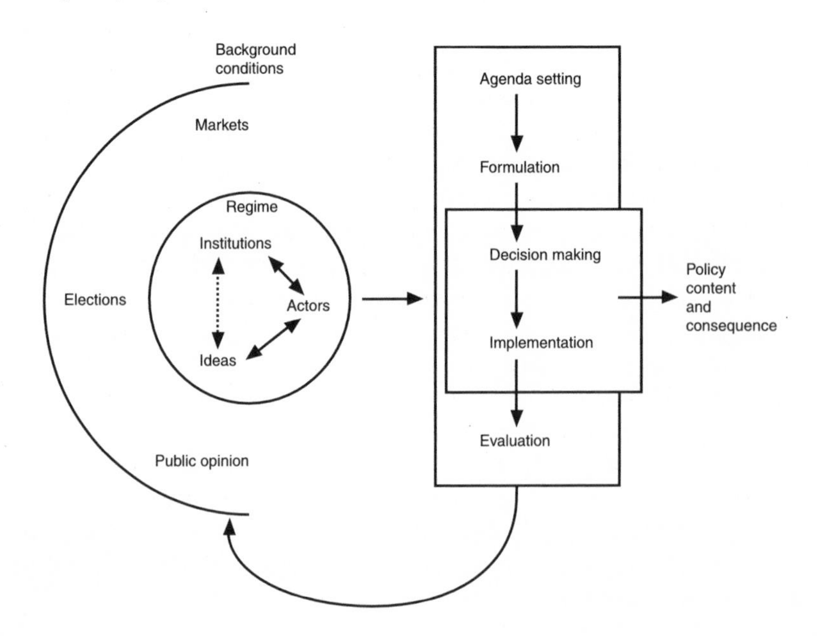

Actors interact with each other and are situated in networks or webs of relationships. These relationships are usually uneven, for actors differ significantly in their influence and proximity to power and, hence, their ability to directly affect the shape of policy outcomes. Various concepts and typologies have been developed to try to characterize the variety of relations that exist in different policy sectors.[28] The key variables are proximity to power and the number and variety of actors who are influential in the policy domain in question. We will use the term *policy communities* to refer to particular configurations among actors in each policy area. Our analysis will examine how the relationships among key actors change over time and across different subsectors.

Institutions are rules and procedures that allocate authority over policy and structure relations among various actors in the policy process. There is considerable disagreement in the political science literature about how much institutions matter.[29] Across different methodological perspectives, there is widespread agreement that institutions can affect the resources and especially the strategies of political actors. Some of the strongest advocates for an institutional approach also argue that institutions influence how actors define their interests. Most attempts to do so stress the role of institutions in influencing what actors come to believe is possible in terms of policy changes.[30] We think that this effect is more appropriately considered a matter of strategy and that it is better to conceive of interests as prior to and independent of institutional variables.

Institutions can influence policy through two avenues. First, they structure the authority and relations among government actors (for example, among executive, legislative, and judicial actors or between federal and subnational governments). Second, they influence the relations between societal interests and the state. By specifying how interest groups can participate in the policy process, institutional rules shape the resources actors can bring to bear and the strategies they adopt in pursuing those interests.

Ideas are both causal and normative beliefs about the substance and process of public policy. Normative beliefs speak to ends. For example, the idea that species have the inherent right to exist informs many people's views on the merits of wildlife protection. Causal beliefs speak to means. For example, the emerging science of conservation biology informs our understanding of the relation between a particular forest practice and the maintenance of wildlife habitat. According to Goldstein and Keohane, ideas can influence policy in three ways. First, they can act as road maps that help actors understand the relationship between their goals and strategies to reach those goals. Second, ideas can serve as a focus or glue that narrows the range of alternatives in complex situations. And third, ideas become embodied in policies and institutional rules, and their impact is felt long after their enactment in law.[31]

For a specific policy area and a particular time, these three components – actors, institutions, and ideas – combine to form a distinctive policy regime. There are significant interactions among the regime components. As described, institutions shape the resources and strategies of actors. Ideas inform the interests and strategies of actors, and they can provide valuable political resources for actors. Because of the importance of institutions and ideas, strategic actors attempt to alter institutions and reshape ideas to advance their interests. This interactive effect is described in more detail below.

As with the case of the advocacy coalition framework, policy regimes exist in the context of certain background conditions. There is a wide variety of potentially relevant background conditions, but we focus in this book on five in particular: public opinion, elections, economic conditions, the macropolitical system, and other policy sectors. Perhaps the most important effect of public opinion is in shaping the incentives of politicians who are seeking reelection. Fluctuations in markets – resulting from shifts in commodity prices, exchange rates, consumer tastes, or other factors – can have profound impacts on policy regimes in economic sectors like forestry. Elements of the macropolitical system such as the organization of political parties, electoral rules, and constitutional design can also have a significant impact on regime dynamics. Although the macropolitical system tends to be more stable than public opinion and economic conditions,[32] it is not immune to significant change. The characteristics of regimes vary from policy sector to policy sector. Another potential source of policy change is thus spillovers from one policy sector to another.[33]

The interactions of the regime components, in the context of particular background conditions, produce distinctive policy outcomes. In this framework, policy contents and consequences are the ultimate dependent variable. As described in our discussion of the policy cycle model, these policy outcomes emerge from the decision-making and implementation stages. However, as in most systemic models, there is powerful feedback between policy outcomes and real-world impacts, with consequences for different actors. First, policy-issue characteristics, such as the level of technical complexity or the distribution of costs and benefits associated with the issue, can influence the distribution of interests surrounding an issue[34] and the relative tractability of a problem.[35] Regime characteristics therefore differ across sectors and even within different aspects of a particular sector, as we will see in our cases. Second, actors interpret the consequences of previous and existing policies, creating the potential for policy learning, defined as "relatively enduring alterations of thought or behavioral intentions which result from experience and are concerned with the attainment or revision of policy objectives."[36]

Whatever their source, pressures for policy change always confront considerable obstacles that create a significant bias toward the status quo. Indeed, early policy theories seemed more oriented toward explaining inertia than change.[37] One of the most important sources of inertia is the path dependence created by the legacy of past policy choices. This is another aspect of the feedback between the policy outcomes and regime components discussed earlier. Policy legacies structure political coalitions, conceptualizations of the policy problems, and appropriate solutions to those problems.[38] The inertia resulting from a long-institutionalized course of action can significantly increase the costs of departing from the status quo. Powerful regime actors who benefit from the status quo can be important obstacles to change, and they can be expected to employ all the political resources at their disposal to thwart changes that adversely affect their interests. Institutional designs that fragment power, such as federalism or the American system of checks and balances, can also provide an obstacle to change by increasing the level of consent required before new policies can be adopted.[39] Each of these forces will be developed in more detail throughout this book. The general shape of policy outcomes is determined by the relative balance between these forces for change and for stability at any given time.

Generating Hypotheses

The movement from framework to theory requires more careful specification of causal relations between variables. In a framework explicitly emphasizing multiple causality that task is inherently difficult. As described, the pressure for policy change and the resistance to change can come from a number of sources, and it is impossible to develop a generally applicable sequence of causal events. However, it is possible to generate hypotheses about patterns of relations and to specify circumstances that would be more likely to lead to policy change.

One important distinction is whether the change comes from within the regime or results from exogenous changes in background conditions. The regime approach starts with strategic actors as the core variable. At the very heart of the regime are policy makers, government actors with authority, the most important of whom are elected officials and appointed officials. Although elected officials have a number of goals, including the pursuit of their own visions of good public policy, they are frequently constrained in the choice of policies by their desire to be reelected. Because of their own authority and the wide variety of controls they exercise over bureaucrats, elected officials are typically the most important actors determining policy choices, at least in terms of the general objectives and design of policy.

Although the rational actor approach is useful in reminding us of the central role of elected officials, it is virtually silent on the role of ideas and learning.[40] In the regime framework, policy is determined by the interaction of policy makers with other strategic actors, both public and private, all pursuing their interests within a particular institutional and ideational context. That context frames issues, structures incentives, and allocates advantages and disadvantages to different actors. Here, the inclusion of ideas separates this approach from rational choice perspectives. Within a given institutional context, actors adopt the strategies most likely to advance their interests. In framing their arguments, actors appeal to widely shared values and expert authority as much as possible. When knowledge is contested, actors select those arguments most consistent with their interests. Learning occurs when, based on their interpretation of experience, actors adopt new beliefs about their interests or the best strategies to pursue those interests.

Cognizant of the structural biases of particular contexts, actors frequently adopt strategies to alter the institutional arena or ideational context of decisions to promote their interests. Here the regime approach is inspired by Baumgartner and Jones's important work on punctuated equilibrium.[41] There are three types of these second-order strategies. First, actors pursue changes in the rules of the game within a particular institutional arena (for example, a change in administrative procedures). Second, actors pursue a strategy of shifting venues, atttempting to move the locus of policy making to an arena in which their perceived advantages are greater (for example, from the executive branch to the courts).[42] This strategy might go beyond shifting from one institutional venue to another to moving outside government altogether (for example, to the international marketplace). Finally, actors pursue idea-based strategies of reframing problem definitions in a manner more amenable to their interests.[43] Strategic actors, elected and unelected, actively attempt to shape elite and public opinion. In this case, the mechanisms of influence are not notice and comment periods, cabinet directives, or judicial review but rather symbolic manipulation, advertising campaigns, the dissemination of research, and focusing events. The media become crucial actors here as well because they influence which messages become dominant.

Thus, the regime approach is centred on strategic actors acting within and on institutional and ideational contexts. As a result, some of the major pressures for change emerge from within the policy regime. Recall, however, that the three regime components (actors, institutions, and ideas) operate within a larger environment of background conditions, including economic conditions, elections, public opinion, and the broader macropolitical system. Powerful pressures for change emanate from these background conditions, which Sabatier refers to as exogenous factors. For example, changes in markets or politics, domestic and international, provide the impetus for change by shifting the resources and strategic opportunities of regime actors. These background changes are typically the most powerful forces for significant change in policy. Indeed, the regime approach wholeheartedly endorses the advocacy coalition framework's hypothesis that "the core (basic attributes) of a governmental action program is unlikely to be changed in the absence of significant perturbations external to the subsystem, that is changes in socio-economic conditions, system-wide governing coalitions, or policy outputs from other subsystems."[44] Restated in the language of the regime framework, significant policy change is unlikely without significant change in background conditions.

The challenge is to take this fundamental proposition, seemingly anomalous within the advocacy coalition framework, given its focus on learning, and specify it in more detail. The problem is that precisely because the background conditions are exogenous to the subsystem, they are the least predictable elements in the framework. It is possible, however, to specify hypotheses about the relations between directional changes in background conditions and some types of pressures for policy changes.

Regarding economic conditions, there tends to be an inverse relation between profitability and the power resources of industry groups in a particular sector. Where policy change involves overcoming the opposition of business groups, this proposition can help identify the conditions for when to expect policy change. As Lindblom and many others have noted, the most important power resource of business is its control over investment.[45] To get reelected, the government needs a satisfied public, and that depends on the availability of jobs. In a market economy, the driving force behind jobs is business investment. Thus, governments have a definite incentive to create and maintain a healthy business climate, which requires they remain attuned to business interests. This basic logic applies at all times, but the relative force of these pressures varies with the business cycle. When business is poor and profitability declines, business threats of shutdowns and layoffs seem very credible, and governments tend to be highly responsive. Governments are particularly vulnerable to industry arguments that any regulatory action taken to increase the cost of production will cost jobs and further threaten government approval ratings. When business is booming, industry is less likely to make these threats, and, when it does, the threats are taken less seriously. Industry finds it harder to make compelling arguments that it cannot afford the additional costs, governments are less sympathetic to such arguments, and environmentalists – sensing the vulnerability of their opponents – press their case with renewed vigour. In other words, the worse off a business sector is economically, the better off it is politically.

One of the biggest distinctions between the advocacy coalition framework and the policy regime framework is the emphasis of the latter on public opinion and politicians. One potential test of importance of this distinction is the following hypothesis about the relation between public opinion and policy changes: Significant changes in policy that go against the interests of business groups (or other dominant actors) are unlikely without a burst in public salience of new values.[46] This hypothesis builds on Baumgartner and Jones's argument that "most issue change occurs during periods of heightened general attention to policy."[47] Big changes in policy are unlikely without the intervention of elected officials. Elected officials are usually willing to defer to specialized policy subsystems, but when broader publics get interested, electoral threats and opportunities emerge, creating incentives for elected officials to get involved. The difference between elected officials and other actors in the process is that when they develop an incentive to get involved, so long as they can construct a winning governing coalition, they have the power to change policy to advance their interests.

The other major difference between the regime framework and the advocacy coalition framework is how it conceptualizes ideas. The regime approach constructs ideas as a separate conceptual sphere distinct from actors and institutions, whereas the advocacy coalition embeds ideas within the concept of the belief systems of actors. Unquestionably, in the regime approach the influence of ideas is manifested through interpretation-based actions of actors. However, ideas, both causal and normative, have dynamics that are relatively autonomous from strategic actors, and they thus are best considered as an analytically distinct sphere. Ideas can be conceived of as a power resource for actors. The legitimacy of certain arguments contributes to the influence of the actors promoting those arguments. Causal arguments are given credence by the views of acknowledged experts in the field, whereas normative arguments get their credibility from their compatibility with the interests or values of the relevant public.

Integrating Policy Cycles and Policy Regimes

Very little work has been done to integrate the policy cycle model with broader explanatory theories of public policy. One notable exception is Howlett and Ramesh, but they focus their linkage on the policy formulation stage and to a lesser degree on the policy evaluation stage.[48] In our view, the regime framework can be applied to illuminate each stage of the cycle. For example, agenda setting involves the interaction of actors and ideas in the struggle of shaping how problems are constructed, and institutions can influence the ability of different actors to force their issues on the agenda. Implementation, as Mazmanian and Sabatier emphasize, depends on the allocation of institutional authority, an appropriate causal theory, and the support of relevant political actors.

Table 1.1 suggests some of the most important ways that different aspects of the regime framework are influential at different stages of the policy cycle. It is relatively straightforward to fill in each cell of the matrix, demonstrating that there are important links between each of the regime elements and background conditions on the one hand and cycle stages on the other hand. However, it is likely that some of the explanatory factors will be more important at some stages than others. For example, ideas and public opinion are most likely to be influential in the agenda-setting and formulation stages. Technical knowledge is likely to be most important at the formulation and implementation stages. Our case studies will explore the relative importance of these linkages, and we will return to this issue in the conclusion.

Figure 1.1 shows the basic causal relationships linking the policy regime and policy cycle model. Arguably, there is feedback between each element depicted in the figure, but we have chosen to highlight only those relationships we view as most important. Policy outcomes are the core dependent variable to be explained. We characterize them as emerging from a combination of the decision-making stage, which produces policy content consisting of goals, objectives, instruments, and settings, and the implementation stage, which produces real-world consequences. The interaction of regime elements, in the context of particular background conditions, sets a policy cycle in motion and yields distinctive policy outcomes. Although we focus on policy outcomes as the dependent variable, we recognize the feedback that occurs between policy consequences and various regime elements. In addition to the legacy of past policies, there is also an important intellectual aspect to this feedback. Actors interpret the consequences of policy, and learning may occur. Of course, learning is neither easy nor inevitable, as actors with conflicting interests will contest the meaning and significance of various consequences.

The BC Forest Policy Regime and Change in the 1990s

Strategic actors are at the core of the regime framework. The major actors in BC forestry are the forest industry, workers, environmentalists, First Nations, elected politicians who form the government, and the major bureaucratic organizations: the Ministry of Forests and the Ministry of Environment, Lands and Parks.[49] Although the motivations of actors and the sources of influence for these actors are all complex, it is possible to outline a crude but generally accurate characterization of actor interests and resources. The forest industry is interested in profits. It has considerable economic clout as a result of its dominant position in the provincial economy. It controls a wide range of decisions that produce jobs and provincial tax revenues fundamental to the political survival of the government in power.[50] Forest workers are interested in secure jobs. Forest

Table 1.1

Policy cycle

	Policy regime				
	Agenda setting	Formulation	Decision making	Implementation	Evaluation
Actors	Struggle to put their issues on the agenda; shape problem definition	Shape options to suit their interests	State actors make decisions; private actors constrain choices	Participation narrows, though actors continue struggle for influence	Emphasize consequences affecting their interests; frame problem to restart cycle
Ideas	Social construction, problem definition	Influence range of alternatives considered, causal theories linking alternatives with consequences	Become embedded in decisions	Causal theory influences success	Learning; techniques of analysis; causal understanding
Institutions	Create differential agenda access	Influence which actors get involved in analysis	Shape the influence of actors	Influence access, oversight	Influence who gets involved and how

Background conditions

	Agenda setting	Formulation	Decision making	Implementation	Evaluation
Public opinion	Shape what politicians want; constrain problem definitions	Constrains range of alternatives	Constrains choice	Change in public support can undermine program	Influences which consequences receive emphasis
Elections	Produce mandates	Mandate can preclude alternatives	Determines legislators and executives making decisions	Change in government can undermine program	Fodder for campaigns
Markets	Determine salience of economic issue, space for others	Constrains options from undermining business climate	Bias against decisions that threaten business climate	Can undercut conditions for success	Relative salience of economic consequences

workers have the power of the vote and organization through their unions, particularly through their ties to the New Democratic Party. On many aspects of forest policy, especially those involving environmental issues, forest workers find themselves allied with the forestry companies.

The core interests of environmentalists are in preserving wilderness and reducing the negative environmental consequences of forestry. Environmentalists derive their influence primarily from their ability to represent significant parts of the electorate, and, as we will see, they have developed increased economic power through their influence on consumers in the international marketplace.

The government in power holds the overriding objective of being reelected. This is not the only objective of the government. Its members may also be concerned about pursuing their vision of good public policy or in helping their political support base. But their desire for reelection means that their actions are constrained by public opinion. The magnitude of that constraint depends on how salient the issue is to the public. If the public is indifferent, opinion matters little. But if the issue is a high priority, it can be a driving force behind policy. In terms of resources, politicians, particularly in the cabinet, have the fundamental resource of authority that grants them the power to allocate costs and benefits to the others.[51] This authority involves control not just over the substance of policy but also over the institutional structures of government in terms of administrative procedures and organizational arrangements.

Unelected government officials, bureaucrats, seek influence and prestige, whether for their own sake or to pursue their policy interests. Scholars differ on whether bureaucrats are primarily interested in maximizing their budgets or in enhancing their autonomy.[52] For resources, bureaucrats rely on a combination of authority and expertise.

First Nations are principally interested in Aboriginal title – to reclaim what they view as their land to use as they see fit. Unlike environmental groups, First Nations groups have difficulty using public opinion as a resource to motivate policy makers to address their concerns. Instead, their major political resource has proven to be the power of law, as courts have increasingly expanded doctrines of Aboriginal title.[53]

Depending on the issue, different actors can form coalitions. On environmental issues, industry and labour are usually closely allied. This coalition is particularly evident in the BC Forest Alliance, a lobby group created in 1991 to represent industry concerns. For its first nine years, the group was led by Jack Munro, a long-time president of the IWA. In responding to accusations that he crossed an important political line, he responded: "It bothered me then and it bothers me today, that I supposedly jumped the fence. I say to the people who invented the fence, where is the god-damn fence?"[54] At various points over the past several decades, First Nations and environmentalists have formed alliances. Indeed, BC forest politics is best characterized by a shifting conflict between two major coalitions: a development coalition representing industry, labour, and some parts of the government; and an environmental coalition in opposition.[55] This concept of competing coalitions can be illuminating as long as the crucial distinctions between governmental actors, who wield real political authority, and societal actors are recognized.

The BC forest policy regime has three significant institutional features: provincial dominance, government ownership, and executive dominance. Under Canada's Constitution, jurisdiction over forests in Canada belongs almost exclusively with the provinces. The federal government does have some overlapping jurisdiction in three areas:

- fisheries, because of the consequences of forest practices for anadromous fish habitat
- First Nations, because of their responsibility for addressing land claims
- international trade, because of pricing issues underlying disputes with the United States over softwood lumber.

In the past, the federal government sought leverage over forest policy through its spending power, in particular by contributing to reforestation funds, but those cost-sharing agreements have ended. Despite these areas of federal jurisdiction, forest policy is overwhelmingly dominated by the provincial government.

The government of British Columbia owns virtually all (95 percent) of BC's forest land. Small areas of land were alienated in the late nineteenth century, usually through railway grants, but the province has chosen to retain ownership of the remainder. This fraction of government-owned land is distinctive in comparative terms. Table 1.2 shows the fraction of public ownership of forest land in various jurisdictions. BC's forest land is managed through a complex system of tenures, the most common forms of which involve the government delegating certain forest management responsibilities to private timber companies in exchange for long-term guarantees of timber supply (see Chapter 4).

The third major institutional characteristic of the forest policy regime is executive dominance. Rather than separating the legislative and executive branches, as in the American model, parliamentary systems combine them. The cabinet and the prime minister and provincial premiers are selected from the party with the most seats in the legislature. This fact, combined with exceptionally strong party discipline, means that the executive (that is, cabinet) of the majority party dominates policy making, and there is virtually no significant role played by the legislature independent of the cabinet. A striking illustration of this fact is that throughout the

Table 1.2

Publicly owned forest land by selected areas

Area	Publicly owned forest land (%)
British Columbia	96
Alberta	96
Canada (average)	94
Russia	90
US Pacific Northwest	56
US (average)	45
Finland	29
Sweden	13
New Zealand	4

Source: G. Cornelis van Kooten and Ilan Vertinsky, "Introduction: Framework for Forest Policy Comparisons," in Bill Wilson, G. Cornelis van Kooten, Ilan Vertinsky, Louise Arthur, eds., *Forest Policy: International Case Studies* (Wallingford, UK: CABI, 1999), 13.

history of BC forest policy, the major changes occurred following the deliberations of cabinet-appointed royal commissions. A brief history of forest policy in BC is presented in Table 1.3.[56]

The absence of an independent legislature means the government has no incentives to create the types of nondiscretionary duties that dominate the American policy landscape. Legislation generally provides broad grants of authority but almost never binds the Crown to perform any particular task. Because of this large amount of government discretion, the role of the courts in policy making is also limited.

Forest policy in BC centres on the Ministry of Forests. Before the major reforms of the 1990s, BC forest management was governed by the Ministry of Forests Act and the Forest Act, both of which set out general standards for forest management that did little to constrain the discretion of the minister or the chief forester, who is the ministry official responsible for the establishment of harvest levels. As a result of this enabling structure of statutory law, policy tends to result from bargaining between the executive and the relevant societal interest groups. Before the late 1980s, environmental groups in Canada tended to play a small role in the policy process, and the dominant mode of policy making was therefore bipartite bargaining between government and business.[57] Since then, however, the policy process has become more pluralistic, with the bargaining process expanded to include environmentalists and other relevant stakeholders.[58]

Since the 1991 election of the New Democratic Party, there have been two major pushes for institutional change. The first, initiated by the government, was a highly innovative effort to use consensus-based negotiation to resolve land use disputes. These developments will be discussed in detail

Table 1.3

BC forest policy, pre-1912 to 1989

Pre-1912: Era of unregulated exploitation. Some outright land sales (mostly involving railroads); some timber leases.

1912: Following recommendations of the Fulton Commission, the Forest Act was introduced. Focus on timber allocation, revenues, and economic development. Created Forest Service. Timber Sale Licences awarded on competitive bids. Also contained objective of "protecting the water supply."

1947: Amendments to the Forest Act. Followed Sloan Royal Commission report of 1945. Initiated sustained yield policy to regulate the rate of cut. Created a new form of area-based tenure, in which more management responsibilities delegated to private companies (eventually known as Tree Farm Licences or TFLs).

1956: Report of second Sloan Royal Commission.

1958: "Sommers Affair," Minister of Forests convicted for taking bribes in issuing TFLs.

Late 1950s to early 1960s: Move away from competitive bidding in awarding tenures.

1967: Volume-based Timber Sale Licences given greater security in exchange for licensees assuming responsibility for basic silviculture.

1976: Pearse Royal Commission report recommends more security for tenure holders, more effective markets in timber products, and greater attention to environmental concerns.

1978: Forest Act amendments (and Ministry of Forests Act). Clarified allowable annual cut determination process, created small business program. Explicit incorporation of environmental values in this legislation for the first time.

Early 1980s: Deep staff cuts, "sympathetic administration."

1987: Forest Act amendments required private sector to bear full financial responsibility for reforestation, in response to American countervailing duty pressures.

1988: Proposal to "rollover" tenures into more secure TFLs fails in the wake of public pressure.

1989: Forest Resources Commission appointed to assess public participation, tenure reform, and forest practices, especially clearcutting.

in Chapter 2. The second, initiated by the environmental community, was an effort to import American-style legalism into BC forest policy. Unlike the environmental litigation campaign in the US Pacific Northwest that revolutionized forest policy there, the impact of the BC litigation campaign has been relatively limited.[59] This outcome is not surprising, given the structure of the law that courts are asked to apply: it is enabling but almost never mandatory. The bread and butter of so-called action- forcing statutes in the United States – the requirement for certain administrative actions within a specific period of time – is simply unavailable to BC environmentalists. The few deadlines in the Forest Act can be easily circumvented by its remarkable Section 152, which gives blanket authority to the minister to "extend a time required for doing anything under this Act."[60]

The litigation campaign – in combination with other political pressures – does seem to have had some effect on legislative and procedural design, resulting in a more open and formal policy process. For example, the process of determining allowable annual cuts has changed fundamentally, in part in response to litigation. Rather than a quiet negotiation between the chief forester and the timber company, there is now a multistage process of analysis, explicitly calling for and incorporating public comment. When announcing the final decision, the chief forester publishes an extensive rationale. Another major change to BC forest policy has been the enactment of the Forest Practices Code, a legal codification of many of the diverse guidelines and regulations that previously were applied on a contractual basis to timber companies. The extent of openness and formalization that emerged in this area will be discussed in Chapter 3.

Even after all these changes, BC Minister of Forests David Zirnhelt still proclaimed, "Don't forget the government can do anything."[61] Of course, there is hyperbole in his statement, but it has become notorious for its political tactlessness, not for its inaccuracy.

Ideas, about both the institutional process and the substance of policy, have supported the traditional forest policy regime in British Columbia. Traditional norms of Westminster parliamentary government justified extraordinary grants of discretion to the cabinet and its delegates, with minimal mechanisms for accountability other than periodic elections. Policies of this earlier regime were legitimized by the concept of sustained yield forestry, which in British Columbia was used to justify the rapid conversion of old-growth forests to more routinely managed second-growth forests, what Jeremy Wilson refers to as the liquidation-conversion project.[62]

Gradually, the ideas used to legitimate the traditional regime were increasingly contested by actors clamouring for access to the core of the regime. Disenfranchised groups, as well as many members of the public, became dissatisfied with their opportunities for participation under the old system and demanded a greater role in decision making. These changes contributed to institutional reforms, leading to a more pluralist process. Changes in science, particularly the understanding of forest ecosystems, also undermined traditional practices. Greater understanding of and attention to the values associated with old-growth ecosystems promoted and strengthened demands to protect these forests from conversion to more industrially oriented forestry.

The early 1990s witnessed a remarkable combination of circumstances that created expectations of significant policy change. The altered political landscape resulted both from the emergence of new actors and from significant changes in background conditions that transformed the resources and strategic opportunities of forest policy actors.[63] The emergence of new environmental groups energized the environmental movement with new resources and strategies. Ecotrust and BC Wild brought in financial resources from wealthy American foundations.[64] The Sierra Legal Defence Fund, also supported by American foundations, injected legal expertise, served as a watchdog on implementation, and launched an aggressive litigation campaign that, while short on legal victories, intensified pressures for government reforms. Greenpeace International, the World Wildlife Fund, and the Natural Resources Defense Council brought BC forestry into the international spotlight and, as highlighted below, fundamentally altered the incentives of both the government and the corporate sectors.[65]

Three changes in background conditions were extremely important: public opinion, elections, and markets. Perhaps the most important change has been in public opinion. Beginning in the late 1960s, British Columbians, along with residents of other advanced industrial nations, began to place increasing importance on environmental values.[66] In the context of forest policy, this change marked a shift from a nearly exclusive focus on the economic benefits derived from logging to a greater general concern with preserving wilderness areas and regulating the environmental side effects of logging. Although this value change has been an enduring one, several marked fluctuations in the salience of environmental issues have occurred, with one peak around 1970 and another around 1990. The general effect of these changes has been to force politicians to give greater attention to environmental interests in policy making.

Elections have also led to fundamental shifts in resources. The historical electoral dominance of the proindustry Social Credit party in BC was interrupted briefly in the early 1970s, and the NDP took over again in the wake of the collapse of the Socred party in 1991. The NDP was narrowly reelected in 1996 and clung to power for the remainder of the decade. Forest policy creates interesting dilemmas for the NDP because it has both labour and environmentalists within its coalition. However, socioeconomic and demographic changes in the province – the decline of forest industry employment as a fraction of total employment, the rise of the service sector, and the dramatic growth of urban areas removed from any direct economic dependence on the forest sector – have served to increase the political importance of environmental interests to the NDP at the expense of forest workers. In both periods of NDP rule, the government was controlled by officials more inclined to represent environmental interests than were their probusiness predecessors.[67]

The third fundamental change relates to the market within which the BC forest industry operates. As a commodity producer dependent on international exports, the forest industry has been highly cyclical. Economic conditions of the industry have a significant impact on the distribution of political resources among various actors, with industry power being inversely related to industry health. Structural changes in the market for softwood lumber in the early 1990s seemed to have created a sustained period of high prices and expanded markets for BC wood products. The changes were caused by significant reductions in timber supply in the US Pacific Northwest as a result of environmental restrictions imposed there in the late 1980s and early 1990s. These bull-market conditions were shattered by the Asian economic crisis in 1997, when the industry went into a tailspin. Figure 1.2 provides an overview of the business cycle in BC forestry over the past two decades.

These three significant changes in background conditions altered the power resources of various actors in BC forest policy and contributed to much of the policy change that occurred. Although the extent of change was limited until the 1990s, the reforms of the mid-1970s were largely the result of the NDP becoming the government in the early 1970s and the general shift toward environmental values. The pace of change increased considerably again in the late 1980s. The burst of public interest in the environment in the late 1980s forced the recalcitrant Socred government to undertake initiatives, such as the Old Growth Strategy, to improve the environmental image of the sector. But as long as a probusiness party remained in power, such efforts were likely to be largely symbolic or limited in magnitude.

In 1991, policy change began to accelerate as a result of a critical combination of factors. The most obvious was the election of the proenvironment NDP government. The party's platform contained a number of environmental initiatives, including doubling the amount of protected wilderness areas in the province from 6 percent to 12 percent. The NDP was forced to honour its environmental commitments because of its interest in reelection. Although the recession in the early 1990s resulted in a general decline in the salience of environmental issues across Canada, BC forests continued to be an intense political issue, with support for preservation high in the urban areas of the province. As the most important environmental issue in the province, a "greener" forest policy was essential to the party's electoral strategy of moving beyond its core support in labour to capture white-collar, middle-class urbanites.

Because of the institutional structure of BC policy making, the strategic alternatives of environmentalists were relatively limited. The design of

Figure 1.2

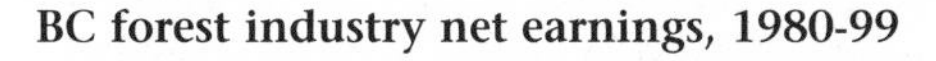

BC forest industry net earnings, 1980-99

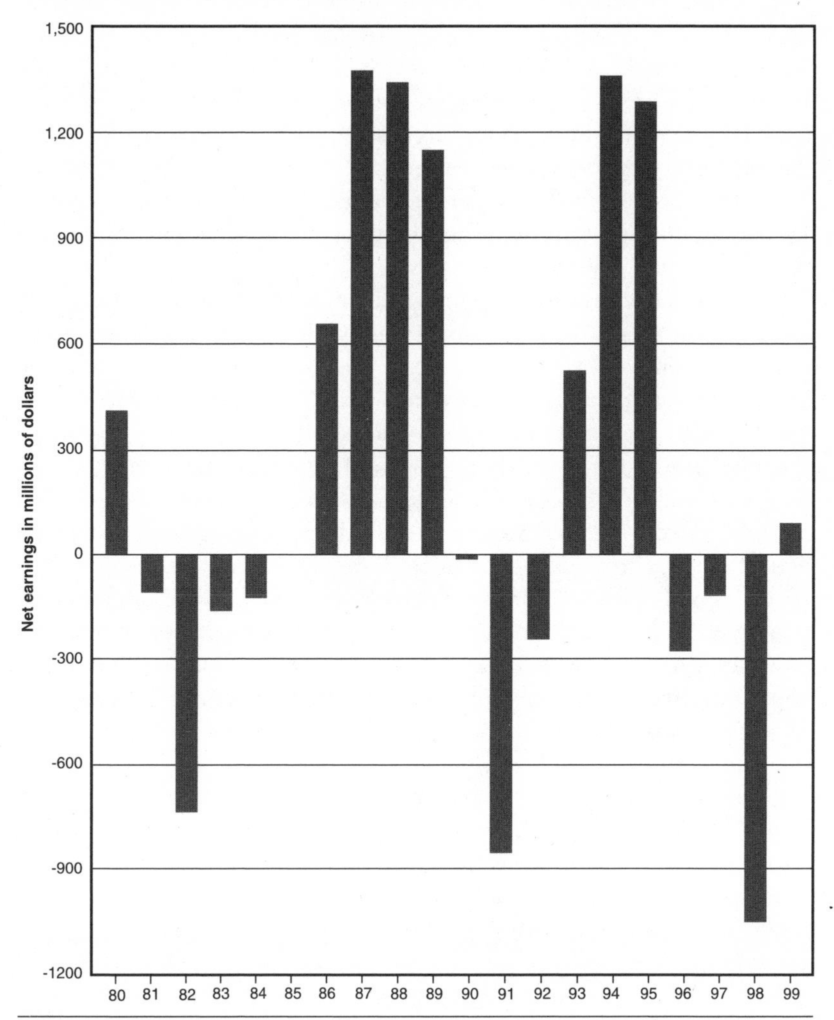

Source: COFI Fact Book, 2000.

statutes made litigation relatively ineffective. The provincial dominance over jurisdiction also made a nationalization strategy difficult.[68] Lacking judicial or national alternatives, environmentalists embarked on an innovative international campaign that met with considerable success. Moving beyond traditional strategies to influence governments by shaping public opinion, environmentalists began taking advantage of market forces to alter the incentives of corporations. Led by Greenpeace International, environmentalists began targeting industrial consumers of BC forest products, initially in Europe and then in the United States, threatening to promote boycotts of their products if they did not stop purchasing BC forest products that environmentalists claimed were being produced in environmentally destructive ways.[69] Although few contracts were cancelled, this campaign succeeded brilliantly by giving BC forest companies and the province an economic interest in improving their environmental record. Despite increased production costs, the companies came to accept that additional regulation was essential for maintaining their market share. This strategic opportunity for environmentalists is linked to changing background conditions, in this case the salience of the environmental issue in international public opinion.

Corporate practices were thus under assault both by a government seeking a greener image to garner votes and by customers demanding more environmentally sound forest practices. This fact encouraged the industry to move from a defensive posture to one more accepting of policy change. The potential blow to the bottom line of swallowing this bitter pill was softened greatly by the structural changes in the markets for wood products discussed earlier. Because prices were high and demand robust, with both predicted to remain so, the forest industry was far less resistant to dramatic policy changes than it would have been otherwise.

This combination of forces created a potentially powerful wave of change in BC forest policy, but to succeed it had to overcome formidable obstacles. Although the boom years of the early 1990s softened industry opposition, the need to maintain a healthy investment climate limited how far the government could go. And when the industry experienced a sharp downturn in 1997, the political dynamics reversed. Government became far more receptive to industry arguments about the need to relax regulatory restraints and reduce costs, and environmentalists were put on the defensive. Premier Glen Clark, who took over from Mike Harcourt in 1996, reflected his far greater sympathy for the union wing of the party when he denounced environmentalists as "enemies of BC" for promoting a boycott campaign.

Dividing the Problem

Thus, the struggle over forest policy in the 1990s was about the outcome of this clash between the factors promoting and those resisting change. Although many of these forces were at work throughout the forest policy sector, their particular manifestation varies from subsector to subsector. We have chosen to divide the forest policy sector into seven issue areas:

- land use (how the government decides which areas to set aside for protection and in which areas to allow logging)
- forest practices (how the government regulates logging)

- tenure (how the government allocates harvesting rights on government-controlled land)
- Aboriginal title (how the government deals with the claims of First Nations peoples for control over their traditional lands)
- timber supply (how the government determines the rate at which to harvest timber)
- pricing (how the government decides how much to charge for Crown timber)
- forest jobs (how the government promotes jobs and sustainable communities in timber-dependent regions of the province).

In selecting the aspects of the forest policy issue on which to focus, we have tried to address all the major categories of issues within the sector in the 1990s. A review of the major forest policy commissions over the past century reveals that five of the issues have been enduring concerns at least since the mid-1970s.[70] The exceptions are forest jobs and Aboriginal title. These two issues emerged in the 1990s and have become undeniably central to forest policy in BC. Although government concerns with employment levels have always played a role in policy making across the sector, the jobs issue took on special importance in the 1990s because so many developments converged to raise uncertainty about the industry's future. Aboriginal title is perhaps the most controversial choice. This issue is justifiably considered a policy sector on its own, but given that developments there have spilled over into the forest sector, any serious discussion of BC forest policy would be profoundly flawed without a careful discussion of Aboriginal issues.[71]

There are close relations between many of these issues, but each has its own particular history and to some extent its own characteristic configuration of actors, ideas, and institutions. Indeed, there is some uncertainty about the appropriate level of aggregation for the study of public policy.[72] The case studies will reveal how different the arrangements are at the subsectoral level and thus the most appropriate level of aggregation. As we will see, there are complex interrelationships among the subsectors. In some cases, spillovers from some subsectors facilitated change in others. In other cases, the dynamics of some subsectors, which we will call critical subsectors, placed severe constraints on changes in other subsectors. We will return to this issue in the conclusion.

Plan of the Book

This chapter has presented the analytical framework and provides an overview of the forces influencing forest policy in the province in recent decades. Chapters 2 to 8 will examine each of the major ingredients of BC forest policy, focusing on the major developments in the 1990s. Each of the policy chapters follows a common structure: an outline of the policy problem in that subsector; an overview of the policy indicators used to assess change; the particular subsectoral characteristics of the regime framework (that is, the specific actors, institutional context, or ideas that dominate in the subsector); a historical overview, reviewing developments before the 1990s; an analysis of the issue as it went through each stage of the policy cycle, focusing on how different elements of the regime framework influenced different stages of the cycle; and a conclusion, addressing key questions about how much policy change and why. Chapter 9 will provide a conclusion to the analysis.

As mentioned, one of the major challenges of this book is to determine whether the policy changes of the 1990s reflect a fundamental change in direction for provincial forest policy or whether they are better conceived of as relatively minor departures from the status quo. In this determination, we need to distinguish policy outcomes as our dependent variable (what we are trying to measure and explain) from a variety of independent (explanatory) variables developed in the regime framework.

Determining how fundamental the policy change in question is, is necessarily a question of interpretation, frequently tantamount to deciding whether a glass is half empty or half full. In our interpretations, we use several devices. First, we distinguish the various components of policy content: general policy goals, more specific policy objectives, policy instruments, and the settings on those instruments. There may be change in some but not all of these components.[73] Our ultimate concern is with the relative balance struck between the fundamentally competing interests at stake. In some cases, this balance involves competition between environmental and economic objectives, but in others, the relevant competing objectives will be different. Second, we try to conceptualize for each case what might be considered a fundamental change in paradigm. One example is the recommendations of the Clayoquot Sound Scientific Panel. The panel argued that, for that region, ecosystem values should dominate and industrial forest values should be derivative. If that model were to be applied provincewide, it would certainly represent a fundamental change.

We will return to reflections about these issues in the concluding chapter. As we will see, paradigm change did not occur in any of these seven issue areas. Overall, the forest sector in BC experienced only modest policy change.

2
Experimentation on a Leash: Forest Land Use Planning in the 1990s

The 1990s cycle of forest land use planning resulted in some striking changes. After two decades of piecemeal attempts to resolve land use issues, the provincial government moved decisively toward a comprehensive planning approach based on regional zoning processes. Although not universally acclaimed, the new processes allowed fuller application of ecological knowledge and provided enhanced opportunities for ordinary British Columbians wanting a say on regional land use issues. The hundreds who took the opportunity to work alongside government officials at or near land use planning "tables" helped craft a lengthy list of regional agreements, most of which were highlighted by significant additions to the protected areas system. Over the decade, nearly 5 million hectares were added, increasing the proportion of the province in the parks system from under 6 percent to about 11.5 percent. The list of new protected areas included some large expanses of old-growth forest, including a number of areas that had been the object of intense conflict between the forest industry and environmentalists during the preceding decades.

This record is not all that surprising. Indeed, given the number of change-promoting and -facilitating regime alterations during the five years before the NDP takeover, it would have been surprising if the 1990s had failed to produce significant change. From the outset, there was little question that the new government would move aggressively on its goal of doubling the protected areas total to 12 percent by the end of the decade. Public support for enlargement of the protected areas system was generally high. Many industry and bureaucratic actors who had resisted the idea of expansion over the previous decades had come to believe that park additions were inevitable. The issue was salient, particularly to the large contingent of green voters who had helped elect the NDP. Premier Mike Harcourt and his advisors recognized the importance of retaining the loyalty of the green constituency, and they knew that, for many members of this constituency, the government's response on protected areas would symbolize its attitudes toward the environment in general. The 12 percent target was a hard objective that would serve as the basis for an unambiguous evaluation of whether the government had made good on its promise.

Still, it was far from inevitable that straightforward government commitments and a receptive public would translate into subsector policy change as extensive or rapid as what occurred. The NDP government knew that forest companies and their workers were strongly opposed to erosion of the forest land base. It soon realized that its commitment to seeking land use peace through regional consensus-seeking processes would provide these actors with rich opportunities to derail the protected areas agenda. Here, as during the forest practices policy process described in the next chapter, the new government was reminded of the points underlined in Chapter 6 concerning the critical timber supply subsector. Development of protected areas policy and other forest environment initiatives could not be shielded from the powerful forces shaping determination of allowable annual cut (AAC) levels.

These features of the subsector regime and its relationship to the timber supply subsector meant that those trying to translate the NDP land use planning objectives into policy were faced with some difficult challenges. Perhaps most important, the new government had to confront the dilemma of how to move toward its politically important protected area goal while at the same time making good on its commitment to disperse significant land use planning power to local communities dominated by forest industry actors. Its response to this dilemma reflected successful learning on the job: by the midpoint of the first term, the government was well into implementation of an approach that can be referred to as regionalized experimentation on a leash. It relied heavily on advice generated by the regional and subregional stakeholder tables set up under the auspices of its flagship planning initiatives, the Commission on Resources and Environment (CORE) and the Land and Resource Management Plan (LRMP) process. As these experiments unfolded, however, it became clear that cabinet – or, more specifically, the premier and his close advisors – would be maintaining a tight rein. Any doubts about this being a cabinet-dominated system, or about this being a government with the capacity to steer the processes it had set in motion, were removed once policy implementation commenced. Harcourt's key ministers and advisors were everywhere, adjusting policy instruments, leading the additional policy development processes needed to fill in critical "policy blanks," and manipulating relationships among the actors involved in the regional experiments.

Experimentation generated learning. In terms of the model of policy cycle stages used to frame the following analysis, the monitoring and evaluation activities that commenced at the outset of the implementation phase generated dense patterns of feedback. These patterns shaped further rounds of agenda setting, problem definition, policy formulation, and decision making involved in the government's moves to extend and adjust its policy package. Like the treatments offered in other chapters, then, this one illustrates the blurred lines separating the five stages identified in that model. We will see, for example, that jousting over problem definitions continued long after the focus shifted to decision making and implementation.

We will also see that the overall change process here is best understood as constituting nested cycles. Regional cycles interacted with interconnected provincial cycles, some of which unfolded within this subsector and some of which did not.

The second section of this chapter will relate the NDP's objectives to the political and ideational currents that evolved during the decade before its 1991 victory. The third and fourth sections will examine the Harcourt government's choice of objectives and instruments, and then consider the policy settings that emerged as it began to implement policy. The fifth section will focus on events after 1996, exploring the Glen Clark government's manipulation of the now mature LRMP process along with its response to issues avoided during the Harcourt years. We begin with an introduction to the subsector policy regime.

Characteristics of the Forest Land Use Policy Regime

The policy regime here closely resembles those described in other chapters. As elsewhere, outcomes were shaped by multidimensional conflict between the development and environmental coalitions. And, as elsewhere, this conflict played out in an institutional context marked by cabinet control and fairly consistent Ministry of Forests (MOF) dominance, and in an ideational context shaped by the ascendancy of the biodiversity discourse and growing doubts about the liquidation-conversion project, the package of post-second World War forest policies aimed at "rationing" the liquidation of old-growth forests and ensuring their conversion to productive second-growth forests.

Broad similarities between the land use regime and other regimes do, however, mask some differences. First, in this subsector, the cleavage between the development and environmental coalitions has been little leavened by intracoalition tensions. In each camp, some differences over approaches and tactics can be observed, but overall, there is clear unity of purpose. Forest workers and companies big and small coalesce to resist threats to the "forest land base," whereas diverse environmental groups join to push for a larger and more ecologically representative protected areas system. Second, the concept of ecological representation and others at the core of the biodiversity discourse play a particularly important role, providing an increasingly potent foundation for wilderness movement arguments about the importance of preserving remaining old-growth valleys and forcing the development coalition to respond to new tests of good stewardship.

The institutional setting has continued to be marked by Ministry of Forests control over forest land use decision processes. MOF dominance was most evident before 1985, when the agencies responsible for parks and the environment were able to exert only a weak counterweight force.[1] Increasing public concern about forest stewardship caused the MOF's star to decline in the late 1980s, setting the stage for some loss of status and control during the Harcourt years. The new government briefly elevated the standing of the Ministry of Environment, Lands and Parks (MOELP) and granted important responsibilities to CORE and another new instrument, the Land Use Coordination Office (LUCO). By the beginning of the Clark regime, however, it was clear that the MOF's early 1990s problems had been only temporary. CORE, which had guided an important stream of developments between 1992 and 1994, was dissolved by 1996. The MOELP's star waned, and the MOF-designed LRMP processes became the chosen vehicle for developing regional land use plans.

From a wider perspective, any ups and downs in MOF fortunes are less important than the constants of the institutional context. The arrival of new agencies and associated institutional changes did little to alter two central features of the situation. First, even during its difficult years, the MOF continued to control most of the land base, using the discretionary latitude granted under the Forest Act to put its stamp on a wide assortment of outcomes. Second, as noted in Chapter 6, the chief forester continued to set the all-important allowable harvest levels.

Turning to societal actors, the wilderness movement includes a loose alliance of recreational, naturalist, and advocacy groups. The movement's major strengths are its diversity, broad public support, large and intensely committed pool of activists, and well-cultivated connections to important allies both inside and outside the province.[2] Its diversity has manifested itself in resilience, adaptability, and tactical ingenuity. It has mixed direct and indirect lobbying, using an assortment of approaches to raise public concern and galvanize its supporters into communicating their views to government officials. By the early 1990s, it had widened its list of allies to include powerful European and American environmental groups, preparing the ground for the internationalization of BC forest environment issues that played a key role in the determination of protected areas and forest practices policy outcomes.

The forest industry alliance is dominated by the twenty to twenty-five companies that together control over three-quarters of the total provincial timber harvest. As others chapters note, the industry's political power derives from its structural advantages and from its possession of the financial resources needed to support lobbying and public relations work. In the 1990s, the industry's stance on forest land issues was articulated by the Forest Alliance of British Columbia as well as by the individual companies threatened by land use changes in particular regions. On protected areas issues, the industry received strong support from the major union representing forest workers, the Industrial, Wood and Allied Workers of Canada (IWA-Canada), as well as from an array of local "share" groups linking forest workers and their families to others in forest-dependent communities. Industry arguments about the importance of resisting erosion of the forest land base have generally been seconded by MOF spokespersons. Thanks to the efforts of a green faction within the ministry, however, the MOF's hostility to the preservation of old-growth forests had somewhat lessened by the time the NDP was elected. This and related shifts in outlook during the final years of Social Credit helped position the MOF to rebound from difficulties experienced in the few years on either side of 1990.

All of these state and societal actors have had to respond to major shifts in ideas and background conditions. During the 1980s, public opinion favouring expansion of the protected areas system grew stronger with increases in societal concern about the environment and with the diminution in MOF-industry legitimacy resulting from failed attempts to implement promised sustained yield and integrated management reforms. The rise of the biodiversity discourse helped move the public focus onto old-growth forests, complicating MOF-industry efforts to appease environmentalists. Each of these shifts, of course, resulted from complex causal chains, some of which had exogenous roots. For example, developments in the US Pacific Northwest had a major impact on the discourse, heightening awareness of the biological diversity present in old-growth forests. Or to take another example, Social Credit's mid-1980s policy of "sympathetic administration," which contributed so significantly to the delegitimation of the sustained yield-integrated management paradigm, was a response to the recession that engulfed the industry in the early 1980s. Key shifts in the political and ideational contexts can also be linked to preparatory work done by the competing coalitions. For example, in the late 1980s, the environmental coalition did an effective job of heightening public concern about the forest environment, of promoting the ascendant biodiversity discourse, and of parlaying the report of the World Commission on Environment and Development (Brundtland report) into an effective campaign for the 12 percent goal. As the next section shows, these initiatives were critical in the skirmishes over problem definition that set the stage for developments in the 1990s.

Historical Background: Setting the Agenda for the 1990s Policy Cycle

All of the land use changes initiated by the NDP could be linked to major shifts in the problem definition in the years before it took over. As Chapter 1 showed, these and the shifts taking place in other subsectors resulted from the interaction of actors and ideas. In the period 1985-91, the environmental coalition effectively challenged the way the wilderness conservation problem had been constructed, in the process pushing onto the agenda the closely related issues of old-growth preservation and the ecological representativeness of the parks system. As well, by directly confronting companies intent on logging areas such as the Carmanah, the Walbran, and the Tsitika, the movement precipitated the "wars in the woods," which the NDP was forced to address in its deliberations on the problems the government would have to address in the 1990s.

The industry-MOF problem definition that had reigned supreme before 1985 was fairly straightforward: there was no wilderness conservation problem, but those campaigning for large increases in the protected areas system would create one if they succeeded. The existing system provided adequate protection; preservation of additional large amounts of forest land would have major negative economic consequences. In the case of major "withdrawals" from what the development coalition began to refer to as the forest land base, these consequences would include the bill for compensating forest companies losing appreciable amounts of their timber supply. Given these potential costs, environmentalists should be content with the present parks system along with the additional protection of wildlife, fisheries, and recreational opportunities that would result as forest managers put increased emphasis on integrated resource management.

During the 1970s and early 1980s, the challenge to this "no problem" definition came mainly from organizations (or coalitions) trying to preserve specific wilderness areas such as the Stein, the Valhalla, and South Moresby. By the mid-1980s, the limited potential of this piecemeal approach had become apparent to environmentalists. Frustrated over how much time was being poured into these campaigns and by the lack of results, leaders of the environmental coalition began to advance alternative approaches to wilderness protection. Some called for American-style wilderness legislation, and others assembled and promoted provincewide lists of prime wilderness candidates. A number presented arguments for a comprehensive approach to parks system expansion to the 1985-6 Wilderness Advisory Committee (WAC), a group reluctantly set up by the Social Credit government in the hopes of finding a package solution to the increasingly nettlesome set of wilderness issues. Federation of Mountain Clubs activist Stephan Fuller synopsized the arguments for a new approach, telling the WAC that existing planning structures allowed continued "incremental destruction of wilderness ... before it is properly evaluated as a resource ... Claim staking, resource development road construction, and other development are permitted with limited consideration of the alternative land use of the area. A 'no development' option is rarely if ever considered."[3]

The movement's efforts received major boosts from externally induced shifts in the climate of ideas. The Brundtland Commission's suggestion that the world's protected areas total, which at that point stood at about 4 percent, should be tripled was quickly seized upon by the World Wildlife Fund Canada (WWF).[4] In a move that nicely illustrates adept policy entrepreneurship, the WWF launched its national endangered spaces campaign in 1989, arguing that each Canadian jurisdiction should develop an action plan for achieving the goal of 12 percent protected areas.[5] And drawing on an important element of the emerging biodiversity discourse, it said governments should aim to protect areas representative of each of Canada's major ecosystems.

The BC wilderness movement had long pointed out that the parks system overrepresented high-elevation rocks-and-ice landscapes at the expense of low-lying forested ones. It began to articulate these concerns more forcefully after a series of analyses of the system's ecological representativeness. In the most important of these, a WWF offshoot – the Earthlife Canada Foundation – teamed up with the American organization Ecotrust to assess the development and preservation status of valleys up and down the BC coast.[6] This inventory classified the 354 coastal watersheds larger than 5,000 hectares as pristine ("virtually no evidence of past human or industrial activities"), modified ("slightly affected by a limited amount of industrial activity"), or developed ("significantly affected by logging activities, highways or other industrial development such as powerlines or pipelines"). Just one-third of these watersheds were found to be undeveloped (about 20 percent pristine and 13 percent modified), with this percentage falling to less than 20 percent for valleys over 20,000 hectares. Only 9 of 354 watersheds were fully protected. Other groups and individuals deployed rapidly improving geographical information systems (GIS) computer-mapping techniques to reinforce arguments about gaps in the parks system's representation of the province's diverse ecosystems.

These analyses illustrate how evaluation of one set of policies works through feedback loops to influence the agenda-setting and problem definition activities that inaugurate a new policy cycle. This work also reflects the shift toward greater focus on old-growth forests that coincided with increased attention to biodiversity and related concepts. These concepts quickly gained currency within the policy community in the two or three years before the NDP takeover. Learning about how they were being incorporated into preservationist arguments in the US Pacific Northwest was advanced by an MOF-sponsored exercise known as the Old Growth Strategy Project, launched in early 1990.[7]

The Social Credit government's willingness to support the Old Growth Strategy was one indication of the development coalition's loss of control over the agenda and problem definition in the late 1980s. Another was the government's decision to soften its line on parks system expansion. After digesting the WAC recommendations, it allowed the Parks Branch and the MOF to undertake public processes aimed at developing and evaluating lists of preservation candidates. Parks Branch officials used the opportunity to air a landscape representation model that, because of cabinet opposition, had previously been kept under wraps. In 1990, the two agencies combined their wilderness review process into a public review process known as Parks and Wilderness for the 90s.[8]

Launching the 1990s Policy Cycles: Redefining the Problem, Establishing Objectives, and Choosing Instruments

By 1991, the old industry-MOF problem definition was under attack from various sides. According to the environmental coalition's ascendant arguments, the real problem was that the protected areas system was not adequately protecting the province's ecological diversity, particularly its forest ecosystems. The system, it said, had to be radically enlarged, and expansion had to be based on comprehensive planning, with special emphasis put on boosting the representation of threatened ecosystems. Environmentalists linked these arguments to ones debunking industry concerns about negative economic impacts. Any negative consequences of an expanded parks system, they stressed, could be easily mitigated by better forest management and timber utilization. Although these arguments certainly did not win widespread support among forest executives and workers, more and more members of the development coalition did begin to consider seriously the proposition that some expansion of the parks system might be a relatively painless way of relieving the environmental pressure.

These developments within the competing coalitions presented the NDP government-in-waiting with its own problem-definition challenges. During the period when Social Credit's accumulating political problems made it more and more likely that the NDP would soon be taking over, Harcourt and his advisors were forced to search for a way of defining the problem that would satisfy both the green and the IWA components of the party's support base. They did so by stressing the serious costs associated with the war in the woods. Clearly, the old system was not working. Increased strife between workers and environmentalists jeopardized social peace and increased uncertainty, undermining the economies of forest-dependent communities.

When the fall 1991 election campaign began, the NDP had crafted the ideas advanced by various policy entrepreneurs into a compromise package of goals and objectives for the policy subsectors most germane to environmentalists.[9] An NDP government's primary goal would be "peace in the woods." This core commitment spawned the closely connected derivatives needed to maintain the allegiance of the IWA and green sides of the party: stronger environmental measures and protection of forest industry jobs. In the land use planning area, these goals translated into a shortlist of key objectives. Environmentalists were offered the concrete pledge that the size of the protected areas system would be doubled to 12 percent of the province, along with a more abstract commitment to improved ecosystem representation. Forest workers were promised measures to protect jobs threatened by creation of new parks or other "conservation activity." The protected areas objectives would be designed and implemented through new consensus-seeking approaches.

As John Kingdon's model would predict,[10] the convergence of the politics, problems, and solutions streams pushed the protected areas issue up the agenda. In the years before the NDP takeover, interest group pressures and shifts in public opinion altered the political context, convincing actors throughout the policy community that the hardline stance on protected areas had to be softened. Environmentalists popularized an alternative problem definition and, by vigorously pressing their case at various sites across the province, helped create the war in the woods that figured so prominently in the NDP's conception of the problem. The collapse of Social Credit created an opportunity for the NDP to take control of the issue. Its response drew on solutions suggested by a variety of policy entrepreneurs. Confident that it had a win-win solution, the NDP gave its pledge for peace in the woods prominence in the 1991 election campaign, ensuring that the protected areas issue would be at the top of the political agenda as it settled into office.

In considering the Harcourt government's choice of instruments, it is again helpful to think in terms of a constellation of interconnected elements arrayed along a continuum from the abstract to the concrete. Procedurally, the peace in the woods goal would be achieved through shared decision making. On a substantive but still relatively abstract level, the package of instruments would revolve around commitments to a comprehensive assessment of land use and to a zoning approach based on counterbalancing protected areas additions with zones designated for high-intensity resource development.

Over the first two years of the Harcourt regime, these ideas were translated into a number of organizational and programmatic initiatives, including CORE, the LRMP processes, the Protected Areas Strategy (PAS), LUCO, and Forest Renewal BC (FRBC). These combined to create a complicated stew. All were instruments with roles in pursuit of the NDP's objectives, but they varied on a number of dimensions. CORE, LUCO, and FRBC, for example, were new organizations that immediately became key actors. The PAS and the LRMPs, on the other hand, were processes that depended for support and guidance on various organizational actors. In addition to LUCO, these actors included the cabinet along with committees of ministers, deputy ministers, and regional officials from the natural resource and economic development ministries. It is also important to note that one of these organizational instruments, FRBC, was a product of decisions in the jobs subsector, described in more detail in Chapter 8. Despite its extrasubsector origins, FRBC had a key facilitating role in efforts to achieve land use peace.[11]

This picture was further complicated because some of the primary instruments quickly spawned series of regional arrangements known as land use planning "tables." CORE had a large central office staff along with a mandate to pursue a provincial land use strategy, but it will be best remembered for its creation and guidance of four regional land use planning processes. These and the various LRMP processes took on lives of their own, drawing disparate sets of local stakeholders and agency officials into the analytic and consensus-building processes that led to regional land use recommendations.

We hesitate to clutter the account with talk of subsubsector cycles and the like, but it is important to recognize that this mix of central and regional organizations and processes meant that the overall policy process in this subsector had a complex, triple-nested character. Here, as in several other chapters, we describe lateral nesting of two types: there is spillover from subsector to subsector as well as from cycle to cycle within the subsector. In addition, to take account of concurrent provincial and subprovincial processes, we need to introduce the notion of vertical nesting: forces operating within particular regional land use cycles interact with those shaping provincial-level policy change. Thus, for example, the land use planning cycles in the CORE and LRMP regions interacted with the provincial level policy cycles shaping the evolution of CORE, LUCO, and the zonation system as well as with cycles unfolding in the jobs, forest practices, and timber supply subsectors.

As the remainder of this chapter shows, this assortment of developments did not unfold sequentially. Different cycles were launched at different times, with developments in one cycle constantly feeding back to influence decisions on how to adjust the instruments and settings involved in other cycles. The provincial policy framework evolved as CORE, LUCO, and the LRMP initiative got off the ground and matured, meaning that later regional processes received different cues regarding provincial policies than had their predecessors.

Given this complexity, the development of concrete policy instruments was a surprisingly straightforward process. New governments in Canada and elsewhere often find that linkage between the stages of objective setting and instrument choice is complicated by bureaucratic resistance. As is frequently pointed out, permanent officials possess a powerful arsenal of techniques for diverting or stalling the reform aspirations of parties newly arrived in power. In this instance, however, the bureaucracy was generally sympathetic to the NDP's parks expansion and land use planning proposals. Indeed, in various parts of the bureaucracy, policy development work during the final years of Social Credit had anticipated or paralleled that taking place within the NDP. As noted, the Parks and MOF bureaucracies had combined forces to begin a comprehensive review of protected areas candidates under the rubric of Parks and Wilderness for the 90s. The MOF had started to develop the new subregional land use planning approach that became the LRMP process, while an interagency team of officials called the Provincial Land Use Strategy (PLUS) working group had started to develop proposals for process reform.

Those trying to design new advisory mechanisms could draw on the province's experience with regional and local land use decision-making processes. By 1991, the base of experience was lengthy and varied. For more than three decades, governments had experimented with structures for facilitating interagency consultation at the provincial and regional levels. And for at least two decades, governments seeking to resolve conflicts over particular wilderness areas had explored ways of giving representatives of the public a role in these advice-generation processes. The history of forest land use decision making in the 1970s and 1980s features a long list of task forces, study teams, public advisory committees, planning teams, public liaison committees, and the like.[12] These were constituted in different ways, but they usually combined citizens representing different interests with teams of officials from the MOF and other resource agencies.

As noted, multifaceted dissatisfaction with these experiences led after 1985 to increased calls for a more comprehensive approach. Various individuals and organizations tried their hand at sketching blueprints for development of a provincial land use strategy. Drawing on these models, the Harcourt team moved quickly to design the instruments needed to pursue its objectives. By mid-1992, it had assigned the lead role to the newly created Commission on Resources and Environment. The mandating statute directed the commissioner to "advise [cabinet] in an independent manner on land use and related resource and environmental issues."[13] CORE was asked to formulate recommendations for a provincial land use strategy and to develop and implement dispute resolution mechanisms as well as regional and community planning processes. Former provincial ombudsman Stephen Owen was named as first commissioner.

By the end of 1991, CORE had begun to develop land use negotiation processes in the regions that had been grappling with particularly difficult land use issues: Vancouver Island, the Cariboo-Chilcotin, and the Kootenays. In designing these processes, CORE officials sought to implement the concept of shared decision making, an approach featuring "structured, collaborative negotiation" and aimed at ensuring that "by participating directly in the decision making process, all legitimate interests have an opportunity to be heard and to 'own' the decision."[14] "As a result," CORE said, "confrontation, with its unavoidable social and economic costs, is replaced by negotiation leading to a decision acceptable to all parts of the community." Each regional negotiating table would develop recommendations regarding the allocation of the land base to different zones and would also provide advice on "transition and mitigation strategies for those affected by re-allocation."

In the CORE regions and elsewhere, advice on protected areas would be funnelled through the Protected Areas Strategy (PAS) process. This process would extend the work of Parks and Wilderness for the 90s (and to a lesser extent the Old Growth Strategy) and would help CORE and the regional tables "evaluate regional protected area candidates within an overall provincial context."[15] The strategy, directed by a committee of assistant deputy ministers, would pursue the 12 percent goal, in the process trying, among other things, to "protect viable, representative examples of the natural diversity of the province ... the major terrestrial, marine and freshwater ecosystems."[16]

Another key set of parameters framing the regional land use decisions was fleshed in by early decisions on the zonation model. In its 1993 decision on a land use plan for the contentious Clayoquot Sound area of Vancouver Island, the government laid out the template. The land use designation process would involve assignment of territory to zones of differing land use intensity. Protected areas additions would be counterbalanced by zones designated for high-intensity resource development, thus providing some succour for development interests. In addition, the prospects of regional land use consensus would be facilitated by enabling the land use tables to place at least some contested territory in a vaguely defined "special management" category. In these Special Management Zones (SMZs), as they came to be called, resource extraction at lower-intensity levels would be allowed. Crucial decisions about just how much lower would be put off to some future date. In the new land use system, then, the key policy settings would result from decisions on how much land would be placed in each of the zones. A crucial second layer of settings outcomes would be filled in with decisions about the rules governing practices (and particularly the intensity of extractive activities) in the various zones.

Policy Implementation, Evaluation, and Adjustment during the Harcourt Years

The "feedback-intense, nested cycles" nature of land use policy making in the 1990s makes it difficult to draw a clear distinction between the decision-making and implementation stages, but it is not unreasonable to think of implementation as having begun as the CORE regional processes started to consider substantive zoning proposals. CORE's brief but remarkable history centred on the regional land use planning processes carried out under its auspices in four regions: Vancouver Island, Cariboo-Chilcotin, West Kootenay-Boundary, and East Kootenay. In each, CORE staff tried to herd long lists of regional stakeholders toward consensus. Although the accomplishments of these tables varied, none was able to reach full agreement on land use designations. The task of recommending how each of the regions should be zoned thus fell to Commissioner Owen and his staff. In an impressive series of reports issued during 1994, Owen extended what had been accomplished at the tables into detailed land use prescriptions. In each case, he applied the multicategory land classification scheme first applied in the Clayoquot Sound decision. The Special Management Zones (which were labelled in that way in two of Owen's reports, as sensitive development areas in another, and as regionally significant lands in the fourth) were defined in vague terms, thus intentionally or otherwise providing the government with a means of bridging the gulf between the competing coalitions. Each side, that is, was allowed to read what it wanted into the special management concept.

It was soon apparent that more than a few vague bridging concepts would be required to achieve consensus in the CORE regions. Although Owen took pains to recommend measures to mitigate negative economic impacts of land use changes, forest industry workers and their supporters reacted very negatively. The responses to his Vancouver Island and Cariboo-Chilcotin reports were particularly hostile. A crowd estimated at between 15,000 and 30,000 demonstrated against the Vancouver Island report on the lawns of the provincial legislature in Victoria, and Owen encountered belligerent audiences when he tried to sell the Cariboo-Chilcotin plan to area residents.

These events increased the urgency of cabinet-level deliberations on how to adjust the evolving package of instruments and settings. In fact, evaluation, learning, and debate about necessary adjustments had begun even before Owen's reports appeared. By the end of 1994, the cabinet had drawn – and begun to apply – several interconnected lessons from the early experiences. First, the system lacked the capacity needed to coordinate flows of information, advice, and directives among cabinet, the ministries, committees of officials, CORE, and the regional land use advisory bodies. Second, progress also required that cabinet move promptly to fill in key policy gaps, particularly those highlighted by Harcourt's promise to provide help to forest workers whose jobs were threatened by land use decisions. Third, the gap-filling exercise could ignore the issue of what compensation was owed to companies losing cutting rights as a result of land use decisions. Fourth, although broad-based consensus-seeking advisory approaches would continue to play an important part in achieving land use peace, the CORE approach could be improved upon. Finally, cabinet and its closest advisors would have to take a more direct, hands-on role than envisaged by the architects of the initial phase of the experiment.

Let us consider each of these lessons and the "feedback loop" changes that followed.

Addressing the Deficiency in Integrative Capacity

The realization that the cabinet information-processing system lacked the capacity to effectively manage the flows of directives and advice among different parts of the system led to the establishment of the Land Use Coordination Office (LUCO) in early 1994. Since then, LUCO has performed numerous important integrative roles. It coordinated relationships between CORE and the ministries, and organized advice to senior officials and cabinet on what to do with CORE recommendations. It supervised the start-up stages of the Land and Resource Management Plan (LRMP) exercises, oversees and supports the committees of senior regional officials (Interagency Management Committees) that guide these exercises, and advises cabinet on a range of protected areas issues, including the critical matter of how the 12 percent "budget" should be parcelled among different regions.

Filling in the Policy Gaps

From the outset, participants at the CORE regional tables made it clear that the chances of their coming to agreement were undermined by uncertainty concerning the government's plans on related issues. The IWA and other spokespersons for forest-dependent communities wanted the government to clarify its intentions regarding financial support for workers who stood to suffer as a result of land use changes. The government's April 1994 response is described in detail in Chapter 8. In brief, the government announced that, over the following five years, about $2 billion of revenue raised through stumpage increases would be used to build a Forest Renewal fund. The money would be disbursed by a new Crown corporation, Forest Renewal BC. Its programs would support reforestation and silviculture, the reclamation of marginal agricultural land, research on silviculture and environmentally sound forest practices, environmental restoration, rehabilitation of watersheds, and various initiatives aimed at increasing value-added processing of BC timber. Although it was not clear what proportion of Forest Renewal expenditures would flow toward displaced forest workers, the government left little doubt that those workers could expect to be major beneficiaries of a number of the new programs. The origins of Forest Renewal were complex, and it was intended to serve a number of functions. Among other things, though, it was an attempt to make good on Harcourt's pledge that "not one forest worker will be left without the option to work in the forest as a result of land-use decisions."[17]

Environmentalists at the CORE tables had expressed a different set of concerns about policy gaps, making it clear that their willingness to support land use plans in the regions would depend in part on their receiving the right signals concerning government moves to toughen forest practices rules. The movement's willingness to compromise on its protected areas goals would obviously be greater if it had some assurance that the protection of biodiversity across the remainder of the land base was to be improved. As noted in Chapter 3, the government responded to these concerns during 1993. Its Forest Practices Code was unveiled in mid-1994.

Putting Off the Compensation Issue

To the surprise of many, the Harcourt government decided to avoid the compensation issue. Here the learning process persuaded cabinet that it could get away with maintaining its deliberately vague policy on reparations for forest companies adversely affected by land use decisions. This issue had lurked in the background throughout the 1980s, with forest companies and their supporters periodically pointing out that, given the applicable provisions of the Forest Act, the compensation costs of implementing the environmental coalition's wilderness agenda would be extremely high. These warnings took on greater immediacy after the BC and federal governments agreed to create the South Moresby national park reserve in 1986. Negotiations over compensation for the company that lost the lion's share of cutting rights in the area (Doman Industries) continued until late 1991, when the new NDP government agreed on a $37 million package.

At the outset, the Harcourt team seemed to agree with those who contended that it could not move ahead with its protected areas plans until the compensation issue was clarified. Shortly after taking over, it established the Commission of Inquiry on Compensation for the Taking of Resource Interests. The commissioner, SFU economist Richard Schwindt, was asked "to inquire into the principles and processes for determining whether, in what circumstances and how much, if any, compensation should be paid, to the holders of resource interests that under an enactment have been or are taken, for public purposes and without the consent of the holder, by the Crown or another authority."[18] After considering submissions from a variety of interested parties, Schwindt rejected a comprehensive approach, recommending instead that the right to compensation should be determined case by case and governed by the principle that licensees are entitled to compensation related to "the financial harm done to investments made with the expectation of an uninterrupted supply of a specific volume of fibre."[19]

A few months deliberation on Schwindt's report convinced the cabinet that it should avoid the issue. Its decision was no doubt influenced by arguments that if the Doman-South Moresby agreement was taken as a precedent, the government would have to shell out about $40 for every metre of compensable wood removed from licensees. Knowing that the industry would respond with hostility to any attempt to decree easier withdrawal provisions, the government dropped the idea of developing a new policy. Protected areas deliberations were left to unfold without any guidance on possible compensation implications, and Harcourt's ministers spent the remainder of the NDP's first term dodging questions about the eventual compensation bill that might have to be faced. As we will see, however, the issue could only be avoided for so long.

LRMP Processes Become the Chosen Means of Regional Land Use Planning

By responding to these policy gaps and taking charge of negotiations, the government was able to win acceptance of land use plans in the CORE regions. The whole experience, however, raised doubts about the CORE approach. The search for alternatives, as it turned out, did not have to travel very far. While CORE's regional processes were unfolding, the MOF had quietly begun to implement its newly designed Land and Resource Management Plan concept. Based on design work begun before the NDP's election, LRMPs were originally conceived as subregional exercises that would kick into gear once a regional land use strategy was agreed to. This notion, however, was quickly set aside, and the first of the LRMP processes began deliberating on a region-sized chunk of territory in the Kamloops area in late 1992. While the CORE processes lurched unsteadily and publicly toward deadlock, the government officials and citizen representatives on the Kamloops LRMP table quietly achieved consensus on a detailed land use plan. This group's relative success was, no doubt, owed in part to the fact that it began without the baggage that had accumulated during decades of land use conflict in at least three of the CORE regions.[20] CORE suffered in the comparison with this and other early LRMP processes simply because it was asked to tackle the tough parts of the province. In addition, it seemed that the LRMP model provided a more promising basis for integrating the work of government officials and stakeholders. Whereas officials from the resource ministries stayed in the background during the CORE table processes, the LRMP model assumed they would be centrally involved. Standard interagency planning mechanisms would be wedded to a strong citizen advisory structure, with each table given some latitude as to how to shape this linkage and respond to other key start-up issues.

Whatever the reasons for the CORE-LRMP differences, the comparison was not lost on a cabinet looking to avoid more conflagrations of the sort precipitated by CORE's Vancouver Island and Cariboo-Chilcotin reports. CORE was wound down and then shut down, and the LRMP model was anointed as the chosen vehicle for further delivery of the government's land use planning objectives. In a variant of this response, the government in 1995 turned a nest of land issues in the southwest mainland corner of the province over to a new multistakeholder advisory committee, the Lower Mainland Regional Public Advisory Committee.

Cabinet-Led Bargaining

The CORE tables' failure to reach consensus led the cabinet to reevaluate the potential of the shared decision-making philosophy that had guided Owen and his team. Harcourt and his advisors concluded that, although the tables' work and Owen's reports provided valuable bases for land use consensus in the regions, cabinet would have to play an active role in leading development of the final agreements. Cabinet had given the CORE experiments plenty of leash; it was now time to tighten control and engage in some traditional brokerage politics.[21] Because it had the authority to impose land use plans, cabinet knew it would be taken seriously when it engaged key stakeholders in bargaining over ways of improving what had been suggested by Owen.

This switch in approaches paid off. With the premier's deputy taking a lead role, the government's emissaries set out to determine what it would take to convince the key stakeholders in each of the CORE regions to buy into a plan.[22] After a complex series of closed-door negotiations, consensus was reached in each region. These final plans were similar to what Owen had recommended. In each region, a comparison of the CORE proposal and the final plan reveals some variation in the number and boundaries of protected areas and Special Management Zones, but shows similar allocations of land to the different categories.

A number of factors helped convert harsh CORE critics into deal-making compromisers. First, as Owen later put it, the adversaries went through the inevitable transitions from venting to dealing.[23] Second, in each region, those who led the complex post-CORE negotiations made full use of the leverage provided by the argument that the cabinet was determined to act, regarded Owen's report as a credible default option, and would not hesitate to impose it if the region's stakeholders could not agree on ways of improving it.[24] Third, the cabinet assuaged forest industry opponents by stressing that the Special Management Zones were not parks-in-waiting. Finally, as noted, Harcourt's pledge to provide special assistance to forest workers threatened by land use changes firmed up considerably in 1994, giving the government's agents some important new chips to use in the negotiations.

Cabinet approval of final land use plans in the four CORE regions added about 1 million hectares to the protected areas system, boosting the protected areas percentages to 13 percent on Vancouver Island, 12 percent in Cariboo-Chilcotin, 11.3 percent in West Kootenay-Boundary, and 16.5 percent in East Kootenay. Each of the plans also assigned sizable areas to the Special Management Zone category.[25] By 1996, 147 areas encompassing more than 4 million hectares had been so designated.[26]

The Harcourt government's increasingly confident use of the levers of power was also reflected in its readiness to make a number of stand-alone decisions on wilderness areas. It finished off some old business by announcing preservation of the Khutzeymateen Valley north of Prince Rupert, and the Stein in the southern Interior. Two large areas – the Tatshenshini-Alsek (946,000 hectares) in the remote northwest corner of the province and the Kitlope watershed (317,000 hectares) on the Central Coast – were preserved as a result of one-of-a-kind, cabinet-led evaluation and negotiation processes. In each case, particular combinations of factors convinced Harcourt and his advisors that it would be politically profitable to move in advance of regional planning. Cabinet also decided to preserve the 236,000-hectare Ts'yl-os (Chilko Lake) area in the Cariboo-Chilcotin, moving ahead on the advice of a local stakeholder group set up before the CORE regional process began.

In all, over 2.7 million hectares were added to the protected areas system during the Harcourt years, increasing the proportion of the province protected from about 6 percent to over 9 percent.[27] As the next section shows, the additions – and the lesson learning – continued after Harcourt stepped aside.

Policy Implementation, Evaluation, and Adjustment after 1996

The policy decisions just reviewed reflected monitoring, evaluation, and learning processes that kicked into gear as the NDP's land use planning experiments entered the implementation phase. These processes were guided by political considerations. The cabinet drew lessons concerning what would work, what it could get away with, and which investments of political capital would pay the best returns. Its reflections on these and related questions were strongly influenced by its concern to hold both the IWA and the green components of its support base, as well as by increasing worries about the threat to the industry's health posed by the environmental coalition's boycott campaign.

The government's assessment of the political situation continued to change after 1996. During the Harcourt years, worries about losing the support of forest workers increased after the vociferous anti-CORE demonstrations of 1994 as well as in response to the rise of the Reform Party in rural BC. Rather than causing these fears to dissipate, the decline of the Reform threat after 1996 created a new source of anxiety – the prospect that there would be no splitting of the "free enterprise" vote the next time British Columbians went to the polls. This worry, combined with those resulting from a prolonged forest industry slump and the government's general unpopularity, had a major impact on the political context shaping the implementation and adjustment of land use initiatives during the second half of the decade.

For environmentalists, the political context after 1996 was discouraging. The movement continued to work on a variety of fronts. Groups such as the Sierra Club and Greenpeace devoted considerable effort to preserving large, intact areas of temperate rain forest on the Central Coast. The Western Canada Wilderness Committee (WCWC) and others worked doggedly to force reconsideration of the decision to allow logging in key parts of the 260,000-hectare Randy Stoltmann Wilderness proposal area northwest of Whistler.[28] By early 2000, one of these areas – the Elaho Valley – had become the focus of a particularly bitter battle. On more general fronts, the movement tried to focus attention on the issues of ecosystem representation, the inadequacy of the 12 percent target, and the meaning of "special" in the Special Management Zone category.

In these and other areas, however, members of the environmental coalition found it increasingly difficult to match the electoral leverage of forest workers. The perception in the environmental camp that concerns about the economic health of the industry would dominate the Clark team's calculations led to a pronounced tactical shift from domestic politics to external, market campaigning. A central environmental actor may have somewhat exaggerated the magnitude and impact of this shift when she told a reporter in late 1999 that "the government is irrelevant; it is the marketplace. We give Home Depot 25,000 post cards. Home Depot responds."[29] But with such major customers deciding to phase out sales of wood from forests labelled endangered by environmentalists, threats of consumer boycotts were certainly figuring prominently in industry and government deliberations by the end of the decade.[30]

Although the Clark government ignored environmentalists' calls for it to revisit the 12 percent target or other parameters shaping land use decisions, it did not abandon the land use planning and protected areas expansion processes set in motion before 1996. The next section will review the NDP's continued experimentation on these fronts and then consider its handling of the Special Management Zone, ecosystem representation, and compensation issues.

The LRMP Process Matures

By April 2000, eleven reports from LRMP tables had been finalized and approved by cabinet. Another seven LRMP processes were in motion.[31] Taking into account these processes and the CORE area plans already approved, the government estimated that plans had been approved or were being developed for more than 80 percent of British Columbia.[32]

Although LRMP processes have been far from routine for the many communities that have been involved, these processes have attracted little notice in the provincial media. Being out of the spotlight has helped ensure an atmosphere conducive to local autonomy and compromises. Questions such as which interests should be represented at the table and which approaches best promote the search for consensus have been answered in different ways in different regions.

Not surprisingly, the plans generated by the tables have been shaped by different ideas about what spectrum of resource use possibilities should frame zonation decisions. The Bulkley plan, for example, delineates six zones: agriculture/wildlife, settlement, integrated resource management ("allows for a full range of resource uses and activities"), special management 1 ("exclude[s] all industrial activities except mineral exploration, mining and related infrastructure needs"), special management 2 ("contains key non-industrial values such as visual quality, wildlife habitat, recreation, and sensitive soils ... The impact of industrial activity on these key values must be minimized in order to receive approval"), and protected areas. To take another example, the plan for the Vanderhoof area sets out five zones: settlement/agriculture, resource development emphasis, multi-value emphasis ("manages with specific strategies to integrate a wide array of resource values"), special ("manages for the conservation of one or more resource values such as habitat, scenery and recreation, while still enabling resource development activities"), and protected areas.

LRMP deliberations on protected areas are constrained by LUCO-vetted lists of candidate areas and directives on targets. The candidate lists bear the imprint of preparatory work done by several ministries as far back as the Parks and Wilderness for the 90s exercise, and they have been refined over the years since the late 1980s through interagency negotiation. The protected areas targets assigned to each LRMP area are negotiable, but they do reflect deliberations by LUCO and the deputy ministers committee on how the total 12 percent "budget" should be apportioned across regions. The protected areas totals for the LRMPs completed by the year 2000 vary considerably. For example, the Vanderhoof plan allots 6.8 percent to protected areas, and the Bulkley plan 5 percent. At the other extreme, the Fort Nelson table designated 10 percent as protected area, combining with its Fort St. John counterpart to recommend creation of the 1.17-million-hectare Northern Rockies (Muskwa-Kechika) protected area.[33] By the end of the 1990s, cabinet-approved LRMP plans had boosted the protected area total past 10.75 million hectares, or about 11.4 percent of British Columbia.[34]

The LRMP processes still in operation are grappling with a number of important protected areas issues. The most difficult of these are under consideration on the Central Coast. Environmentalists, as noted, have put particular emphasis on this 4.75-million-hectare region, which includes the large temperate rain-forest area they have dubbed the Great Bear Rainforest. The Sierra Club, Greenpeace, and other organizations initially boycotted the Central Coast LRMP process, arguing that logging should not be allowed to proceed while the table deliberated and that, because 11 percent of the area is already protected, the fate of the remaining old-growth valleys would be predetermined by a 12 percent cap. The process began in 1997 without environmental group representatives, but Greenpeace and the Sierra Club relented after receiving assurances from the premier that protected areas in the planning area would not be capped at 12 percent and from forest companies that they would delay plans to log key valleys.[35] By spring 2000, events at the Central Coast LRMP table were eclipsed by news that three major environmental organizations and six large companies operating in the area had been trying to negotiate a deal that would give the companies respite from international market campaigns in return for an eighteen-month halt to logging and further talks on environmentally sensitive logging approaches.[36] The two sides were forced to postpone these negotiations and undertake some fence-mending when the provincial government and excluded stakeholders (including the area's First Nations peoples) reacted negatively to being left on the sidelines.

Some protected areas advocates have emerged from their LRMP experiences feeling quite positive. For example, a key player in the movement's Northern Rockies victory describes one facet of the Fort St. John LRMP process, noting that "after numerous philosophical discussions, table poundings, and eventual off table meetings," the group had reached a consensus on the difficult issues posed by the Graham River watershed. The upper portion of the watershed, though high in timber values, would be placed in a protected area, while the lower portion, though high in wilderness and guide outfitting values, would be placed in a Special Management Area that would allow logging under special guildelines. "This solution," he said, "was truly innovative, focusing on the concept of 'sequential logging.' This system was arrived at with the realization that the greatest threat to wilderness and wildlife values was often the roading that accompanies logging, rather than the logging itself ... These roads continue to act as an open wound, from an environmental point of view, as they admit the negative effects of access, essentially forever ... [With sequential logging], after each 'cluster' is logged, all roads are reclaimed and re-countered so no motorized access is created."[37]

Not all LRMPs exercises have generated such positive reviews from environmentalists. In 1998, for example, local conservation groups protested what had transpired in the process of deciding the fate of the Robson Valley LRMP area, a 1.3-million-hectare expanse southeast of Prince George. According to spokespersons for the newly formed Fraser Headwaters Alliance, the draft plan failed to protect the area's scenic, wildlife, salmon, and old-growth values. In their estimation:

> The Robson Valley LRMP process failed because of the flawed nature of the process and subsequent changes to the document made behind closed doors ... The Table was started after an assessment from an independent facilitator determined that there was strong community consensus for a "bottom-up" approach. However, the process gradually became more and more "top-down," as representatives of government ministries continued to intervene in the process, and change the rules from month to month ... Rather than waiting for Table recommendations regarding protected areas, the government set percentage targets for the LRMP, which became upper limits ("caps"), although the Table did not agree to this ... The government further restricted the amount of area that could be protected by including Mt. Robson Provincial Park in the Robson Valley's percentage target – this despite the fact that Mt. Robson is largely "rock and ice" and does little to contribute to the protection of representative ecosystems and biodiversity ... Even outside of protected areas, the Table was restricted as to areas that it could recommend for special management by a "percentage game" handed down as policy dictated by the Forest Practices Code.[38]

As the next section indicates, issues to do with the meaning of the Special Management Zone designation became a major focus of debate as attention turned to implementation of the land use plans.

Jockeying over the Meaning of "Special" in the Special Management Zones

Taking their cue from the government's 1993 Clayoquot Sound land use plan, all of the regional planning exercises made extensive use of the Special Management Zone (SMZ) designation. In so doing, they illustrated how leaving key issues fuzzy can facilitate consensus. From the outset, those responsible for arranging compromise solutions recognized the advantage of defining this category ambiguously enough to allow the competing camps (or at least optimists in these camps) to reach sanguine readings. The government was happy to avoid, at least temporarily, detailed specifications that would be bound to leave one or other side unhappy. As Ric Careless points out, the SMZs "were critical to achieving success at the negotiation table. Often, the Special Management Zones were the tool that provided the means to break key deadlocks and bridge the gap between industry and conservationists. In many respects these Special Management Zones epitomize the good will and faith of those individuals who had the courage to sit down and negotiate with their adversaries."[39]

By early 2000, the government had designated about 275 SMZs, encompassing about 11 million hectares, or about 21 percent of the total area covered by approved plans.[40] According to the government's projections, SMZs would eventually account for 19 million hectares.

Confusion and disagreement over what practices and planning procedures should apply within SMZs grew as the numbers mounted. Environmentalists' concerns were elaborated in Jim Cooperman's 1998 assessment, *Keeping the "Special" in Special Management Zones: A Citizens' Guide*. After carefully reviewing accounts from environmentalists familiar with early SMZ planning experiences in various regions, Cooperman concluded that management in many SMZs "is too often 'business as usual.' Problems include continued high rates of logging, continued use of clearcutting, inadequate respect for non-timber values, and the dearth of more detailed, long term planning."[41] To put SMZ management on a track more in keeping with environmentalists' expectations, Cooperman said, the government should commit to the use of ecoforestry practices, in particular the variable-retention silvicultural system recommended by the Clayoquot Sound Scientific Panel.[42]

Faced with these arguments and a countervailing set voiced by industry interests concerned about uncertainty and potentially high costs, the government asked a special working group made up of officials and stakeholder representatives to advise it on the implementation of SMZ objectives. In its draft report, released in early 2000, this group recommended several ways of improving planning processes. Although these recommendations may lead to change, even a moderately cynical interpretation would be that the group's main function is to allow the government to put off having to clarify SMZ guidelines. Assuming continuation of the present government's focus on creating jobs, maintaining timber supply, and reducing industry costs, future outcomes will probably be disappointing to those seeking to ensure that the SMZs are governed by strong ecosystem-protection rules. The experience of environmentalists who have carefully tracked implementation of the Vancouver Island Land Use

Plan is instructive. They have ended up dissatisfied not only with the handling of SMZs but also with what they regard as timber-centred management regimes set up to guide operations in the zones designated for "high intensity resource development."[43]

Attempts to "Harden" the Ecosystem Representation Goal

The political realities after 1996 have also worked against the environmental coalition's attempts to focus attention on the issue of ecosystem representation. As noted, the movement had embraced this notion in the late 1980s, using it as a foundation for its arguments about the importance of preserving old-growth valleys. The rocks-and-ice bias of the parks system, environmentalists argued, had to be reversed. Thanks in considerable part to the efforts of sympathetic officials, the goal of improved ecorepresentation was featured in the Protected Areas Strategy endorsed by the Harcourt government. From the outset, though, the government was wary of environmentalists' efforts to have this treated as a hard goal. It kept the focus on the global 12 percent target, adding when called upon that it was making progress toward greater protection of underrepresented ecosystems.

Environmentalists trying to keep the ecosystem representation issue from sliding down the agenda did receive a boost from a comprehensive evaluation conducted by LUCO staff near the end of the NDP's first term and from updated figures released in 1999.[44] Using both of the ecological classification systems customarily employed in the province, the LUCO staffers showed that, despite some improvements, many of the province's ecosystems remained poorly represented in the parks system. Table 2.1 summarizes results for biogeoclimatic zones, indicating, for example, that the proportion of the alpine tundra zone protected increased from 10.9 percent in 1992 to 20.8 percent in 1999, whereas the proportion of the Interior cedar–hemlock zone protected went from 6.1 percent to 8.8 percent. Skewing in favour of the alpine tundra zone and other high elevation zones (mountain hemlock, Engelmann spruce–subalpine fir and spruce–willow–birch) continues; in fact, over 60 percent of all territory designated between 1992 and 1996 was from these zones.[45] Forested ecosystems such as the Interior cedar–hemlock and the Interior Douglas fir zones continue to be underrepresented. In addition, the proportion of subalpine old-growth forest preserved is substantially higher than the proportion of low-elevation old growth. One recent analysis concludes that, although the proportions of nonforested and high-elevation forested areas protected are 19.8 percent and 14.2 percent, respectively, the comparable figure for low-elevation forest areas is only 6.3 percent.[46]

Environmentalists' inability to parlay evidence like this into pressure for stronger ecosystem representation should not be attributed solely to forest industry resistance and government bias toward low-cost protected areas

Table 2.1

Percentages of biogeoclimatic zones protected, 1991-99

Zone	Proportion of zone protected		
	1991	1996	1999
Alpine tundra	10.9	17.4	20.8
Spruce – willow – birch	10.9	11.2	16.2
Boreal white and black spruce	1.9	2.7	4.7
Sub-boreal pine – spruce	2.8	4.8	6.2
Sub-boreal spruce	3.2	3.3	5.3
Mountain hemlock	8.9	10.8	12.1
Engelmann spruce – subalpine fir	9.1	11.5	13.7
Montane spruce	2.5	7.1	7.1
Bunchgrass	0.4	7.8	8.0
Ponderosa pine	0.5	2.4	2.4
Interior Douglas fir	1.8	4.0	4.0
Coastal Douglas fir	0.9	2.1	2.6
Interior cedar – hemlock	6.1	7.6	8.8
Coastal western hemlock	6.0	8.9	9.5

Source: Karen Lewis and Susan Westmacott, *Provincial Overview and Status Report* (Victoria: Land Use Coordination Office, 1996), Table 13. Updated figures supplied by LUCO, June 1999.

options. Slow progress here also reflects the fact that it is easier to mobilize public support for straightforward goals such as the 12 percent target than for more complicated ones like ecorepresentation. As well, the gravitational pull toward higher-elevation territory has a number of roots, including some that highlight the wilderness movement's diverse goals. Environmental groups responding to their constituents cannot be immune to the fact that, as an evening's dose of television, beer, and SUV ads reminds us, high-elevation spaces are cherished within North American culture for their scenic and recreational values. As well, the environmental coalition's different biodiversity preservation goals do sometimes conflict. For example, although its ecological characteristics are already well represented in the parks system, the vast alpine and subalpine terrain of northern BC provides some of the best remaining opportunities on the face of the planet to preserve wilderness areas big enough to allow interactions between large predators and their prey to unfold relatively free from human influences. The Harcourt government's addition of the 946,000-hectare Tatshenshini-Alsek protected area was defended in part on these grounds, even though its addition significantly increased the preexisting overrepresentation of the alpine-tundra biogeoclimatic zone.

These points aside, the wilderness movement will continue trying to focus attention on the underrepresentation of low-elevation ecosystems. Ecorepresentation arguments will be featured prominently in specific campaigns, such as those aimed at preserving the remaining intact old-growth valleys on the Central Coast as well as in general efforts to convince government and the public that adequate biodiversity conservation measures cannot be accomplished if the 12 percent target is treated as a cap.[47] Once again, though, those pushing on these fronts could take little encouragement from the political winds prevailing during the final years of the decade.

Facing the Compensation Issue

Another set of worries for environmentalists was unleashed in 1999 when the other shoe dropped on the issue of what compensation was owed companies that had lost cutting rights as a result of protected areas decisions. As noted, the Harcourt government had briefly considered the issue and then decided that it could be put off. After mothballing Schwindt's 1992 report, the government gingerly engaged the industry in exploratory talks. The issue emerged from the backrooms in September 1997, when MacMillan Bloedel petitioned the Supreme Court of British Columbia, seeking a declaration that the province was obliged to pay compensation for harvesting rights it had removed when it created several new parks on Vancouver Island. According to the petition and accompanying documents, the company had believed that negotiations were proceeding satisfactorily until August 1997, when it received a letter from an official in the attorney general's ministry stating the province's position that the company was not entitled to compensation.[48]

Because it was wary of enunciating this position in legislation, the government had no option but to return to the bargaining table. Sixteen months of complicated negotiations ended in April 1999 with the announcement that the government had agreed to compensate MacMillan Bloedel for harvesting right reductions in excess of the 5 percent allowed under the Forest Act.[49] The company would receive $84 million for "foregone net income," with this amount to be paid in resource rights and land and/or cash. MacMillan Bloedel expressed a willingness to accept a noncash package that would give it fee simple ownership of some Crown land (32,000 hectares) and allow it to remove some land it already owned from Tree Farm Licences (about 91,000 hectares). Clearly, the company put considerable value on the opportunity to free some of its operations from the profit-limiting rules governing forest practices and stumpage on Crown timber.

Although its budgetary problems disposed it toward the land-package option, the government said that no decision on how to compensate the company would be made without public consultation. A government emissary would consult representatives of First Nations, local governments, interest groups, and others as well as listen to reaction at public meetings in several communities. In responses that were no doubt influenced by Deputy Premier Dan Miller's public musings about the time having come to consider privatizing more forest land,[50] environmental groups condemned the land-package option. It would, said the former executive director of the Sierra Legal Defence Fund, open "the floodgates to privatization."[51] The news that the government was also negotiating compensation settlements with other companies raised more general concerns among environmentalists about the chilling effects on old-growth preservation campaigns.[52] According to the MOF, however, the total bill for compensating forest companies (including MacMillan Bloedel) was expected to be $150 million, a figure much lower than estimates bandied about earlier.[53]

In August 1999, the government's emissary reported strong public opposition to the idea of a land swap: "People were horrified at the idea of Crown land being used to settle this claim."[54] The minister of forests said he had gotten the message and would recommend the cash settlement route to his cabinet colleagues.[55]

It is unfortunate that the decade passed without a full public debate of the issues so expertly introduced in the Schwindt report, particularly the fundamental question of whether the province should be obliged to compensate.[56] On the other hand, if it turns out that the total compensation bill is "only" in the $150 million range, the NDP will be justified in claiming that the Harcourt government's decision to adopt a backroom bargaining approach was sensible. This approach enabled it to circumvent what had, in the years before its election, seemed to represent a major obstacle to expansion of the protected areas system.

Conclusion

The package of outcomes described in this chapter does not meet our test of paradigm change. The values and ideas at the centre of the biodiversity discourse have attained a prominent place in forest land use planning deliberations, but decisions continue to be heavily influenced by traditional economic considerations. To cite some hypothetical examples of changes that might be deemed paradigmatic, the power to rezone the forest land base has not been turned over to a committee of conservation biologists or to committees of local citizens. The Clayoquot Sound model has not been extended across the coast. The cabinet still controls the decision-making process, and the Ministry of Forests still dominates the advice-generation process.

The scope and speed of the changes in the land use planning subsector are, nonetheless, striking. The NDP governments of Mike Harcourt and Glen Clark designed and implemented a sweeping reform of the land use decision-making structure. The regional planning system put in place represents a significant response to demands for greater community input. As this system matured, hundreds of ordinary British Columbians took advantage of new opportunities to influence decisions on how the land base of their regions should be zoned. Few of those who poured time into these exercises would agree totally with the results or express full satisfaction with the process. Most, however, would agree that the CORE and LRMP tables gave them much more meaningful opportunities to influence outcomes than had existed in the preceding decades. Ongoing events in places such as the Elaho and the Slocan Valley remind us that it is too early to proclaim the arrival of "peace in the woods," but in many areas the level of acrimony does seem to have dropped.

The new land use planning processes also reflect the arrival of a more rational approach. Piecemeal consideration of protected area candidates driven by levels of political noise has been replaced by comprehensive examination of the full land base, involving more thorough application of scientific knowledge. In most cases, for example, members of land use planning tables have been able to draw on the results of sophisticated "gap analyses" to assess the potential contributions of different protected area candidates to ecosystem representation goals.

Substantial process change was matched by significant substantive change. Nearly 5 million hectares were added to the parks system in the 1990s, bringing the total area protected to over 10.75 million hectares or 11.4 percent of the province. By early 2000, another 11 million hectares had been set aside in Special Management Zones, with projections suggesting that the completion of land use planning across the province would bring this total to 19 million hectares, or 20 percent of the province.

The limitations of the NDP's response to the environmental coalition's protected areas aspirations are evident in the figures on ecosystem representation shown in Table 2.1, as well as in guidelines governing timber supply impacts that were included in the SMZ components of various land use plans. Although some significant expanses of old-growth forest were protected, low-lying forest ecosystems continue to be underrepresented. Completion of the Central Coast planning process will likely improve the situation, but it is clear that accomplishment of environmentalists' goals for old-growth protection would require an upward adjustment of the global 12 percent target. Meanwhile, environmentalists looking for some consolation from the Special Management Zones have encountered something akin to the 6 percent cap so crucial to the story related in the next chapter. Here, the nature of the "numbers games" has varied from region to region. Some plans set resource targets for SMZs, whereas others include provisos that the timber impacts of SMZ management prescriptions must not exceed a specified percentage.[57] For example, the Cariboo-Chilcotin plan says that 70 percent of productive forest land in SMZs must be available for harvesting. The Vancouver Island plan says that AAC levels in SMZs are not to be reduced by more than 10 percent. Such provisions have naturally disappointed those hoping that logging in SMZs would be in accordance with low-intensity, ecosystem management approaches such as those proposed in the Clayoquot Sound Scientific Panel recommendations on variable retention.

The policy changes that did occur in the land use subsector can largely be attributed to change-promoting regime alterations in the 1980s and to the Harcourt government's deft management of a number of key regime elements. The years preceding the NDP's election were marked by important changes to the ideational context, the institutional setting, and background political conditions. Even striking packages of favourable preconditions, however, do not translate easily into policy change. The 1990s initiatives were accomplished despite some powerful countervailing forces, the neutralization of which required careful manipulation of various dimensions of the subsector regime.

The environmental coalition can be credited with a central role in bringing about the key regime alterations. Its campaigns left a mark on background conditions, institutions, and ideas. It effectively challenged the sustained yield-integrated management paradigm, undermining public faith in MOF-industry stewardship. It promoted the ideas at the heart of the biodiversity discourse, expanding public support for the notion that the protected areas system should be both much larger and much more representative of the province's ecological diversity. The environmental coalition helped convince actors throughout the forest policy community that piecemeal consideration of wilderness issues had to be replaced by a comprehensive consideration of how best to zone the land base. And along with various allies, it raised public doubts about the credibility and competence of the actors who had traditionally dominated forest policy decision making. These and associated changes put MOF and industry leaders on the defensive, convincing more and more of those who had led resistance in the 1970s and early 1980s that concessions were now required.

Those who feel the NDP's land use planning accomplishments failed to sufficiently expand old-growth protection will, of course, be more interested in the factors that limited the degree of change. Our interpretation here, as at the end of the next chapter, returns to the central point made in Chapter 6 concerning the critical timber supply subsector. The forces described there spilled over into other subsectors, manifesting themselves as obstacles that constrained possibilities wherever the NDP grappled with how to make good on its forest environment objectives. These obstacles can be tied directly to the industry's power to minimize reductions in its access to high-value timber. Across the decade, proposals for protection of prime old-growth timber (and for low-intensity logging in the SMZs) continued to be met by resistance from the industry's workers, investors, and community supporters. This resistance was generally tempered by the fact that, in at least some corners of the development coalition, protected areas concessions were seen as providing a relatively painless way of containing environmentalism and relegitimating MOF-industry control. Nevertheless, as environmentalists participating at the various land use tables have invariably been reminded, forest industry interests are strongly committed to continuation of old-growth liquidation policies and can be counted on to oppose reductions to the scope and intensity of logging operations. And, as environmentalists do not need to be reminded, the structural advantages of the forest industry continue to ensure that government listens very carefully to these concerns.

All things considered, the NDP government responded effectively to the challenge of resolving conflict over the forest land base. It designed its instruments well, giving the regional land use advisory processes enough latitude to ensure that they would retain credibility and be able to identify whatever potential for community consensus existed. This, however, remained experimentation on a leash. The cabinet and its close advisors maintained a careful watch over the regional tables as policy implementation kicked into gear. Learning quickly from the early results, these officials were soon confidently intervening to fill policy gaps and adjust settings. Members of the Harcourt team recognized that the path to land use consensus would have to be greased with dollars; forest workers threatened by change would have to be appeased. Most important, they recognized that, more often than not, the results generated by advisory processes would have to be massaged into final outcomes through some traditional behind-the-scenes bargaining.

Events in this subsector, then, reflect successful efforts to manage a situation that, because of a concatenation of developments in the previous decade, was ripe for change when the NDP took over in 1991. The NDP came to power with clearly defined protected areas objectives and a strong belief that it could win political points by making good on its promise to bring peace to the woods. In their pursuit of these objectives, the Harcourt and Clark governments took an experimental approach. They devolved considerable authority to regional land use processes, and they then quickly adjusted instruments and settings in response to lessons learned as these unfolded. This approach provided a striking illustration of the importance of one feature of the institutional setting emphasized in our introduction to the policy regime. At least in this subsector, cabinet exercised tight control.

3
The 6 Percent Solution: The Forest Practices Code

Forest practices are those activities designed to mitigate the undesirable consequences of timber harvesting. Harvesting can disrupt fish and wildlife habitat, cause soil erosion, and reduce the attractiveness of areas for tourism. If inadequate attention is paid to reforestation, harvested areas may not regenerate adequately. In BC, the practice of clearcutting – harvesting all the trees in a given area at a time – has been a lightning rod for controversy for more than a decade. This chapter examines the changes in forest practices regulation resulting from the enactment and implementation of the 1994 Forest Practices Code, one of the most substantial and far-reaching changes to BC forest policy in the 1990s.

To assess the magnitude of change in forest practices, we will examine policy goals and objectives, the nature of instruments used, and the settings of those instruments. To the extent possible, we will also examine changes in the various consequences of implementing forest practice regulations. Although information on consequences is relatively limited, there are data on costs and some indicators of on-the-ground performance. Forest practices are complex. To simplify the analysis, we will focus here on two extremely important areas – the regulation of clearcutting and the rules for protecting fish habitat – and examine how they were changed by the introduction of the Code.

This chapter shows that there have been substantial changes in the regulation of forest practices in British Columbia. There has been a pronounced shift in goals and objectives toward greater consideration of environmental values; for example, the principal goal of the Code is "balancing economic, productive, spiritual, ecological and recreational value of forests." The instruments for regulation have been formalized. It is harder to judge how much settings have changed, but in at least some cases they have been made more stringent. Although there has been a shift toward greener forest practices, the government capped the magnitude of change, deciding to limit the impact of the Forest Practices Code on the allowable annual cut to no greater than 6 percent. That percentage is a very appropriate indicator of the magnitude of change in this subsector. In terms of meeting its objective of finding an appropriate balance of the various interests in forests, the Forest Practices Code has failed. Indeed, the Code has contributed to the essential dilemma of BC forest policy: it has escalated costs so that the industry has trouble competing economically, yet its environmental performance has not been satisfactory to influential political interests.

After reviewing some of the regime characteristics distinctive to this policy subsector and a brief history of the policy area, the chapter reviews each stage of the policy cycle in some detail. Throughout this process, the fundamental regime forces at work are revealed. The chapter also chronicles the emergence of a powerful new force: private certification organizations that are coming to play a significant role in corporate forest practices. The conclusion examines the magnitude and type of policy change, and the relative importance of various regime forces in bringing about that change.

Subsectoral Regime Characteristics

Forest practices are subject to many of the general forest policy regime characteristics outlined in Chapter 1. There are a significant number of actors with a keen interest in forest practices.[1] Elected officials play an important role, especially when significant issues are at stake as they were in this case. Because of the technically complex nature of the area, bureaucrats are extremely influential in both the development and the implementation of forest practices. As a result of the Code, one of the most important changes was a reduction in the monopoly position of the Ministry of Forests and an expanded role for the Ministry of Environment, Lands and Parks. Industry groups have intense interests and play an exceptionally active role. Environmentalists, both domestic and international, have been very influential, particularly on evocative issues such as clearcutting. Unlike some other forest policy sectors, neither labour nor First Nations have assumed a very active role.

Like other areas of forest policy, provincial jurisdiction is paramount. However, the links between forest practices and fish habitat, including anadromous fish such as salmon, gives the federal government some jurisdictional clout through the federal Fisheries Act.[2] The provincial cabinet and bureaucracy are the dominant institutions; neither the legislature nor the courts have played a meaningful role.

One extraordinary feature of this policy subsector is the strategic innovations of environmental groups. Frustrated with limited channels of influence within the system of executive-centred bargaining, environmentalists developed the strategy of appealing to the consumers of BC forest products in the international marketplace to cease purchasing products that they argued derived from unsustainable practices.[3] This international campaign originated with the Clayoquot Sound controversy in 1993, and it has escalated significantly since then. This strategy was highly effective: it gave business an economic interest in improving forest practices. Later in the decade, environmentalists supplemented the boycott pressures with demands for independent certification of forest products, a development that, as described later, may have profound implications for the structure and operation of the policy regime in this subsector.

Because of its relatively technical nature, the policy regime in this area tends to be more narrow and closed than the one in land use planning.[4] Although interest groups, including environmental groups, have participated and influenced consultation processes, policy formulation has been dominated by expert bureaucrats.

In terms of ideas, forest policy has evolved from a nearly exclusive emphasis on timber values to a more complex weighing of timber with other values such as recreation and wildlife protection. Since the 1970s, the dominant idea has been the central model guiding forest practices regulation: integrated resource management. Under this model, multiple values of the forest – timber, wildlife, recreation, and so on – are pursued in the same area. The key task for managers is to strike an optimum balance among these competing values. More recently, several alternatives to this dominant model have emerged.

One alternative approach is intensive zoning. This approach differs from the historical model, in which all areas were managed almost exclusively for timber. The modern version argues that integrated resource management is inherently limited in its ability to provide multiple values; a superior approach would be more intensive zoning where different areas of the forest are zoned for different resources. For example, intensive industrial timber harvesting would be promoted in one large area, and other large areas would be set aside for wilderness and recreation.[5] A second alternative approach, advocated by environmentalists, is ecosystem management. In this model, multiple uses, including logging, would still be allowed, but the dominant value in all areas would be the maintenance of ecosystem function.[6] Although integrated resource management is still the dominant model, it is increasingly contested by these two alternatives.

Historical Background[7]

Before the 1990s, the dominant goal of forest policy in the province was unquestionably timber harvesting. As a result, most of the concern over forest practices after the Second World War focused on reforestation; that is, the application of harvesting and silvicultural techniques to produce the desired regeneration of future stands.[8] Despite the increase in environmental pressures beginning in the late 1960s, environmental values did not explicitly get incorporated into governing provincial statutes until 1978.[9] The core statute, the Forest Act, embodied only an oblique reference to environmental values, allowing them to be considered among the "purposes other than timber production" in the determination of the allowable annual cut (Section 7[3][a][v]). The companion Ministry of Forests Act was more explicit. In the five objectives describing the "purposes and functions" of the Ministry of Forests, only one directly mentioned environmental values, and even then it did so in the context of balancing them with industrial values. Section 4(c) directs the ministry to "plan the use of the forest and range resources of the Crown, so that the production of timber and forage, the harvesting of timber, the grazing of livestock and the realization of fisheries, wildlife, water, outdoor recreation and other natural resource values are coordinated and integrated."[10]

Before the Code was enacted, the principal instrument for the regulation of forest practices was the licence agreements between forest companies and the government. Unfortunately, there are no systematic studies of forest practices in this period, and any assessment is greatly complicated because practices were governed by a bewildering array of statutes, regulations, guidelines, and specific licence agreements. In some areas and for some resources, regional guidelines were used as the basis for provisions in licences. In the case of clearcutting, there were no limits in statute or regulation on clearcut size. As a result, limitations on clearcutting varied significantly from area to area, in both the average and the maximum size of clearcut allowed, and in the regulatory force of the limits. For instance, the Okanagan Timber Supply Area plan contained a limit on the maximum average cutblock size of less than 30 hectares but allowed clearcuts up to 50 hectares.[11] On the coast, the initial 1972 Coast Planning Guidelines established a maximum cutblock size of 80 hectares, but, as a result of "sympathetic administration," this limit was not enforced after 1980.[12] The 1992 Coast Harvest Planning Guidelines stated that new cut blocks "should average 40 hectares." Before 1992, there were also no limits on adjacency; that is, how soon an area next to an existing clearcut could be logged. The 1972 guidelines required "leave areas" between cutblocks, but that guideline was not enforced after 1980. The 1992 guidelines for the Vancouver region required that clearcut blocks reach "free growing stage" before adjacent blocks could be logged.[13]

The protection for rivers and streams, or "riparian areas," was similarly governed by nonbinding guidelines with substantial regional variation. Coastal areas were governed by the British Columbia Coastal Fisheries Forestry Guidelines, initially established in 1988 and updated in 1993. These guidelines were established in a classic bargaining process between the federal Department of Fisheries and Oceans; the BC Ministry of Forests; the BC Ministry of Environment; and the industry trade group, the Council of Forest Industries (COFI). The guidelines defined four classes of streams, each class being associated with specific fisheries values and forest management objectives. The guidelines set out "streamside management zones" that are 10 to 30 metres wide, depending on the stream width. Within the first 10 metres adjacent to the stream, harvesting was not permitted, though exceptions to prevent destructive windthrow were allowed. Partial cutting was permitted in the rest of the zone, subject to the relatively restrictive requirement of maintaining "original stand characteristics." There were no regionwide guidelines for the BC Interior, but some specific areas had them. For example, the Okanagan Timber Supply Area had its own guidelines. The minimum width of a streamside management zone was greater (20 metres), but the limits on harvesting within the zone were far less stringent: there was no "no harvest strip," and harvesting within the zone could remove up to 50 percent of the preharvest closure.[14] Although these guidelines were nonbinding, they did attain greater legal force if they were included in stand-specific cutting plans.

Historically, timber-oriented goals dominated forest policy. The dominant mode of policy instruments was a complex and confusing mix of nonbinding guidelines and site-specific provisions contained in permits or licences. As a result of this mode of regulation, forest practices regulations were characterized by a limited or uncertain legal basis, substantial regional variation, and weak enforcement.

Policy Cycle

Agenda

It is frequently difficult to establish exactly when an issue gets on the governmental agenda. With the Forest Practices Code, however, there is evidence that the Code was on the BC government's agenda by July 1991. Before the election, Socred Minister of Forests Claude Richmond issued a "discussion paper" requesting input into which approach should be pursued. In the overview to the six-page brochure, Richmond stated in a letter, "I believe the question is not whether we need a code ... but how we can put a code into place." The letter concludes, "Based on your response, the Ministry of Forests will develop a forest practices code."[15]

In his model of agenda setting, Kingdon described three streams: politics, problems, and policies.[16] The first two streams are the most important in influencing the agenda stage; the third stream is more important in the formulation stage. The agenda-setting process for the Forest Practices Code was dominated by the politics stream. First, the jump in salience of environmental issues around 1990 gave significant momentum to environmental criticisms of BC forest practices. These pressures were strong enough to lead even a recalcitrant Socred government to initiate development of a code. Although the salience of environmental issues generally declined in the early 1990s, concerns over the environmental consequences of BC forestry persisted. Government-commissioned surveys of the attitudes of British Columbians revealed very strong support for tightening forest practice regulations.[17]

Second, the NDP election in late 1991 brought to power a government far greener in orientation than its predecessor. Indeed, the NDP platform explicitly called for the enactment of a forest practices act as part of its environmental agenda.[18] The existence of a new party in power lessened the chances that the government would reverse its plans to develop a code.

The third, and ultimately most influential, channel of the politics stream was the change in strategy by the environmental movement to shift the venue of forest politics from the province to international markets and politics. Environmentalists began laying the groundwork for the campaign in the late 1980s and early 1990s. One of the tactics was to frame BC forestry as the "Brazil of the North" to take advantage of the strong public sentiment against deforestation of tropical rain forests.[19] The campaign intensified dramatically in the wake of the Clayoquot Sound controversy during summer 1993, attracting considerable attention from prominent environmental groups and celebrities. Visits by the Australian rock band Midnight Oil and Robert Kennedy Jr., a representative of the American group the Natural Resources Defense Council, attracted a great deal of attention to the cause.

As described in Chapter 1, environmentalists pursued this strategy because the institutional structure of the BC forest policy regime limited their opportunities for influence. In the United States, environmentalists pursued a very effective campaign against the forest industry in the Pacific Northwest by relying on the courts and enlarging the scope of conflict to the national level. In BC, the discretionary statutory structure and provincial jurisdiction precluded those strategies, and instead environmentalists chose to "go international."

While the politics stream dominated the agenda process, development in the problem stream added fuel to the fire. In particular, the publication of two reports (the so-called Tripp reports) investigating the impacts of forest practices on salmon habitat embarrassed forest companies and the government and prompted calls for reform. An investigation of compliance with the fisheries guidelines on Vancouver Island exposed practices that led even the minister of forests to claim that he was "absolutely appalled" and that the practices were "completely unacceptable."[20] The report found that, when the guidelines were followed, they were reasonably effective, but compliance with the guidelines was generally poor, including those for streamside management zones. The report concluded: "There was, on average, one major or moderate impact on one stream for every cut block inspected."[21] A follow-up report examining the entire coastal region found similar levels of noncompliance.[22]

These fisheries reports were important not only because they contributed to the view that there were significant problems with the existing policy framework but also because they promoted a particular definition of the policy problems. The reports found that when the guidelines were implemented, they worked. The problem was that the guidelines weren't being followed in enough situations. The recommended solution was twofold: strengthen the legal force behind the standards and step up enforcement.

The politics and problem streams pushed changes to forest practices onto the BC government agenda in the early 1990s. The policies, or solutions, stream was dominated by one idea: consolidate the existing patchwork of guidelines and regulations into an overarching piece of legislation. The official genesis of the proposal seems to be the 1991 report of the Forest Resource Commission, appointed by the Socred government in 1989 in the wake of the "rollover" disaster (see Chapters 4 and 6). The report recommended "a single, all-encompassing code of forest practices through the introduction of a Forest Practices Act."[23] The Ministry of Forests July 1991 discussion paper explicitly mentions the report as the genesis for the idea. The NDP picked up on this idea and included it in its platform for the 1991 election.[24]

The proposal for a comprehensive legislative code emerged from a particular definition of the policy problem in the early 1990s. In a 1993 public document, the government defined the problems as follows:

- "Insufficient legal powers – lack of a single, consistently applied forest practices act."[25] At the time, forest and range management activities were said to be governed by 26 statutes, 700 regulations, and 3,000 guidelines. This bewildering patchwork needed to be consolidated and rationalized.
- "Lack of strong, up-to-date rules governing all areas of forest and range practices." Standards have been applied inconsistently across regions and in some cases do not exist at all.
- "Occurrences of poor and inconsistent industry performance": Basic stewardship requirements were therefore not met.
- "Inadequate monitoring and enforcement." Staff cutbacks in the 1980s reduced monitoring and placed greater responsibilities on companies for monitoring and reporting.
- "Weak penalties." In some areas the government lacked authority to fine noncompliance; in areas where fines were authorized, the maximum penalty of $2,000 was considered inadequate to deter violations.

- "Insufficient auditing": The government failed to sufficiently review company forest practices.[26]

In the context of heightened domestic and international political scrutiny of BC forest practices, the pre-Code framework created two specific problems. First, to the extent that the government did have standards in place, the exceptionally complex nature of the regulatory framework made it difficult to present and explain them to the world. Second, as the Tripp reports revealed, the nonbinding nature of the standards was undermining their effectiveness. The solution that emerged was a formalization of the regulatory framework into a comprehensive, statutory code of practice.[27]

Formulation

The process of formulating the Code began in the waning days of the Socred regime. In July 1991, the Ministry of Forests issued a forest practices code discussion paper. The Forest Resources Commission, which had already issued its major report, was charged with consulting relevant interests and recommending a framework for a code, which it did in July 1992.[28] At that point, the bureaucracy took charge, led by an interministerial Forest Practices Code steering committee established in May 1992.[29]

In November 1993, the government released a discussion paper along with proposed rules. The basic process was one of assembling and improving existing standards, and the government proposed to codify many of them in law. No great break from the past was envisioned, such as a move to an incentives or results-based approach.[30]

During the winter, consultations with interested groups were conducted, and the most formal was conducted by Gordon Baskerville, head of the Forest Resources Management Department of the University of British Columbia. One of the striking aspects of Baskerville's review is the level of support, from many quarters, for the basic concept of a code. For example, part of Baskerville's summary of the forest industry's comments states: "The Code is an essential element towards management of the forests, and maintenance of market share."[31] The reference to market share makes clear the industry's preoccupation with the rising pressures of green consumerism. Environmental groups making presentations criticized the proposed code for not going far enough in reducing or banning clearcutting and fully adopting an ecosystem perspective.

Despite these opportunities for input, interest groups viewed the process as dominated overwhelmingly by bureaucrats and the two lead ministers, Andrew Petter of the Ministry of Forests and Moe Sihota of the Ministry of Environment, Lands and Parks. Environmentalists were pleased that the Ministry of Environment was so involved in the process, but they were concerned – especially having become familiar with consensus-based processes in CORE and other environmental initiatives – that they were not given a broader role. Industry officials were taken aback when their requests to meet with ministers at various times were summarily rejected.

Decision Making

In applying the policy cycle model, it is frequently difficult to identify where one stage ends and another begins. This is perhaps most challenging in the case of decision making and implementation. For our purposes, we will consider decision making to consist of the establishment of the most important aspects of policy content. This approach requires going beyond enabling legislation to include the development of regulations. Without the latter, one can identify the policy objectives and legal framework but not the specific instruments or their settings.

Promising a "new regime of accountability," the government introduced the Forest Practices Code Act of British Columbia to the legislature in mid-May 1994. In doing so, it emphasized the enactment of "world class forest practices" and stiff enforcement penalties of up to $1 million per day. The government estimated the Code would cost between $250 and $300 million per year.[32] Draft regulations were issued later that month. In releasing the proposed standards, Minister of Forests Andrew Petter stated, "Unlike many current standards, the proposed new standards are mandatory and enforceable. They limit the size of cutblocks, protect community watersheds and will ban clearcuts where alternative harvesting systems are more appropriate. Ecological requirements will drive decisions on appropriate harvesting methods."[33] The Code was enacted by the legislature on 7 July 1994. On 12 April 1995, eighteen regulations and sixteen high-priority guidebooks were released, followed by others over the next month. The Code provided for a transitional period of implementation that began in June 1995 and was completed in June 1997.

The objectives of the Code are enunciated in the preamble:

> WHEREAS British Columbians desire sustainable use of the forests they hold in trust for future generations;
> AND WHEREAS sustainable use includes
> (a) managing forests to meet present needs without compromising the needs of future generations,
> (b) providing stewardship of forests based on an ethic of respect for the land,
> (c) balancing productive, spiritual, ecological and recreational values of forests to meet the economic and cultural needs of peoples and communities, including First Nations,
> (d) conserving biological diversity, soil, water, fish, wildlife, scenic diversity and other forest resources, and
> (e) restoring damaged ecologies.

Compared to the Ministry of Forests Act, far more weight is given to environmental values in the text, suggesting that the Code represents an effort to embody a stronger commitment to environmental values than had existed in BC forest policy up to that point.

Structurally, the Forest Practices Code has three layers: the act passed by the BC legislature, a series of binding regulations issued through order-in-council, and a set of nonbinding guidebooks meant to guide implementation of specific provisions. Functionally, the Code has three components: planning, forest practices, and enforcement.

Planning. At its heart, the Forest Practices Code is a framework for planning. Indeed, planning is a much greater focus of the Code than the regulations of forest practices as such. Higher-level plans are essentially zoning decisions about the emphasis given to values, and they include resource management zones, landscape unit objectives, and sensitive area designations. These plans guide the development of operational plans oriented toward more specific aspects of forestry. Originally, the Code provided for six operational plans: Forest Development Plans, Logging Plans, Silviculture Prescriptions, Stand Management Prescription, Five-Year Silviculture Plans, and Access Management Plans.[34] The central planning document was the Forest Development Plan (FDP), a five-year plan setting out all the proposed management activities for the area. The most important site-specific plan is the Silviculture Prescription (SP). One of the most important features of the Code Act was that once higher-level plans were in place, operational plans had to be consistent with them. The planning framework invests an enormous amount of discretionary authority in the district managers (there are forty-three in the province) of the Ministry of Forests, who are responsible for approving operational plans.

Forest Practices. Given the complexity and variability of the pre-Code provisions, it is difficult to assess precisely how the Code changed policies in specific areas of forest practices. Following through on our two specific areas of focus, we can see that there has been substantial change in both the formality of the instrument and the setting of the instrument. In the case of clearcutting, as described above, there were no clearly defined rules prior to the Code. In the Code's Operational Planning Regulation, specific legal limits were placed on clearcut size: 40 hectares in the coastal and southern Interior regions and 60 hectares in the northern Interior regions. In coastal regions, this regulation changed a guideline average of 40 hectares to a legal maximum of 40 hectares. In the Interior, legal limits were imposed where none existed before.

Stream protection follows a similar pattern. In coastal regions, nonbinding guidelines on required buffer strips and other features were transformed into binding regulations. In addition to formalizing the instrument, the settings on the instrument were also changed to be more protective of the resource. The Coastal Fisheries/Forestry Guidelines (CFFG) required streamside management zones for fish-bearing streams to be as long as channel widths on both sides of the stream. The guidelines stated: "Generally, all trees within 10 metres of the streambanks should be retained," but they allowed for exemptions with the approval of fisheries departments. As shown in Table 3.1, the Code increased the buffer strips significantly for streams greater than 1.5 metres in width. (Fish streams less than 1.5 metres in width – category S4 under the Code and often important coho habitat – get less protection under the Code than under the CFFG.) In addition to the no-harvest buffer strip, the Code provides for "management zones" beyond the "reserve zone" where limited logging was intended. Thus, the Code significantly increased stream protection on the coast. In the Interior, the provincewide standards in the Code meant even more dramatic change because formerly no regional guidelines were in place.

Although the stream protection rules apply provincewide, the *Riparian Management Area Guidebook* developed to supplement the regulations does contain recommendations that vary regionally. For instance, the "best management practice" suggested for small fish streams on the coast (S4) is to retain 50 percent of trees within 10 metres of the bank. In the Interior, the recommendation is to retain all trees within 10 metres of the bank.[35]

Enforcement. In addition to planning and practices, the Code also introduced a new system for enforcement and compliance, including a dramatic increase in the maximum fine, from $2,000 to $1 million. In its public relations on the Code, the government gave considerable emphasis to the stringent fines available as an indicator to their tough, new enforcement approach. A compliance and enforcement branch was created in the Ministry of Forests, and the Code Act also provided for an appeals commission to hear the complaints of licensees about enforcement actions.[36] The Code did not give other members of the public, such as environmentalists, the right to appeal regulatory or enforcement decisions (or nondecisions).

The Code Act does contain an innovative method of regulatory oversight, the Forest Practices Board. This new watchdog body does not have any legal or regulatory power, but it has the authority to conduct investigations into the Ministry of Forests' implementation of the Code and to audit corporate forest practices. It is also established as the organization to hear complaints from the public, including environmentalists, about ministry decisions under the Code or corporate forest practices. The board does not have authority to take action on a complaint, but it can investigate a complaint and make recommendations to the ministry. It can also launch an appeal to the Forest Appeal Commission on behalf of the public. Although ministry decisions are still insulated from direct public challenge,

Table 3.1

Guidelines for stream protection under the Forest Practices Code

Riparian class	Average channel width (m)	Reserve zone width (m)	Management zone width (m)	Total RMA width (m)
S1 large rivers	≥100	0	100	100
S1 (except large rivers)	>20	50	20	70
S2	>5≤20	30	20	50
S3	1.5≤5	20	20	40
S4	<1.5	0	30	30
S5	>3	0	30	30
S6	≤3	0	20	20

[shaded] Fish stream or community watershed
[unshaded] Not fish stream and not in community watershed

Source: British Columbia, Ministry of Forests and Ministry of Environment, Lands and Parks, "Riparian Management Area Guidebook" (Victoria: Queen's Printer, 1995), 14.

the Forest Practices Board – through its investigatory and auditing roles – grants the public much more access to information about ministry decisions.

Constraining Trade-offs: The 6 Percent Solution. During the development of the regulations and guidebooks to implement the legislation, concerns emerged about the impact of the Code on provincial harvest levels. To address these concerns, Minister of Forests Andrew Petter established a 6 percent cap on the impacts of the Code on the allowable annual cut. The origins and rationale for this decision are mysterious – if the decision has ever been written down, it has not been made public. Industry and labour expressed their alarm about the timber supply implications of draft Code regulations on November 1994.[37] In February 1995, newspaper reports showed that Petter argued the range of impacts would be 4 percent to 6 percent but that by April 1995 he had settled on 6 percent.[38] Government documents suggest this decision was made as early as November 1994.[39]

The most complete justification for the decision is the February 1996 *Forest Practices Code Timber Supply Analysis.*[40] But the report is presented in a very curious fashion. It does not claim to be a justification for a prior policy decision but rather a technical report estimating the impacts of the Code on timber supply. Yet in a remarkable coincidence, the estimated impact turns out to be exactly 6 percent, as shown in Table 3.2.[41] This outcome is particularly striking given the tremendous amount of uncertainty in any forecasting exercise and the large number of assumptions that need to be made when conducting such an analysis. For example, no Identified Wildlife Strategy had been issued yet, but the report estimated its impact as 1 percent. The total effect of the various provisions estimated in the report added up to 8.2 percent, but the figure was reduced by assuming that Visual Quality Objectives (VQOs) could be relaxed to produce a 2.2 percent gain. This precise number was included in the report despite the fact that no revisions to the VQO standards had been proposed or made.

At several points, the text of the report contains statements that reveal the larger purpose of the exercise. For example, the conclusion states: "Operational interpretation of the guidebooks will affect to a large degree, the impacts that the FPC requirements have on provincial timber supplies ... Implementation of the FPC consistent with these assumptions should result in AAC impacts outlined in this report."[42] The report, presented as a forecasting exercise, quickly evolved into a justification for fundamental policy decisions. It has been used to justify the 6 percent cap as well as the specific apportionment of 6 percent to particular aspects of the Code, such as biodiversity at 4.1 percent.

Interviews with senior government officials reveal that the development of the 6 percent cap was not as haphazard as it may seem. When serious concerns emerged about potential Code impacts in fall 1994, government experts began the modelling exercise that was eventually published in February 1996. According to Chief Forester Larry Pedersen, "We were able to say 6 percent with some reasonable certainty to ministers, under a set of assumptions," before the legislation had been finalized.[43] The bureaucratic analysis was able to show that a Code could be delivered at around 6 percent as long as two conditions were met. First, negative timber supply

Table 3.2

Estimate of the impact of the Forest Practices Code on harvest levels

	Provincial harvest impacts and (benefits) (%)	
	Short term	Long term
Riparian	2.1	6.0
Biodiversity		
stand	1.8[a]	
landscape	2.3	
stand and landscape	4.1	4.3
Watershed assessments	1.0[b]	1.0
Identified wildlife	1.0	1.0
Soil conservation	0	-2.3
Visual quality objectives	-2.2	-2.0
Watershed restoration		-0.3 to -1.0
Total impact	**6.0**	**7.0 to 7.7**[c]

[a] Increases by 1 percent in absence of landscape unit biodiversity objectives.
[b] Increases by 1.5 percent in absence of landscape unit biodiversity objectives.
[c] Does not include further reductions resulting from silviculture investments and other strategies.
Source: British Columbia, Ministry of Forests and Ministry of Environment, Lands and Parks, "Forest Practices Code Timber Supply Analysis" (Victoria: Queen's Printer, 1996).

impacts had to be compensated in part by a relaxation of the Visual Quality Objective, an initiative pushed by Minister Andrew Petter. Second, the various aspects of the Code had to be developed and implemented in a way that was consistent with the analysis.

What emerged over the intervening year and a half was an iterative process between the analysis and the implementing regulations and guidebooks. Riparian policies had been established early on by ministerial direction. The real wildcard, given the potential magnitude of its impact, was the *Biodiversity Guidebook*. An August 1995 memorandum from the deputy ministers of forests and environment shows how earlier versions of the impact analysis report influenced the development of policy.[44] In several critical areas of interpretation, the memorandum states what the analysis assumed in order to arrive at the 6 percent figure, and it then turns and transforms those assumptions into policy. For example, the analysis assumed that "high emphasis" biodiversity requirements were applied to only 10 percent of the landscape. The memorandum (and subsequently the guidebook itself) transformed this assumption into a policy limitation.

This 6 percent cap is a fascinating illustration of both the limits to comprehensive decision making and the extent of policy change embodied in the Code. The Code involves a difficult balancing act between timber and environmental objectives. A rational-comprehensive decision-making process – the type that forest planners are fond of – would have struggled to derive some optimal balance of these factors, given the combination of objectives of the government and consequences of alternative policies. Yet, just as Lindblom and other decision theorists would predict, policy makers are unwilling or unable to rank their priorities systematically.[45] Moreover, a combination of inherent uncertainty, and limits to the cognitive capacities or the resources available to devote to gathering the relevant knowledge, means that the analysis of alternatives was very limited. This chaotic situation was partially resolved by a political decision to limit timber supply impacts to 6 percent. Although it seems arbitrary, the decision has reduced the uncertainty for decision makers and other interests, and it has helped structure the implementation process. In a 1998 interview, Chief Forester Larry Pedersen stated: "To this day we still fall back on this report for policy direction."[46]

The 6 percent cap provides an important signal of the policy objectives of the government and the extent of change involved in the Code. The government did not choose to set the cap at zero; it shifted policy in a proenvironment direction. But it was also limited in how far it was willing to go in that direction – not 10, 20, or 50 percent. We will return to a discussion of the magnitude of change in the conclusion.

This analysis of the decision stage shows a change in policy content – goals and objectives that reflect a movement toward greater concern with environmental values, limited by the 6 percent dictate. The most dramatic change was in the formalization of instruments. Whether those changes amount to a mere change in process or real change in forest practices awaits a review of the implementation stage. The push for more environmental objectives was clearly motivated by a combination of new elected officials, provincial public opinion, and the environmental group shift in strategy to emphasize the international arena. The limits on the extent of that shift were motivated by the desire to contain the impacts of this new initiative on the critical timber supply subsector.

Implementation

The Forest Practices Code took effect in June 1995, with a transition period for full implementation ending in June 1997. It is too early to provide an appropriate evaluation of implementation, but we can point to certain patterns and pressures. From a political perspective, the Code seems to have solved some of the government's problems but not others. The good news is that polls of provincial public opinion show the government has received credit for introducing the Code (although the public remains sceptical about enforcement). The bad news comes on several fronts. The Code has done nothing to chill the criticisms of the environmental community, and international market pressures still loom large. At the same time, the costs of implementing the Code have been burdensome and have contributed to an economic crisis in the industry. In that the Code's primary objective is "balancing productive, spiritual, ecological and recreational values of forests," the Code must be deemed a failure. It has not provided ecological benefits that have satisfied powerful political interests, and, at the same time, it has been a major cost burden.

Higher-Level Plans. One of the principal objectives of the Code was to develop a more comprehensive and integrated system of planning. Planning at the operational level appears to be working relatively well, but very significant delays have been experienced in the establishment of higher-level plans. As of mid-July 1999, only three regional and subregional higher-level plans have been put in place for the entire province, one based on the regional land use plan for the Cariboo-Chilcotin and the other two based on Land and Resource Management Plans established in Kispiox and Kamloops. As for Landscape Units, the key unit for the management of biodiversity, they have been established in only three of the province's forty districts.[47] This lack of progress leaves the overwhelming majority of the provincial forest land base without the guidance of higher-level plans.

The absence of higher-level plans does not vitiate the Code planning framework because district managers have a great deal of guidance in planning regulations and guidebooks. However, with so few Landscape Units established, it does create particular problems for the province's approach to protecting biodiversity. In the absence of these plans, biodiversity is not going without protection, but the default provisions to be used in the absence of Landscape Units is the lowest form of biodiversity emphasis provided in the *Biodiversity Guidebook.*[48] It is expected that the March 1999 publication of *Landscape Unit Planning Guide* will facilitate the designation of other units.

The new *Landscape Unit Planning Guide,* which has replaced the *Biodiversity Guidebook* as the key document for managing biodiversity under the Code, clearly shows the force of the 6 percent cap of impacts on harvest levels. The letter introducing the guide states: "The impact of landscape unit biodiversity objectives on provincial timber supply is not permitted to exceed 4.1 percent in the short-term and 4.3 percent over the long-term. We will continue to develop direction and management controls in addition to those outlined in the guide to ensure that this provincial commitment is met."[49]

Forest Practices. The government has invested little effort in tracking implementation of forest practices, at least in a manner that has been communicated to the public. Statistics on clearcut size do exist, and they are displayed in Table 3.3. Between 1988 and 1998, average clearcut size across the province dropped from 43 to 26 hectares, a decline of 40 percent, but virtually all of this decline occurred before 1995, when the Code came into effect.[50] Greater attention to environmental concerns and clearcutting in particular in the late 1980s and early 1990s seems to have had more direct impact on clearcut size than has the Code.

On stream protection, the government has been criticized for its implementation of the new regulations by the Sierra Legal Defence Fund (SLDF)

Table 3.3

Average cutblock size, 1988-98, in hectares

1988	43.4
1989	45.5
1990	39.9
1991	35.7
1992	35.3
1993	34.6
1994	26.3
1995	31.1
1996	28.3
1997	26.2
1998	26.4

Sources: Government of Canada, BC Ministry of Forests. *Just the Facts: A Review of Silviculture and Other Forestry Statistics*, 2000. (Available at www.for.gov.bc.ca/hfp/forsite/jtfacts/11-harv-cutblock.htm.)

and the Forest Practices Board. In an apparent attempt to replicate the methods and influence of the Tripp reports of the first half of the 1990s, the environmental group issued a report blasting government and industry performance. They argued that some streams were not identified at all in plans and that some streams were misclassified (for example, as nonfish-bearing streams when fish were in fact present). Their most striking finding was that 83 percent of the streams they surveyed were clearcut to their banks. For small fish-bearing streams (for example, S4 streams that can serve as important coho habitat), 79 percent were clearcut to the bank. The report did note that these impacts were legal under the Code. Although no-harvest reserve zones exist for larger fish-bearing streams, the smaller streams are protected only by "riparian management zones" where the harvesting prescription is up to the discretion of the company recommending the plan and the district manager of the Ministry of Forests approving the plan.[51]

The SLDF report strongly suggested that the best practices recommended in the *Riparian Management Area Guidebook* were not being followed. Recall that the guidebook suggests that, for small fish streams on the coast (S4), 50 percent of trees within 10 metres of the bank should be retained. The report said that 79 percent of the streams are clearcut to their banks. Although this aspect of the report has received little attention, it does undermine one of the strongest industry criticisms of the Code: that the nonbinding guidebooks become de facto regulations because company foresters fear that the government officials will approve nothing less. If the SLDF figures are correct, they suggest that company foresters routinely recommend, and that government officials routinely approve, riparian protection far less stringent than what is suggested in the guidelines.

This report received considerable press attention and sent off alarm bells in the government. The government ordered its own review, which, although not nearly as critical as that of the SLDF, did raise some significant concerns, particularly about the proper identification and classification of streams. Unlike some of the industry critics of the SLDF report, the government review emphasized the importance of small streams for fish habitat. At this point, the government took the extraordinary step of requesting that the Forest Practices Board conduct its own independent investigation.

After considerable delay, the Forest Practices Board issued its report in June 1998. The board was far more positive about riparian area protection than was the Sierra Legal Defence Fund, but it did note some significant problems with policy implementation. It found relatively high compliance with planning and practice requirements, and it also found that the amount of alteration to streams has decreased significantly since the introduction of the Code. The biggest problem the board found was with the classification of streams, particularly smaller streams. Nearly half the two classes of small fish streams (S3 and S4 under the Code) audited were "underclassified," leading to "inappropriate practices in a number of cases."[52] The board noted that the recommendations in the guidebooks for retention in riparian areas were frequently not followed. Among other recommendations, the board suggested increased training to improve classification along with the enactment of more specific requirements for retention.

Enforcement and Compliance. Another major thrust of the Code was to strengthen enforcement and compliance. The Code seems to have been relatively successful in this regard. The compliance record is good. For June 1995 to June 1998, provincial officials conducted 113,000 inspections, about 94 percent of which were found to be in compliance with the Code.[53] In those instances where compliance was not found, many problems were resolved in the field. The remainder resulted in some type of contravention decision.

Although the rate of compliance is impressive, the pattern of enforcement actions applied to contravention reflects overwhelming reliance on the "softer" end of the remedy spectrum. Of a total of 1,283 contravention decisions over three years, there have been only nine prosecutions under the tougher criminal provisions of the statute. Despite the much-ballyhooed $1 million fines emphasized by the government when it introduced the Code, no fine has come anywhere close to that maximum. The highest fine issued has been $265,000, and that went to the federal government for unauthorized harvesting of Crown timber in the construction on the Alaska Highway. No other penalty has been greater than $100,000. Forty percent of all monetary penalties have been less than $500, 60 percent have been less than $1,000, and 89 percent have been less than $5,000.

Logging Costs. While environmentalists have denounced the Code as ineffective, industry has complained bitterly about the costs of complying with it. At a December 1996 meeting of the Forest Sector Strategy Committee, a high-level advisory body of key stakeholders, industry representatives presented an alarming report, entitled *Industry on the Brink,* about the state of the industry. The report raised concerns about profitability in the wake of cost pressures resulting from the Code and stumpage increases. In response, the government agreed to commission an objective study of costs. The result, a study performed by the accounting firm KPMG and published in April 1997, suggested that the costs of the Code were indeed significant. The study compared "delivered wood costs" in 1992 and 1996, and it then divided the increase according to a highly refined set of categories.[54]

The results state that delivered wood costs increased from $50 per cubic metre in 1992 to $87 per cubic metre in 1996, an increase of 75 percent. Increases in stumpage accounted for 45 percent of the increased costs, the

Code accounted for 33 percent of the increase, and a miscellaneous category of "non-Code-related cost drivers" accounted for the remaining 22 percent. The fallout from this study was confirmation of the industry claims of dramatically increased costs. Despite the fact that stumpage hikes were shown to be a more important driver behind the cost increases than the Code, the added $12 per cubic metre of Code-related costs – 14 percent of total delivered wood costs – was still considered unquestionably significant in an industry with small (or negative) profit margins.[55] A more recent report, prepared by PricewaterhouseCoopers and issued by COFI in January 2000, stated that from 1989 to 1998 logging costs had increased by $12 per cubic metre, "almost entirely the result of the Forest Practices Code."[56]

Explaining Implementation Patterns. How can we account for this pattern of implementation? One of the most influential discussions of the contributions to implementation success and failure can be found in Mazmanian and Sabatier. They have six conditions for effective implementation.[57] This analysis states their conditions and then briefly highlights how those conditions relate to the Code.

(1) *The enabling legislation or other legal directives mandate policy objectives that are clear and consistent or at least provide substantive criteria for resolving goal conflicts.* The Forest Practices Code Act of British Columbia sets out a series of objectives in its preamble, but they are clearly not consistent with each other, and no substantive criteria for goal conflict are established. Petter's "6 percent solution" does constrain how the competing values are balanced: it allows a move toward more environmental sensitivity but also limits it. Environmentalists would argue that achieving the environmental goals of the Code Act are not possible given that constraint.

(2) *The enabling legislation incorporates a sound theory identifying the principal factors and causal linkages affecting policy objectives and gives implementing officials sufficient jurisdiction over target groups and other points of leverage to attain, at least potentially, the desired goals.* The question of causal theory is arguably the most challenging issue for the implementation of the Code. Two important causal theories underlie the Code, both of which have been called into question. The first is the assumption that rigorous planning and government approval is necessary and desirable to ensure forest resources are properly managed. The process-based planning requirements may be needlessly complex and be diverting government and corporate resources from more desirable purposes. The second crucial assumption is that "integrated resource management" is the best way to balance competing objectives in forestry. Many argue that this approach has, among other things, led to perverse outcomes where measures originally designed

with the intention of preserving environmental values are having the effect of fragmenting habitat and increasing environmental impacts. Clearcut size is the most dramatic example. There are serious, ecologically based arguments that reducing clearcut size has negative, not positive, environmental consequences. Given a particular harvest level, small-opening sizes increase edge effects, fragmentation of the landscape, and, most important, the amount of road necessary.[58] The problem with the current causal theory (small clearcuts are better for the environment) is that it reduces the overall net benefits available for forest managers to distribute among the competing values. The clearcut size issue is a microcosm of a larger argument, made from both economic and ecological perspectives, that integrated resource management is inherently less capable of providing a combination of environmental and economic benefits than is more intensive zoning.[59]

(3) *The enabling legislation structures the implementation process to maximize the probability that implementing officials and target groups will perform as desired. This condition involves assignment of authority to sympathetic agencies with adequate hierarchical integration, supportive decision rules, sufficient financial resources, and adequate access to supporters.* The large amount of discretion delegated to district managers raises questions about how effectively they can be influenced by the objectives of government leadership, whether they tend toward the green or brown end of the spectrum. It is clear that sufficient resources have not been made available to implement crucial features of the Code Act, especially higher-level planning. The 6 percent decision rule constrains the ability of some officials to pursue their objectives (for example, MOELP officials and biodiversity). The question of sympathetic public agencies is complex because there is tremendous variation between district managers, but the increased reliance on MOELP officials has increased the representation of environmental interests in the implementation process.

(4) *The leaders of the implementing agency possess substantial managerial and political skills, and they are committed to statutory goals.* Although this condition is less of a problem for agency leaders, one of the problems revealed in implementation studies is that ground-level foresters are inadequately trained in assessing and classifying streams and other resources. The government has committed itself to better training, but we are not yet able to assess whether the situation has improved.

(5) *The program is actively supported by organized constituency groups and by a few key legislators (or a chief executive) throughout the implementation process, with the courts being neutral or supportive.* This condition clearly reflects an American bias – neither the legislature nor the courts have been particularly relevant to the Code at any stage. However, the

replacement of Harcourt with Clark affected the tone and direction of Code changes and implementation.

(6) *The relative priority of statutory objectives is not undermined over time by the emergence of conflicting public policies or by changes in relevant socioeconomic conditions that weaken the statute's causal theory or political support.* The economic swing in BC has unquestionably strengthened the political position of proindustry forces and undermined the domestic political support for the Code, as the later discussion of revisions demonstrates.

Evaluation and Monitoring

As suggested in the implementation section, the evaluation of the monitoring of the implementation of the Code has been performed both inside and outside the government. Inside the government, evaluation has focused on compliance and, to a lesser extent, on costs. Clearly, many resources have been dedicated to enforcement, and public annual reports provide detailed results of fining, including the number and reasons for fines levied against individual named companies. The government sponsored the KPMG study described above. There seems to be remarkably little interest in monitoring the environmental aspects of the Code.

In the early years of Code implementation, environmentalists were extremely active in policy evaluation, producing a series of relentlessly critical reports about the Code and its administration. The Sierra Legal Defence Fund, in a campaign funded by American foundations, issued five reports on the Code. The most prominent report was the one on stream protection described earlier, which stated that a surprising number of streams were incorrectly classified and that 79 percent of small fish-bearing streams were clearcut to the banks.[60] In a report on clearcutting, SLDF argued that, despite the rhetoric of the Code, clearcuts are still the overwhelming harvesting method of choice across the province: they were used in 92 percent of approved cutblocks surveyed by SLDF.[61] A third report highlighted that, as of 1997, the government had yet to introduce any of the major elements of the biodiversity components of the Code, including Landscape Units, Old Growth Management Areas, Identified Wildlife Species, Wildlife Habitat Areas, and Sensitive Areas.[62] A fourth report focused on steep-terrain logging, claiming that the government was not adequately implementing the Code provisions designed to regulate logging on steep slopes so as to prevent landslides and soil erosion.[63] The final report synthesized the findings of these reports into a "government forest report card," stating that, "based upon grades for 12 individual aspects of forest policy, the government's forestry report card contains 8 F's and 4 D's, for a final grade of F."[64]

These reports have contained a great deal of polemical, evocative, proenvironment rhetoric. However, what is distinctive about the Sierra Legal Defence Fund's effort is the detailed, careful documentation underlying the analysis. Although the reports may not meet the standards of a professional audit, they reflect a level of analytical sophistication that is unusual in the BC environmental community. They have helped focus environmental pressures on the government and industry and hammered home the message, lest anyone forget, that members of the public are watching. In its international campaign, Greenpeace has picked up on these reports in an effort to show the world is watching.[65]

In contrast to the aggressive public approach taken by environmentalists, industry initially took a more low-key tack in their evaluation efforts, focusing on costs and communicating their views directly to government. Industry put the cost issue on the premier's agenda. More important, seventeen industry CEOs participated with the premier in the Jobs and Timber Accord process from which the Bill 47 revisions, to be discussed below, arose. When their financial woes continued despite the new legislation, industry adopted a more public campaign. In November 1998, the Council of Forest Industries publicly issued what it called "a plan to stem the bleeding in forest-dependent communities." The plan called, among other things, for "an immediate government ban on all new policy and regulatory changes that may increase costs to the forestry industry."[66] This more aggressive approach continued into early 2000, as COFI worked hard to influence the agenda of the Forest Policy Review of the fading NDP government.[67]

Another major source of evaluation has been the Forest Practices Board, set up as an independent entity by the Code Act to act as a public watchdog. As a result of a combination of start-up problems and limited resources, the board got off to a very slow start. These difficulties were aggravated because of the complexity of the investigation of stream protection in coastal areas that arose after the Sierra Legal Defence Fund report. This report took up much of the board's time and energy in 1997. Initially, the board seemed to adopt a relatively supportive view of the Code's implementation. Its 1997 annual report stated: "The Board is pleased to note that the results of its audits and complaints investigations to date indicate that forestry operations are generally in compliance with Code requirements. There has been general improvement in forest practices in BC and the Board believes this is associated with the introduction of the Code."[68]

In late 1998 and early 1999, the board began to play a more active role in policy evaluation, and it adopted a tone in its communications that was both more confident and somewhat more critical. By early July 1999, the board had published eighteen audits, sixteen complaint investigations, and two special investigations. In public presentations, it has developed a four-part assessment of the Code. First, the board believes "there have been changes and a significant improvement in forest practices since the Code came into effect." Through its activities, it has observed improvements in stream protection and road building, and found a generally high level of compliance. Second, it notes "there is a need for better compliance with Code requirements and a need for more effective opportunities for public involvement." The board is less sanguine about compliance than the Ministry of Forests enforcement reports. It notes that, "although compliance is generally high, we have identified some significant noncompliance with Code requirements in more than half of the operations we have audited."[69] Third, "there are important pieces of the Code that have not yet been implemented," particularly the higher-level plans necessary to protect wildlife and diversity. Finally, it believes that improvements are needed in the quality of plans developed under the Code.[70]

All of these evaluation efforts, especially those of interest groups, were designed to influence the problem definition adhered to by both policy makers and the public. Although environmental groups have struggled mightily to focus environmental concerns on policy makers, there is no question that by early 1997 the dominant problem definition had been transformed from underregulation before the Code to overregulation in the wake of the Code.

Revisions

As discussed in Chapter 1, the power of the industry coalition is inversely proportional to business performance. When the Code was developed in the first half of the decade, the BC economy, and the forest sector in particular, were doing well. By 1996, however, the latest cyclical downturn was well under way, and the industry found itself reeling from a combination of pressures. Although lumber prices remained strong, pulp prices plummeted. Just as competition from lower-cost producers in other parts of the world was intensifying, logging costs in BC increased dramatically as a result of the introduction of the Code and the significant stumpage increases in 1994 as part of the Forest Renewal Plan. Meanwhile, as described in Chapter 7, the 1996 Softwood Lumber Agreement with the United States placed constraints on how much the industry could export south of the border. When the Asian economic crisis emerged in 1997, coastal firms that relied primarily on exports to Asia were left without markets. Production declined dramatically, with accompanying job losses at pulp and sawmills.

The crisis in the industry put pressure on the government to find ways to reduce costs, and revisions to the Code were a fundamental part of the agenda. As a result, despite the strident environmental criticisms of the Code's implementation, the economic context ensured that, when revisions to the Code emerged on the government agenda, the focus was on reducing costs, not expanding environmental protection. In June 1997, the Clark government introduced Bill 47, the Forest Statutes Amendment Act. In announcing the changes, Minister of Forests David Zirnhelt stated: "We introduced the code over two years ago to ensure our forests were managed to meet present needs without compromising the needs of future generations. We made it clear there would be ongoing adjustments, based on experience in the field, to ensure we maintain a balance of world-class sustainable forest practices. We have consulted widely and these initial changes *will make regulations simpler and less costly for government and industry, while preserving our high environmental standards.*"[71] Despite the emphasis on cost savings, the government had seen enough polls showing continued domestic support for strong environmental measures,[72] and it was sufficiently fearful of provoking a renewed international boycott campaign that it took pains to emphasize that streamlining would not compromise environmental standards.

For the better part of a year, the government developed regulations to implement the new legislation; these were finally announced in April 1998. In announcing the changes, Zirnhelt claimed they would "reduce industry's costs by $5 a cubic metre, or about $300 million a year."[73] This cost estimate is striking: it is the same amount the government estimated the entire Code would cost when it was introduced. Although the government had an incentive to downplay costs then, and to exaggerate savings in 1998, its balancing act when it introduced the Code was clearly based on dramatic underestimates of costs. The chief forester openly acknowledged this fact in early 1999: "These studies [KPMG and others] led to a formal understanding that regulation under the Code was having unintended consequences that were placing a much heavier financial burden on the industry than expected."[74]

Unlike when the Code was originally established there was now no opportunity for public consultation and comment on proposed regulations. The government did conduct consultations, but they were by invitation only, and participants had to sign confidentiality agreements pledging that they would not reveal the contents of what they had seen. Environmentalists claimed that they were not treated fairly in this consultation process. It is clear that this round of revisions reflected a tightening of the policy regime and provided a striking contrast to the open, multistakeholder consultations in land use (see Chapter 2) or the extensive opportunities for public input available in the timber supply process (see Chapter 6).

The main thrust of the changes was a streamlining of the planning process. The number of operational plans was reduced from six to three.[75] Companies were given greater stability and certainty through changes in the process for approving plans. Given the instability in the context of higher-level plans, plans would have to meet the requirements in place four months before they were submitted rather than at the time of submission. A process of "gating" was established, so that once a cutblock was approved, it need not be reevaluated in the Forest Development Plan unless dramatic new information emerged. These changes allowed companies to maintain a larger inventory of timber approved for harvest, thus increasing operational certainty and flexibility. Important paperwork requirements were reduced. Assessments of the various resources required to develop plans, such as streams, no longer had to be submitted with them. They did, however, need to be made available to the government (or members of the public) on request.

Another significant change was the move to a more results-based orientation, particularly in the Silviculture Prescription. (Riparian management was explicitly exempted from the move away from process standards.) Although no regulatory changes were involved, the government also promised procedural changes to increase average cutblock size so that it would be closer to the legal limits. According to the government, "Smaller cutblocks significantly increase administrative workload, road requirements, the cost of logging and environmental risk, such as fragmentation of wildlife habitat."[76]

Industry groups offered lukewarm support of the changes. The BC Forest Alliance called the new regulations "a responsible step in the development of BC's framework for sustainable resource management."[77] Industry officials were sceptical of the promised cost saving. Environmentalists denounced the changes, claiming they gutted environmental standards.[78] In their press releases, however, environmental groups struggled to argue this claim. They referred to two changes: the commitment to increasing clearcut size and a procedural change permitting firms to exceed the maximum amount of soil disturbance for road construction if the company commits to remedying afterward.[79] Neither change supports the accusation that Code standards have been gutted. The issue of clearcut size has emerged because de facto clearcut sizes are substantially smaller than regulatory limits (and there are significant ecological concerns related to small clearcuts that environmental groups are hesitant to discuss). The soil disturbance change does reflect a modest relaxation of standards but nothing fundamental.

Environmentalists were far more concerned about the changes in process designed to reduce paperwork and streamline planning. Procedures that industry views as bewildering, unjustified red tape are considered by environmentalists as fundamental to their ability to act as a watchdog on the industry and government. For example, they are very concerned about the increased difficulty in acquiring the assessments performed by companies in support of planning. They are also concerned about reductions in some of the avenues for public review. These concerns have been echoed by the independent Forest Practices Board, which criticized some of the proposed changes in a confidential letter to the government that was leaked to the media.[80]

In enacting the original Code, the government made significant strides toward formalizing the planning and regulatory procedures for forest management as well as expanding environmental protection measures and giving them the force of law. The Bill 47 amendments provide a modest retrenchment in some of the procedural elements, but they leave the bulk of the new policy framework intact, including the most important environmental standards. Given the magnitude of the economic crisis in the forest industry, it is surprising that the industry was unable to attain greater regulatory relief. The modest nature of the changes is testimony to the power of the status quo as well as to the enduring public commitment to environmental values and the continuing pressures from the international marketplace – pressures fostered by international environmentalists.

The Agenda for Further Revisions. Despite the fact that many aspects of the Code have yet to be implemented, fundamental changes to the structure of forest practices regulation appear to remain on the government agenda. While the regulations to implement Bill 47 were being developed, internal government e-mails were leaked to environmentalists suggesting the government had far more fundamental changes in mind. The documents suggested a new approach where only one plan (the Forest Development Plan) was submitted for government approval, and others had to be prepared but would be entrusted to professional foresters. In November 1998, another internal government document was leaked that suggested the government was serious about developing a "results-based code" that "achieves the economic, environmental, and social objectives for the province's forest resources in the most efficient manner possible."[81] The paper questions one of the key assumptions underlying the Code: "The decision to establish in legislation an administrative process for operational planning was based on the view that government review and approval of plans is necessary to ensure that forest resources are adequately managed and conserved. This assumption may not be true and there may be more efficient ways of achieving the environmental and social objectives of the Code."[82] The proposal suggested eliminating the requirement for submitting a Silviculture Prescription for approval, and it opened the possibility of waiving the requirement for Forest Development Plan submission for companies with good performance.

By spring 1999, the government was publicly admitting that its approach to regulation was not working. In a March 1999 speech, Chief Forester Larry Pedersen stated: "But we've also learned that market structure and conditions simply won't support our original plan to regulate the industry closely into conformity with prescribed plans."[83] Rather than launching another major Code revision, however, the government introduced legislation in June 1999 that would authorize pilot projects to test results-based approaches.

Regulating Private Forest Land

In June 1999, the government expanded its regulatory control over the BC forest industry by establishing forest practice regulations to cover privately owned forest land.[84] Although privately owned land makes up only about 2 percent of the provincial land base, it is a significant proportion on Vancouver Island.[85] Before the introduction of these regulations, private forestry was essentially unregulated, except for the general provisions of the federal Fisheries Act. The Forest Practices Code contained provisions (Section 217) for expanding the Code to include private land, but such expansion was strongly resisted by landowners. The government agreed to enact the new regulations outside the framework of the Code. Rather than being implemented by the Ministry of Forests, the regulations would be administered by the small Forest Land Commission, an organization established to administer the province's Forest Land Reserve.

The standards themselves were designed to reflect a different model of regulation, one that is more "results-oriented" than the "process-oriented" Code.[86] (Indeed, the government hoped that the private land regulations might help develop a results-based model that could be used on public lands.) In addition to representing a different approach to regulation, the standards are clearly less stringent than those in the Code. For example, the Code requires no-logging zones around fish-bearing streams larger than 1.5 metres across. The private land rules require only the retention of at least forty trees for every 200 metres of stream length.[87]

The rules for private land clearly involved different regime dynamics than the rules for public land. Landowners (some of whom are very large corporations like Weyerhaeuser and TimberWest) were accorded special status. Rather than following more traditional procedures, the policy was developed as a "memorandum of understanding" with the Private Forest Landowners Association. Environmental groups were permitted to view the proposed standards under strict confidentiality agreements, but, as in the case of the regulations to implement Bill 47, the standards were announced as law without any opportunity for public comment.

Privatizing Forest Practices Regulation

A far more extraordinary development arose in the late 1990s: the emergence of private certification organizations that, because of their market power, have begun to play a powerful role in BC forestry.[88] The most dramatic example of this power was the June 1998 decision of industry giant MacMillan Bloedel to announce that it was abandoning its long-standing practice of clearcutting. In announcing the decision, President Tom Stephens clearly credited the certification movement as an important motivation: "It reflects what our customers are telling us about the need for certified products, but equally important it reflects changing social values and new knowledge about forest ecology."[89] By spring 1999, two other BC companies had followed suit. These corporate decisions reflect a fundamental change in regime dynamics: green consumerism in the marketplace, focused by private certification schemes, has become a lever to influence corporations to adopt forest practices that move significantly beyond those required by government regulations. Although the phenomenon is new, and its impact has only been reflected in part of the industry operating in the province, certification has the potential to create a fundamental shift in the locus of forest practices regulation from the government to private standard-setting bodies.

Forest certification has emerged as part of a larger movement toward increasing the use of voluntary and market-oriented instruments to pursue environmental objectives.[90] There are three competing standard-setting organizations: the International Standards Organization (ISO), the Canadian Standards Association (CSA), and the Forest Stewardship Council (FSC). The most important group in terms of influencing forest policy in BC has been the FSC. The FSC, now based in Oaxaca, Mexico, was created in 1993 by the World Wide Fund for Nature (WWF) after the group became frustrated with the lack of international progress in developing a binding "Global Forest Convention."[91] Its mandate is to "support environmentally appropriate, socially beneficial, and economically viable management of the world's forests." The FSC plays two key roles: it develops a set of "principles and criteria" for the sustainable management of forests, and it accredits certifying organizations to determine whether a particular forest operation complies with them.[92] Once certified, products can be labelled with the registered FSC logo.

The FSC has evolved into a complex governance structure. Internationally, it is governed by a general assembly, consisting of three chambers: social, environmental, and commercial. It operates by a two-thirds majority voting rule, so decision making requires agreement from diverse interests. Despite this appearance of diversity, the organization has still been dominated by a strong environmental perspective, though this trend may change if more industrial forest corporations seek membership in the commercial chamber. The FSC established a set of ten principles and criteria designed to guide certification efforts. However, given the tremendous diversity of forests around the globe, it has also provided for the establishment of regional standards. The FSC has a Canadian affiliate, and regional groups have emerged in the Maritimes, Ontario, and British Columbia. Because all of the standards developed by regional groups are supposed to both reflect distinctive local conditions and be consistent with the ten international principles and criteria, the FSC is evolving into a complex federal organization with considerable management challenges.

The challenges have been particularly vexing in the politically polarized environment of British Columbia. As of September 2000, there are no regional standards in force in BC. The standards development process was initiated in 1996 by a close-knit group of environmentalists with no involvement from major BC forest companies. After much conflict and delay, draft standards were released in June 1999,[93] but they received remarkably little attention. It was understood by many participants that a new process would have to be initiated – one that contained representation from a broader commercial chamber, including representatives of industrial corporations. This process was initiated in June 1999 with the appointment of a new steering committee. Bill Bourgeois, chief forester of Lignum Ltd., a major industrial forest company, was appointed to the commercial chamber. The voting rules require three-quarters of the committee members and a majority of each chamber to be present.

At present, there is a considerable amount of uncertainty about how FSC certification will work in BC. The major question is whether it will be possible to certify any significant industrial forestry operations in the province, particularly those in coastal areas where environmental concerns are most intense. Although several of the principles and criteria may be problematic in BC, the biggest sticking point is Principle 9, which deals with logging in primary forests. Originally, Principle 9 essentially precluded timber harvesting in primary forests, but the text was changed in early 1999 to emphasize the protection of the attributes of "high conservation value forests."[94] The new text opens the door to harvesting of old growth, but it is not certain what types of forest practices would be consistent with the principle.

The FSC permits certification without a regional standard – leaving it up to certifiers to determine whether operations are consistent with international standards. In BC, several major forest companies have retained accredited certifiers to begin the process of seeking FSC certification, including Weyerhaeuser (with extensive coastal holdings), Western Forest Products (which operates in the contested coastal zone), and Lignum (which operates in the Interior). As of August 2000, no major BC forest operation has received FSC certification.

The FSC and the BC forest industry appear to be engaged in a complex strategic game. The FSC is trying to use its credibility in the international marketplace, bolstered in part by the support of major international environmental organizations, including Greenpeace, to reduce dramatically the environmental impact of BC forestry. But to maintain its organizational legitimacy, the FSC must appear to be open to the diversity of stakeholders, which naturally includes industrial forest companies. It also has to be careful not to be seen as unfairly favouring some of the world's forest regions (with their associated industry and national interests) over others. The BC industry understands these tensions, and it has pressed for participation and, in several cases, outright membership. Lignum quietly became a member in early 1999, but a far more significant stir was created in June 1999 when the BC Forest Alliance – since the early 1990s an active defender of traditional BC forestry – announced it too was applying for membership.[95] How far can the FSC go in accommodating the active participation of BC industrial companies without jeopardizing its support from its strident environmental wing? Will the BC industry be able to convince the FSC to certify operations that are profitable on an industrial scale? The fundamental political battles over BC forest practices are shifting from the BC government to a private environmental organization headquartered in Mexico.

As noted, the FSC is only one of three private certification organizations that may come to play a role in provincial forestry, though it is the most environmentally oriented. The International Standards Organization has a certification program under its 14,001 series. This program focuses on an "environmental management system," which requires the establishment of corporate environmental objectives, a management system to implement and monitor those objectives, and a commitment to continuous improvement. The Canadian Standards Association has also developed a forest certification system. This standard was developed by the Canadian forest industry, in collaboration with the Canadian Council of Forest Ministers, explicitly for the purposes of providing an alternative to the FSC approach that would be more compatible with the interests of the Canadian industry. Progress in implementing the standard has been frustratingly slow. In April 1999, MacMillan Bloedel's North Island division was the first to receive CSA certification. Unlike the FSC, neither the ISO program nor the CSA certification scheme provides for a specific product label.[96]

The uncertainty surrounding the various certification schemes has created turmoil in the BC industry and the potential for a dramatic transformation of the policy regime. Once preoccupied with meeting government standards, firms are now struggling to determine whose standards are the most relevant to the emerging market environment. Even the government is getting into the act. In January 2000, the Ministry of Forests announced that it would seek certification for the small business program that it operates. The irony here is remarkable. The government, the regulator of the private sector, is turning around seeking the approval of the private sector for its operations. It is too early to tell what sort of impact certification will have on BC forest practices, but it does have the potential to create a revolution in the governance of forest products, as private standard-setting bodies may become more important in driving forest practices than are government regulators. This development would involve fundamental changes to the forest practices policy regime. The policy makers would change along with the rules for participation of different actors. At its core, the new regime would reflect a dramatic combination of a shift in the strategy of environmental groups to emphasize international markets and the emergence of powerful environmental sentiments among BC forest product customers.

Conclusion

This analysis shows that there was some moderate change in forest policy content in the area of forest practices. Goals and objectives shifted toward the environmental end of the spectrum. Instruments were formalized: nonbinding guidelines were transformed into mandatory regulations, providing the basis for an elaborate new enforcement regime. Settings on the instruments, in the stream protection area at least, were changed to make them more stringent. The magnitude of change was carefully contained, however, by the 6 percent cap on impacts for harvest levels. This cap has been rigidly enforced: the Code has bent to meet its dictates.

What is lacking is a clear sense of policy consequences. There is little information available to evaluate the environmental consequences of the Code. Indicators of clearcut size exist, but they suggest much of the change occurred before the introduction of the Code, and there is widespread agreement that such indicators are not very good proxies for environmental benefits. There are no consistent measures available to assess stream protection or other environmental outcomes, though the Forest Practices Board report on streams does provide evidence that stream protection in coastal areas has improved. Studies of costs suggest that there was a definite imposition of significant additional cost burdens on industry. If policy is defined as the allocation of costs and benefits, there has been a significant increase in costs to industry, with uncertain environmental benefits.

How significant are these changes? The KPMG study estimates the Code accounts for 14 percent of delivered wood costs, with modest cost savings resulting from Code revisions. The government has capped the impact of the AAC at 6 percent. Both of these points indicate change away from the prominence of timber extraction as a policy goal. But they do not suggest any fundamental paradigm shift. The Clayoquot Sound Scientific Panel standards reflect a paradigm shift: their impacts on allowable cut are enormous compared to the Forest Practices Code.[97]

This chapter reveals the complex interactions between regime components and the policy cycle. Although the technical nature of the subject matter privileged expert professionals, the case of the Forest Practices Code demonstrates that broader, more diffuse public forces were felt at each stage of the policy cycle. Public pressures for greater attention to environment goals manifested themselves not just at the agenda-setting stage but also at the decision, implementation, evaluation, and revision stages. Elected officials played a crucial role as well. The election of the NDP virtually ensured there would be a major environmental initiative on forest practices. When it came to making decision about key trade-offs, as in the case of the 6 percent cap, cabinet members made the difficult decisions.

These changes are explained by a combination of factors that are difficult to disentangle. Domestic opinion pressures, a new and more environmentally oriented government, favourable market conditions, and the internationalization of environmental pressure tactics all contributed to the push to adopt more environmentally oriented forest practices. The problem definition dominant at the time – confusing, nonbinding, poorly enforced standards – pushed the solution toward formalization, a comprehensive code of binding regulations. Some of the rhetoric around the Code's introduction, such as the stringent $1 million fines, was clearly exclusively symbolic, given what we now know about the track record for noncompliance penalties. Nonetheless, given the problem definition, selling the Code effectively required it to have real teeth. The 6 percent cap is a powerful indication of how the presence of a critical subsector can constrain the magnitude of change in other subsectors within the same broader policy regime.

The dilemma of formalization, however, means that along with binding regulations come burdensome procedures, the reduction of discretion, and increased costs.[98] When it introduced the Code in 1994, the NDP government claimed the annual cost of the Code would be $250 million to $300 million. Industry disputed those figures at the time, and they have since been proven to be dramatic underestimates. One indication of this fact is that when the government was introducing revisions to the Code in 1998, it claimed that the relatively modest regulatory changes would *save* $300 million a year. What this means is that when the government was introducing the Code, the balancing of costs and benefits it used in designing its provisions was seriously skewed. The government knew it needed to act boldly, but it overshot. The government probably also underestimated environmental benefits, at least in terms of the political measure of easing environmental group criticisms at home and abroad.

The acceptable decision space available to the government seems exceptionally narrow, perhaps nonexistent. Given the existing policy paradigm of integrated resource management, it may not be possible in BC to fulfil the Code's mandate to balance the "productive, spiritual, ecological and recreational values of forests to meet the economic and cultural needs of peoples and communities." Although the alternatives are untested and uncertain, our implementation analysis suggests that the only hope may be to move to an alternative model.

With one case and so many explanatory factors, it is hard to establish how to rank the importance of these different variables. The episode of revision in 1997-8 does provide some insight. When favourable market conditions collapsed, industry power grew. Although the NDP was still in control, the leadership had changed colours from pale green to medium brown. Those changes created significant pressures for regulatory retrenchment. The magnitude of the retrenchment seems small, however. The government claims that it will save the industry $5 per cubic metre, a 40 percent reduction in costs. Industry officials believe that figure is overly optimistic, and initial estimates suggest savings more like $1 to $2 per cubic metre. Fluctuations in party leadership and in the business cycle (from boom to bust) certainly influenced the direction of change on the government agenda, but those pressures were counteracted by the powerful force of international market pressures mobilized by environmentalists.

Some weight should here be given to policy legacies: the inertia built into the status quo, which comprises the laws of political physics and the interests of expert bureaucrats who benefit from a byzantine procedural quagmire that only they fully understand. The Ministry of Forests seems just recently to have concluded, in part because of limited available resources to manage the rules, that it does not have an interest in procedural complexity. It is now searching for reforms that would provide similar or better results with fewer administrative costs. If the government does not manage its political environment effectively, it may be supplanted by private standard-setting organizations that have emerged to rival the influence of government regulators on corporate forest practices.

4
The Politics of Long-Term Policy Stability: Tenure Reform in British Columbia Forest Policy

On 9 June 1999, the New Democratic Party (NDP) government of the province of British Columbia announced the results of an examination into several aspects of the provincial forest tenure system undertaken in 1996-7.[1] In conjunction with the 1998 Forest Statutes Amendments Act, which provided a statutory basis for the initiative, the review resulted in the creation of seven pilot, or experimental, forest tenures in municipalities and reserves in the province. However, despite much attention being paid to this aspect of provincial forest policy over several decades, with numerous proposals having been put forward by various actors for its reform, the announcement covered less than 80,000 hectares of the province's 60.6 million hectares of forest land.[2]

Reform of tenure, or the allocation of rights to harvest timber on Crown and private lands, had been a subject on the provincial government agenda since 18 January 1989, when the previous Social Credit provincial government had temporarily put on hold plans to create 100 new Tree Farm Licences (TFLs) covering half the province.[3] The plan, which had been promoted by the government since the early 1980s, was withdrawn and the entire proposal referred to public hearings.[4] Within weeks, however, the proposal was completely dead and the tenure status quo retained. Although tenure reform had been placed on the back burner by the NDP government following its election in 1991,[5] it was never far from the minds of key government officials and politicians, and it was expected to be a major component of that government's second-term policy agenda.[6]

Why did the 1997-9 and earlier 1989 initiatives fail to result in major reforms to the provincial tenure system? Why was the status quo preferred to policy change? As this chapter will argue, these initiatives were only two in a long series of proposals that failed to substantially alter tenure policy in British Columbia. Despite multiple attempts at substantial reform over the past fifty years, the general contours and structure of tenure policy remain very much the same as when it was first created in the late nineteenth century. Over time, tenure arrangements have become heavily institutionalized and are now very difficult to change without incurring substantial costs. As a result, the system of property rights created by the institutionalization of the tenure system over the first century of the province's history has created an almost insurmountable obstacle to substantial policy change, despite significant changes in recent years in the level and extent of public and policy community support for the existing system.[7] The path-dependent nature of this critical subsector has led to a tendency on the part of policy makers to rebuff proposals for reform and choose to retain the status quo, sometimes for extremely long periods of time.[8]

This chapter will show how the principal elements of British Columbia's system of forest tenure – its goals, objectives, instruments, and settings – were established in the period before and just after the province's entry into Confederation. At that time, the principal goal of this policy subsector was to create the conditions for economic growth in the province, with the objective of supporting the development of a large, stable, and financially secure forest industry in order to do so. The instrument adopted at the outset of the commercial exploitation of the province's vast forests involved government lease of Crown land and timber resources to private corporations for exploitation. The settings for this instrument involved payments made to governments for timber used, with private companies obliged to perform certain tasks related to investment and forest management. This basic system of goals and instruments has remained intact for over 130 years, though the specifics of its operationalization have been altered on many occasions as different instrument settings have been used to match shifts in government objectives.

The record of alterations to the tenure system reflects efforts first to expand the system, then to consolidate it, and finally to rationalize it. This latter stage has occurred in two steps, with rationalization on the coast occurring before that in the Interior. Rationalization remains only partially complete in the Interior today. This is not to say, of course, that the system remains identical to its original formulation in 1865-88. The nature of the obligations placed on leaseholders has changed over the years to include more and more duties and responsibilities, and the size of the areas of forest covered by individual licences has expanded considerably. However, the basic goals of the tenure system, and the basic techniques or instruments used to accomplish these goals, have remained the same.

This chapter examines the history of the evolution of BC forest tenure policy and develops the notion of a "negative decision" in which a policy cycle is abbreviated by a conscious decision on the part of a government to retain most elements of the status quo. The examination of the tenure case suggests how policy community members "frame" an issue and the level of institutionalization of past policy decisions can work together to generate a propensity for arrested policy cycles and long-term policy subsector, and regime, stability.

Subsectoral Regime Characteristics

Ideas

The existing tenure system in British Columbia is a complex mixture of specific licensing arrangements developed over the past century and a half.[9] However, the principal idea lying behind the system is simple: private exploitation of public resources to generate economic wealth. Significantly for the analysis of policy making in this sector, however, alternative ideas have always been present and considered, and rejected, throughout this time. For much of the first century of forest exploitation in the province, for example, many efforts to alter existing arrangements were made, usually toward the public exploitation of public resources or the private exploitation of private ones.

The current system exists very much as a compromise between these two "pure" types of ownership/production regimes, and it has done so for some time. Throughout BC's history, for example, private businesses have urged the alienation (that is, sale) of Crown lands. Although this option was discussed, and rejected, by the 1976 Pearse Royal Commission on Forestry and the 1991 report of the Forest Resources Commission,[10] it nevertheless continues to be endorsed and proposed by leading academic forest economists and business executives.[11] Most recently, the proposals have been oriented toward lengthening the period of the lease toward perpetuity and giving the licensee much greater control over the management of forest lands in areas covered by tenures.

At the other extreme, opposition parties in the distant past, such as the CCF, have urged the removal of private companies and the use of public enterprises to exploit provincial forests.[12] However, the demand for complete nationalization of the industry has not been heard for several decades. In practice, public enterprises have only been used in extraordinary circumstances when governments have temporarily taken over older mills to prevent sudden employment losses in forest-dependent communities, such as occurred in Ocean Falls in the 1970s and in Prince Rupert in the 1970s and 1990s. More modern versions of the proposal have called for a clearer separation of the role of managing forest lands, over which government would play a greater role, and timber harvesting, over which the private sector would have greater control.

Another variation on this public ownership/exploitation theme is apparent in a set of proposals developed by environmentalists. Rather than turn forest management over to the provincial government, they urge that a new system of local, community-based tenures be created. An alliance of environmentalists, forest workers, and First Nations (the Tin-Wis Coalition), for example, argued in the late 1980s that forest management authority should be devolved into the hands of community forest boards.[13] Similar proposals have emanated in more recent years from First Nations for a new system of Aboriginal forest ownership and co-management of forests on public lands located nearby.[14]

Hence a range of ideas, including proposals for 100 percent private control of provincial resources to 100 percent public ownership of the industry, have been apparent in this subsector over the last century. In recent times the range of options put forward has been more limited, but there remain a large number of alternative visions in circulation.

Actors

All of the central actors in this critical subsector share a similar goal structure and generally endorse efforts made to promote the continued health and viability of the forest industry. However, as the brief overview of the major competing proposals above illustrated, each of the major options mooted for tenure reform is backed by a different major actor in the subsectoral policy community, and the preferred sets of objectives, instrument preferences, and choice of instrument settings vary substantially with each proposal and actor.

The existence of a diverse array of proposals accounts for some of the difficulties encountered in tenure reform. It is difficult to see how a government could adopt any of the proposals, or even some combination of them, without alienating at least one of the several major actors now present in the forest sector.[15] This interesting conundrum points to the complex nature of policy stability because, for most of the first century of BC forest policy making, a very different situation prevailed – one in which the two or three major actors in the forest sector agreed on the basic elements of the existing tenure system. That is, as Jeremy Wilson has argued, for almost a century a closed policy monopoly constructed between governments and industry existed in the provincial forest sector. Government interests were sometimes fragmented among various ministries and agencies responsible for various aspects of the forestry, and industry interests among those of small capital, large industry, and labour unions. However, as Wilson noted, "discourse has been dominated by those most intensely involved in forest exploitation, by the interests on the exploitation axis – forest capital, forest labour and the government forest bureaucracy."[16]

Despite their individual preferences for purely private or purely public forest exploitation, forest capital and labour were both more or less satisfied with a compromise public-private situation in which provincial Crown lands were leased to forest companies for extended periods in exchange for commitments to construct and maintain manufacturing facilities. This solution preserved a semblance of government control over the industry while providing the security of raw material supply required for industry to obtain financing for mill construction and development. This latter investment, in turn, generated employment for small businesses and workers and created the economic base for many communities throughout the province, satisfying the primary government goal of economic development. Although demands for additional allowances for small woodlots continued, by and large the compromise on tenure reached at an early stage of provincial history remained satisfactory to all major parties. As discussed above, the provincial debate was punctuated by periodic proposals from business and labour to privatize or nationalize the industry or provincial forests. However, these proposals were never close to being accepted or implemented by any provincial government.

The closed government-industry-labour policy community prevailed for close to a century, and that community has only been upset by the relatively recent emergence of new actors in the sector. Generally, in the forest policy sector as a whole, environmentalists emerged to contest most aspects of the industrial forestry model in the 1970s and 1980s. In the tenure subsector, they were joined by small business and community groups concerned with the concentration of holdings among the larger multinational forest companies. These groups lobbied and agitated for the reform of the tenure system to allocate woodlots and community forests to smaller operators, both private and public – and mainly municipal – in nature. Although these groups did not share all the same goals, because environmentalists generally were concerned less with investment and employment issues and more with those related to ecological imperatives such as wilderness and biodiversity preservation, they could all agree on the objective of moving toward a decentralized, locally controlled system of ecosystem management based on new tenure instruments.[17]

More recently, these groups have been joined by the province's First Nations, who have emerged from a lengthy court battle with the province over Aboriginal title with a set of property and political rights that enable them to potentially affect the existing tenure system by undermining Crown control over the provincial resource base. (This development is discussed at length in Chapter 5.) While the effects of potential First Nations control or influence over the disposition of a substantial amount of the provincial land and resource base are unknown, agreements and arrangements with the province's Aboriginal communities to date have centred on rebuilding Native communities through employment and investment opportunities related to the forest sector. The impact on tenure policy has been minimal because First Nations have seemed content to retain the present leasehold system, with the provision that rents should accrue to their own, rather than to the provincial, governments.[18]

Thus, in this subsector, new actors have brought new ideas forward, and since 1970, they have undermined the historic compromise made between labour-business and government in the early years of the province. As will be discussed below, however, this development has resulted in some instability in the subsector but not yet any major or significant changes to tenure policy. This is largely because the long-term institutionalization of earlier tenure decisions has constrained policy choices.

Institutions

Although the configuration of actors in the subsector has had an important impact, undermining policy stability in this subsector, a large part of the reason that proposals to alter the tenure system have had little effect on altering the status quo is the heavily institutionalized – or path-dependent – nature of the subsector.[19] Tenures are legally binding long-term arrangements that govern many subsidiary institutional arrangements, from mill financing on the part of industry, to government investments in social services and educational facilities in forest-dependent communities, through a wide variety of personal decisions made by individuals involved in or indirectly dependent on the industry.

There is some debate over the ability of governments to alter tenure arrangements regardless of their leaselike nature. That is, it is usually argued that various forms of resource tenures, not only in the forestry sector but also in areas such as mining and fisheries, are privileges that do not in themselves confer property rights.[20] Jurisprudence on the issue is mixed, however, with some court decisions having left the impression that any substantial alteration of a preexisting arrangement requires compensation payments or in-kind equivalents.[21] Also, at present, the entire issue of compensation for regulatory "takings" has been left wide open by Chapter 11 of the North American Free Trade Agreement (NAFTA), ostensibly an open-investment clause but one that has been used with some success by companies against governments instituting regulatory changes of any kind.[22]

Although the general issue of compensation for resource tenure changes is open to interpretation, compensation must be paid for BC forest tenures because this requirement is a part of the provincial Forest Act. Section 50 of the Act allows the government to remove only a maximum of 5 percent of the allowable cut without compensation being paid to the licence holder. In addition, Section 53 states that compensation must be paid if withdrawals for nontimber purposes exceed the 5 percent limit.[23] The Act specifically precludes application of the provincial Expropriation Act, except when forest lands are required for road construction.

As such, tenures are a heavily institutionalized component of the BC forest policy regime. They are legally binding documents, enforceable by the courts, and their unilateral alteration by governments before their scheduled expiry is usually prohibitively expensive, if not impossible. This is because any such action will immediately prompt court actions by companies and leaseholders demanding compensation for both present, and future, revenues and profits.[24]

Although the means and techniques by which compensation is calculated are not specified in law, negotiation is the favoured method, and, until recently, the favoured solution was the exchange of timber lands in the province. That is, when the provincial government has desired to alter one tenure arrangement by way of withdrawals, it has usually compensated licensees by providing an equivalent tenure elsewhere in the province. This system began to break down in the 1990s, however, as most forest land had been allocated and equivalent exchanges proved more difficult to achieve. The result was greater recourse to litigation to settle monetary awards and more public disagreements between the government and licensees over the terms of tenure alteration. A good example of current difficulties is provided by a late 1990s provincial proposal to transfer 30,000 hectares of land to MacMillan Bloedel rather than pay $84 million to the company for compensation over alterations to its licences resulting from park land withdrawals in 1990 and 1995. The land transfer proposal generated intense public debate and forced the government to commission a set of public hearings into the issue. The result was virtually unanimous disapproval and a government decision to compensate the company in cash rather than land.[25]

Historical Background

The specific lease-type arrangements that comprise the British Columbia tenure system have a long institutional history. Shortly before the two colonies of British Columbia and Vancouver Island were united in 1866, a policy of leasing timber, subject to the rents and terms established by the provincial governor, was made official colonial policy through its inclusion in the Land Ordinance of 1865.[26] This approach reflected established practice in the rest of British North America at the time, especially the pattern and practices developed in Quebec and Ontario.

Again in keeping with central Canadian practices, in 1871 the Terms of Union with Canada granted ownership of Crown lands in the province to the provincial government. After Confederation, lease arrangements for timber removal continued to be the norm. The provincial government began a policy of granting large tracts of land to consortia that were planning to build railways in the province, coincidentally placing large amounts of timber into private hands.[27] However, the main thrust of provincial forest tenure policy remained the leasing of public resources to private companies for their exploitation.[28] In 1887, amendments to the Timber Act recognized for the first time the value to the government of the sale of timber leases as a major source of revenue.[29] Outright sales of Crown lands valuable chiefly for their timber resources were forbidden, and timber cutting was made conditional on the issuance of a cutting licence requiring payment of a fixed fee per volume of timber cut.[30]

By 1888, all the basic elements of the modern tenure system were in place. That year, the provincial government brought in a new Land Act that consolidated the provisions of the previous acts.[31] The main thrust of the new legislation was to exchange timber leases for a commitment by lessees to erect lumber mills,[32] and the area of the province covered by the tenure system expanded rapidly as both the government and industrialists began to realize the extraordinary potential for revenue and profit generation contained in the province's forests.[33] Based on this model, the provincial government in 1901 instituted a new system of twenty-one-year pulpwood leases, with low stumpage and ground rent provisions, on the condition that a pulp mill be constructed by the lessee producing a minimum volume of pulp per area leased.[34]

Sensing the increased demand for timber supplies brought about by the development of the pulp and paper industry, and faced with a shortfall in its own revenues, the McBride government moved in 1905 to open up the timber licence system in an effort to secure a higher return from the province's timber lands. Through amendments made to the 1888 Land Act, further timber leases were prohibited and all future sales brought under a new transferable timber licence system.[35]

In a major departure from earlier practice, the provincial government dropped provisions for the establishment of productive mills contained in the previous lease system[36] and opened the door to speculation in timber supplies by making timber licences exchangeable.[37] Although this legislation succeeded in increasing government revenues, the massive speculation and concentration of timber holdings that followed resulted in the government quickly reversing itself and reserving all remaining unalienated land in the province by order-in-council in December 1907. In 1909, the government established a royal commission to determine the effects of past policies and the direction the province would take.[38]

The Royal Commission of Inquiry on Timber and Forestry (the Fulton Commission) was established in 1910 and consulted extensively with conservationists and members of the forestry profession during its brief investigation.[39] The commission endorsed the aims of the special timber licence system, arguing it represented an improvement over previous systems and had successfully brought a larger proportion of the value of the province's forests into the public purse. However, the commission was concerned that as the licences neared expiration they might encourage wasteful harvesting; it therefore urged that the term of the licences be extended in perpetuity or until all the merchantable timber was removed from the lands in question.[40] This last proposal was adopted in 1910 and most of the special licences became perpetual licences, in effect transforming them into multiple small leasehold arrangements.[41]

More significantly, the new Forest Branch created as a result of another Fulton Commission recommendation quickly developed a new form of timber licence, adding a volume-based licence to the existing multiplicity of area-based tenure types. These were the volume-based "timber sales" originally established between the freezing of special licences in 1907 and the adoption of the Forest Act of 1912. The sales of limited stands of timber on the margins of existing leases and licensed areas became an important source of pulpwood for the burgeoning provincial pulp and paper industry.[42] By the beginning of the Second World War, about one-third of the timber cut in the province originated from lands covered by this kind of arrangement.

The introduction of volume-based tenures, the complexity of the area-based tenure system,[43] and the rapid increases in timber removals caused by the creation of the pulp and paper industry led to a major review and reevaluation of provincial forest policy between 1937 and 1945.[44] This review ultimately resulted in the establishment of a new objective for provincial forest policy: "sustained yield." The major goals behind the tenure system – providing public resources to private operators in return for employment, cash, and services – and the actual instruments used to implement this goal – long-term leases – remained the same. However, the instrument settings were altered as contiguous licences and leases of various kinds were combined and consolidated to cover larger areas of forest in an effort to simplify the system and promote tree farming.[45]

The changes in objectives and settings were developed in the second British Columbia royal commission into the forest sector, appointed in 1943 under British Columbia chief justice and former provincial attorney general Gordon Sloan.[46] The commission reported in 1945, and many of its major recommendations were incorporated in revisions to the Forest Act made by the Liberal-Conservative coalition provincial government in 1947.

The report argued that the system of temporary alienations of timber then in place, in which the areas under licence reverted to the Crown once they had been logged, offered "no encouragement" to operators "to treat these lands as permanent tree-farms producing continuous crops." The system instead encouraged the practice of "cutting and getting out."[47] Sloan argued that "responsible operators" realized such behaviour was contrary to their long-term interest in a continuous supply of raw material for conversion plants, and he recommended tenure policy reform to create two types of sustained yield units, the private and public "working circles." In the private working circle, private timber holdings would be consolidated over large areas and supplemented by Crown lands to provide a secure base for harvesting operations. Sloan suggested that the province expropriate "at fair stumpage valuation" the holdings of any private owners who refused to join such circles, which were aimed at providing perpetual supplies for established or new conversion plants.[48] In the public working circles, either Crown Lands would be leased to harvesters or private holdings without associated conversion plants would be combined into larger units to supply open log markets.[49]

In 1947, amendments to the Forest Act allowed the consolidation of private holdings into private working circles through the creation of Forest Management Licences (FMLs) or Tree Farm Licences (TFLs). Public working circles or Public Sustained Yield Units (PSYUs) were created from tracts of Crown land and managed by the revamped Forest Service. The result was the creation of a mosaic of TFLs and PSYUs throughout the province, each with an approved rate of harvest.[50]

In his 1945 report, Sloan had recommended that a new royal commission be established within ten years to assess the progress made by the province toward the sustained yield objective.[51] In 1955, W.A.C. Bennett's Social Credit government obliged, recommissioning Sloan to evaluate the situation of forest resources in the province. The second Sloan Royal Commission received identical terms of reference as did the 1945 Royal Commission, and after extensive public hearings, the commission reported in July 1957.[52]

Most of the recommendations contained in the second Sloan report concerned the fine-tuning of existing policies to keep the balance between public and private interests. Thus, for instance, Sloan devoted many pages to examining the process by which Forest Management Licences were awarded, and he underlined his contention that simply owning existing production facilities should not entitle a company to priority in allocation.[53] Similarly, he recommended that the Forest Management Licences themselves be amended to delete all references to "perpetuity" and to make the term of any licence twenty-one years so that the public would not mistake a licence for an alienation of Crown land or the forfeiture of government control.[54] Following the receipt of the report, several modifications were made to the existing system of forest administration, though the main lines of the system remained those established in 1947.[55]

The system ran into some temporary difficulties after Social Credit Forest Minister Robert Sommers was forced to resign in 1956, then convicted in 1958, for having accepted bribes related to awarding of a forest licence.[56] After 1958, however, the government continued to consolidate and rationalize the patchwork of various old forms of tenure into the new TFL system by enacting various provisions calling for the reversion of logged-over lands to the Crown. The government also attempted to transfer responsibility for several important forest management practices to harvesting companies through the incentive of deductions from Crown stumpage. This was the case with the 1964 changes to the Timber Sale Licences requiring licensees to carry out silvicultural work as well as 1965 changes requiring licensees to construct primary access roads.[57]

Most of these tenure reforms affected areas on the coast, west of the Cascades. However, a second pulp boom occurred in the late 1960s in the Interior, requiring similar consolidation efforts in that region to guarantee resource supplies for the pulp industry.

The boom in the Interior occurred following the introduction in 1966 of the "close utilization," or "third band," timber sale policy by Lands, Forests and Water Resources Minister Ray Williston. Before 1966, cutting rates in TFLs and PSYUs were determined according to "intermediate utilization" standards specifying minimum diameter cuts and maximum tree-stump and top lengths. In 1966, the government offered one-third more than the allocated cut to any firm that agreed to adhere to a "close utilization" standard; that is, one permitting the cut of smaller diameter trees and reducing the amount of timber contained in stumps and tops left lying in the forest after harvesting had been completed.

Because additional cuts were allocated directly to existing licence holders and not bid upon, this offer resulted in an overnight increase of up to 50 percent of existing timber supplies, provided close utilization standards were followed. Most smaller diameter timber was suitable for pulpwood harvesting, though the incentive of free timber also convinced many sawmill operators to invest in the equipment required to process smaller sawlogs. In 1969, the excess material contained in Pulpwood Harvesting Areas as a result of the inclusion of the close utilization timber was sold under Timber Sale Licences as "third band" timber to private companies. These sales were made with the proviso that producing companies agreed to supply chips to holders of the pulpwood harvesting agreements if the need arose. By 1975, over one-half of the timber harvested in the Interior was covered by this form of licence.[58]

By the early 1970s, then, the provincial project of implementing and extending a tree farm tenure system throughout the province was largely complete. However, as a result of the layering of various forms of tenures from the pre- and post-Sloan periods, and the continued expansion of the forest industry from the coast into the Interior and north of the province, many different forms of tenure continued to exist throughout the province and, more significantly, dominated different regions. This development led to several attempts to simplify, consolidate, and rationalize the tenure system, a process that almost succeeded in 1987-9 but ultimately failed as new actors mobilized against the fundamental goal of industrial forestry underpinning the century-old lease system. As a result, the closed government-industry-labour policy community broke down, leading to a period of instability in provincial tenure policy.

Policy Cycle

Several policy cycles have occurred in recent years related to tenure system rationalization, but they have had a limited impact on the configuration or nature of the overall leasehold system. That is, despite the emergence of new ideas and new actors in the subsystem, BC tenure policy continues to provide a good example of long-term policy stability. As noted, the goals and instruments used to implement the policy have remained largely the same for over a century, whereas the objectives and settings of the policy have changed only twice: once during the short-lived timber licence system of 1905-7, when the government desire for increased revenue temporarily trumped other regulatory goals, and again in 1937-47, when the pressures of the pulp and paper industry on the resource base prompted the consolidation of the lease system under the terms of the Sloan Commission report. With these exceptions, all other tenure policy decisions have tended to be negative in that proposals for change have made it onto government agendas and options have been formulated, but the ultimate decision has been to retain virtually all the elements of the status quo.

Cycle I: The Reorganization of PSYUs into TSAs (1972-8)

The contemporary policy debate on the rationalization of the tenure system began in the early 1970s when a policy window opened as a result of the provincial New Democratic Party replacing the long-term Social Credit government of W.A.C. Bennett.[59] Like the Liberal government of the 1930s, the NDP government elected in 1972 inherited a patchwork of timber tenures with a bewildering array of stumpage and royalty rates, tenure types, renewal provisions, and forestry obligations. While coastal areas were consolidated into a few large area-based management units, in the Interior most cutting occurred on volume-based tenures that required extensive public investment to ensure the attainment of the sustained yield policy objective set out by Sloan in 1947 and again in 1956.

Faced with this situation, the David Barrett government stepped up the process of rationalizing and simplifying the existing system of forest regulation. The government appointed the Task Force on Crown Timber Disposal in January 1974 to carry out this work.[60] Chaired by UBC forest economist Peter Pearse, the task force submitted two reports in 1974, the first a review of the old tenures, the second an evaluation of the stumpage appraisal system.[61] The completion of these reports signalled the movement of the issue into the formulation stage of the policy cycle.

Although the issue of tenure reform entered the agenda as a result of the election of the new government, and shifted into the formulation phase following the evaluation of the existing system by the Pearse Task Force, the further movement of the issue through the policy cycle was delayed by several events, including the election in 1975-6 of a second Social Credit government under W.A.C. Bennett's son, William R. Bennett, and by the lack of an articulated alternative to the existing leasehold tenure system.

Initially, however, the NDP government moved in January 1975 to develop some of Pearse's recommendations into legislation, passing the Timber Products Stabilization Act. This legislation authorized the establishment of a Forest Products Board modelled on the Task Force proposal for a new government Timber Authority that would set conditions for chip supply and prices throughout the province.[62] Faced with opposition to its scheme from the forest industry,[63] however, the Barrett government appointed a fourth royal commission into the province's forest resources in June 1975, again under the chair of Peter Pearse.[64]

The Pearse report contained a large number of recommendations pertaining to provincial forest policy. These recommendations ranged from improved fire and disease protection to a renewed emphasis on reforestation and timber management on privately held lands, but its most important recommendations concerned rationalizing the complex provincial forest tenure system and decentralizing the government's forest administrative apparatus.[65] These proposals amounted to only modest modifications to the existing tenure system, however, merely changing the names of the major tenure types and altering only slightly the distribution of publicly managed forest lands throughout the province.[66]

This project of rationalization and consolidation was continued by successive Social Credit governments after 1976. In 1977, following the defeat of the NDP, the new Social Credit government of W.R. Bennett established the Forest Policy Advisory Committee to review the Pearse Commission's recommendations. In 1978, three new acts were passed that established the basis of the current regulatory regime. These were the Ministry of Forests Act, the Forest Act, and, less important, the Range Act.[67] The new acts provided the basis for a rationalization of the patchwork of existing tenures into thirty-four new Timber Supply Areas (TSAs), which included lands from a variety of tenures and especially from the TFLs and PSYUs. Privately owned lands remained concentrated in TFLs, whereas the TSAs became the basic unit of timber allocation in the areas of the province outside the TFLs, each having an allowable annual cut apportioned to the holders of various forms of tenure within the TSA.[68]

Associated with the new tenure forms were a series of obligations imposed on tenure holders to undertake various reforestation and silvicultural activities. Under Sections 88 and 52 of the Act, respectively, licence holders could receive compensation in the form of credits against stumpage for silvicultural work undertaken or in additional allowable annual cuts in anticipation of enhanced future yields.[69]

Cycle II: The Rollover Initiative, or Attempts to Eliminate TSAs and Extend TFLs (1979-89)

These modest reform efforts did not stem criticisms of provincial policy for failing to anticipate future supply shortfalls,[70] and in the late 1970s, the provincial government undertook a large resource inventory exercise to assess the state of the provincial forests and the success of the sustained yield project. This effort culminated in the 1980 Forest and Range Resource Analysis and the five-year Forest and Range Resource Program, which highlighted significant problems remaining in the conversion of the province from a natural to a managed forest.[71] Following the publication of the Forest and Range Resource Analysis in 1979-80, the government published a series of white papers on outstanding forestry issues, such as the role of small business, complaints concerning the appraisal system, and the need for improved, intensive forest management efforts.[72] Proposals for substantial tenure reform remained out of the decision-making process, however, as the government continued to concentrate on the consolidation and rationalization of the provincial system.

A major effort was made to achieve this end in the attempt to extend the provincial TFL system to the Interior in 1987. The so-called TFL Rollover project was put into place by the Social Credit administration of William Vander Zalm, which, beginning in 1985, made several incremental changes to provincial forest policy. This included the creation of a small business timber allocation that provided 5 percent of annual cuts to small business bidders. Proposals for more substantial reforms to the provincial system of forest regulation, however, grew out of a short forest management review carried out in early 1987. The review identified fifty-six issues facing the forest sector and made recommendations for action concerning them. Among the more significant recommendations were proposals to "privatize" small parcels of forests provided purchasers agreed to maintain their status as forest land, replacing government stumpage credits for road construction and silvicultural expenditures with direct contracts, and increasing minimum stumpage rates.[73]

In September 1987, the government released its package of reforms entitled New Directions for Forest Policy in British Columbia.[74] The proposals immediately sparked a large-scale industry outcry against plans to eliminate Section 88 credits and alter the provincial stumpage system.[75] The new system, calculated as a fixed percentage of industry gross sales, was designed to shift reforestation expenses onto the industry at the same time that government annual revenues would increase by an estimated $100 million.[76]

More significantly, at the same time that the government annoyed large business in the sector, it also managed to alienate small business and environmentalists by unveiling a plan to reorganize the tenure system by replacing most volume-based licences in the Interior with area-based Tree Farm Licences. The plan to "roll over" the TSAs with TFLs proposed withdrawing areas covered under new TFLs from government-managed Timber Supply Areas (TSAs) while making new TFL licence holders responsible for forest management activities.[77] Companies would lose as much as 10 percent additional AAC unless they were prepared to meet certain performance standards pertaining to utilization and secondary manufacturing.[78]

Although the industry generally supported the rollover proposals, the entire New Directions package for most major companies represented a significant blow by increasing costs, especially stumpage rates. Small business and environmentalists, ironically, also opposed the package, focusing not on stumpage and costs but on what was perceived as selling off public forests to big business. Although the industry tried to reverse the onerous parts of the package through behind-closed-doors lobbying, the opponents of the rollover scheme went very public. As a result, Forest Minister Dave Parker found himself in the unenviable position of drawing a barrage of criticism both from those who believed he was giving away the province to the major forest companies and from the companies themselves over other aspects of the package.[79]

The government nevertheless introduced the rollover legislation in 1988 and by the end of the year had received about 100 applications to convert Forest Licences into Tree Farm Licences. But mounting opposition, focusing on a scheduled hearing for Fletcher Challenge's application to consolidate several licences in the Mackenzie area of northcentral BC into a six-million-hectare TFL, forced the postponement of the project.[80]

In mid-January 1989, Parker announced that further rollover hearings would be halted until he had had a chance to hear public reaction at several public information sessions he would chair in different parts of the province. The minister said this procedure would give him a chance to dispel certain "myths and fallacies" about the proposed changes.[81] At the meetings, Parker and his senior officials did hear endorsements of the rollover plan from large companies; however, most of the 300 speakers heard at the meetings were environmentalists, loggers, First Nations spokespersons, and small operators who used the opportunity to condemn not only the rollover plan but also other facets of ministry policy.[82] Despite a vigorous government defence,[83] the New Directions plan was shelved after the hearings, with the responsible minister and deputy minister leaving the portfolio shortly afterward.[84]

Cycle III: Tentative Steps towards Reform after the Rollover Defeat (1989-99)

The rollover defeat of 1989 left the province's tenure policy in disarray. The proposal had been much more than a simple reorganization of existing tenures. It was instead the completion of a century-long process of the replacement of the natural forests of the province with tree farms geared toward the production of raw materials for the forest products industry. At the eleventh hour, a coalition of new actors in the subsectoral and sectoral policy communities had coalesced and mobilized to defeat the proposal. However, none of these groups could put forward an alternative policy acceptable to all parties, and none of the long-time players in the sector – including large industry, government, and labour – was content with the half-completed tree farm system.

The government responded to this discontent with the creation in mid-1989 of what was initially constructed to be a permanent Forest Resources Commission (FRC). The FRC was instructed to review all forest issues,[85] though it was also given three priority tasks: to advise on the effectiveness of TFLs, to recommend schemes for improving public participation, and to review ways of improving forest practices. Not surprisingly given the disunity of the opposition and the commitment of the older subsectoral actors to the tree farm project, the FRC ultimately recommended a plan involving conversion of volume-based tenures that was very similar to the recently defeated rollover plan.[86] Unwilling to embark on a second forest reform campaign at the end of a scandal-filled mandate, however, the Vander Zalm government allowed the new proposal, and much of the rest of the FRC report, to languish.

In 1991, the New Democratic Party gained office under Premier Mike Harcourt, ostensibly on a labour-green platform but in fact promising more competent and honest government than that of its immediate Social Credit predecessor. The successful NDP election platform contained a promise to establish a royal commission on forestry to make recommendations on tenure changes;[87] however, tenure reform issues were nowhere to be seen in the party's first term in office. Andrew Petter, who replaced Dan Miller as forest minister in fall 1993, was convinced of the need for tenure reform, but the government understood that to keep its other forestry-related initiatives (such as forest practices and protected areas) on track, industry cooperation was required.[88] Unless the government was willing to accept the industry's position on tenure – essentially, the re-creation of new, area-based, private forest management licences covering the entire province much as the rollover proposal had envisioned – such cooperation would be jeopardized by any alternative tenure reform initiative. As a result, the only change that actually emerged to forest tenure during the first NDP administration involved an expressed willingness to transfer as much as 5 percent of the provincial land base to Aboriginal ownership through the land claims treaty process.[89]

After winning a second mandate in 1996, the NDP government of Glen Clark established a committee to investigate the possibility of a community-based forestry tenure system but continued to avoid any substantial debate on tenure reform. Although several high-profile cabinet ministers spoke out in mid-1999 about the continuing need to reform the system[90] and appeared to endorse elements of the corporate, area-based FML rollover proposal, the government quickly distanced itself from the proposal when it was condemned by both environmentalists and forest unions.[91]

Nevertheless, semiprivatization proposals still circulate in various guises. One such proposal is not to completely sell off the province's forest lands to private companies or individuals but to divide the currently almost 100 percent Crown-owned forests into an ownership pattern of one-third public, one-third small woodlot, and one-third large private company.[92] As early as 1990, for example, groups such as the Truck Loggers Association (TLA) argued the tenure system was an impediment to the development of a diversified, high-value-added manufacturing sector.[93] It proposed separating forestry and manufacturing operations, with processors forced to acquire fibre from timber growers which would be allocated a diverse array of area-based tenures.[94] Woodlots and other small-scale licences would account for over half of the harvest,[95] and the TLA recommended that responsibilities for the other half should be turned over to a new Crown-led company assigned management responsibility for the forest resource. Its primary purposes would be forest renewal and the attainment of the highest possible growth and yield standards.

This type of idea was adopted and put forward by the provincial Forest Resource Commission (FRC) in its April 1991 report. The FRC proposed asking existing licensees to give up rights to some timber in exchange for a more secure hold on the timber they would retain. It also recommended that the proportion of the cut held under secure tenure arrangement by companies with manufacturing facilities should gradually be reduced from about 85 percent to 50 percent.[96] The timber rights taken from present licensees would be reallocated under a diverse array of new tenures. Some would be handed out to communities, Native bands, and woodlot operators in small area-based tenures, and others would be allocated under a new volume-based tenure instrument.[97]

A variation on this private ownership/exploitation theme received some support from prominent provincial foresters in the mid-1990s. In this version, a "zonation" scheme would be developed in which only certain areas of the province would be designated for exclusive forest industry use. In these areas, wilderness, visual, and wildlife requirements would be relaxed and timber maximization would be the preeminent management objective.[98] Although such proposals rest on very shaky grounds related to expectations concerning future silvicultural investment and reforestation efforts, proponents using mathematical models to estimate these parameters have claimed that only 40 percent of the current industrial land base would be required to provide 100 percent of the existing cut.[99]

Within the provincial forest industry, however, tenure reform proposals continued to mirror the contours of the failed 1987-9 rollover project, suggesting that the provincial forest be converted to a 100 percent area-management scheme in which major licence holders would obtain large-scale licences covering virtually all of the province in large, contiguous blocks. Before its 1999 sale to Weyerhaeuser, provincial forest industry giant MacMillan Bloedel, for example, urged the development of such a private sector management system. Harkening back to the 1987-9 rollover proposals, the company suggested that the government rationalize the existing tenure system by converting most major volume-based tenures into area-based Forest Management Licences (FMLs). Governments would pay operators of FMLs to manage the forests, and licensees would be forced to put 30 to 40 percent of the allowable cut up for competitive bids. In this version, a competitive bidding process was intended not only to aid forest management but also to help alleviate concerns about de facto government subsidization of the industry that have led to the establishment of a quota system for BC exports under the terms of the Canada-US Softwood Lumber Agreement.[100]

More recently, a public sector version of the same argument was put forward by academic forest economists who argued that "efficiency is best served by direct public management of the forest resource." They urged the provincial government to enhance the public management of public lands through some form of Crown corporation, with cutting rights clearly separated from forest management activities.[101]

As these different proposals swirled around in the policy community, the Clark government announced a public policy review process in July 1999 as part of its "Forest Action Plan."[102] The inquiry, led by Jobs and Timber Accord advocate Garry Wouters, released several discussion papers, including one that dealt with tenure issues and suggested some form of zonation be adopted throughout the province.[103] The review quickly become embroiled in controversy, however, as the 4,000-member Association of BC Professional Foresters took out full-page newspaper ads criticizing the short timeline established for the review and urging the creation of a royal commission.[104] This development, in turn, prompted the forest industry to launch its own set of public hearings to garner support for its preferred option of longer-term tenures.[105] By the time the Wouters report was released in April 2000, it had been thoroughly discredited, and the caretaker government of interim premier Dan Miller quickly distanced itself from its recommendations, which included a by now familiar plea to complete the post-Sloan rationalization of the existing tenure system by "moving from volume- to area-based tenures through forest stewardship agreements to provide greater diversity in the supply and management of timber."[106]

Hence, despite the continual demands for reform heard from academics, environmentalists, and others outside industry, the only initiative to emerge from the NDP's second term in office was the small pilot project on small-scale, local community resource control developed in 1997-9.[107] Even here, however, the initiative was not only modest in scope but modest in innovation. Community forestry, of course, is not entirely new to BC, having been established through municipal forests created in Mission and North Cowichan in the 1940s. More recently, community-owned TFLs and other types of licences were issued in the west Chilcotin in 1994, in Kaslo, Creston, Lake Cowichan, and Nootka Sound, and in the Revelstoke area in 1995-7.[108]

Conclusion: Conceptualizing Tenure Decisions in Policy Terms – Positive Decisions, Nondecisions, and Negative Decisions

Policies are made in many forms. Besides traditional "positive" decisions in which policy cycles lead to implementation, it is also possible to distinguish between "negative" decisions in which a conscious decision is taken to preserve the status quo and "nondecisions" in which options to deviate from the status quo are systematically excluded from consideration.[109] In both of the later cases, policy cycles are "arrested" and proposals for policy change rejected. Unlike the situation with nondecisions, in which certain options simply never enter into policy deliberations, agenda setting and policy formulation do occur with negative decision making and some decision is actually taken. But unlike the situation with positive decisions, policy implementation does not follow from a decision.

There is a substantial literature on positive decision making and the contours of such decision-making processes are well known; they relate to such factors as how state capacity and subsystem complexity interact to influence the nature of policy outcomes.[110] Nondecision making has also been the subject of many inquiries and studies, beginning with the community power debates in political science in the early 1960s and 1970s[111] and extending into quasi-experimental studies in contemporary psychology.[112] Much contemporary discourse analysis in policy studies also implicitly focuses on nondecisions because it often reflects upon how ideologies operate to restrict the types of options put forward in policy making.[113]

Very little research into negative decisions, however, exists.[114] Nevertheless, the elements that lead to such decisions can be discerned from an examination of the policy cycle operating in the case of BC tenure policy. That is, BC tenure policy after 1970 represents several instances of arrested policy cycles, in which decisions were taken to preserve the status quo rather than to effect major policy change. The key analytical question with negative decision making that this case can help unravel is why a policy process begins but then ends at the decision-making stage.[115]

In the BC tenure case examined above, several elements of this process become clear. First, an important factor affecting policy stability was the entrenchment of previously existing tenure arrangements in law, with the prospects of large penalties and costs being incurred for their alteration in any fashion opposed by the tenure holders. The extensive court-adjudicated institutionalization of policies in this subsector served, and continues to serve, as a major barrier to substantial policy change.

Second, the nature of the relevant subsectoral policy community has also served as a major impediment to change, although for different reasons at different times. That is, before 1987-9, the closed and restricted nature of a tenure subsystem that was essentially limited to large tenure holders and government acted to spur policy reforms but only in a manner consistent with the established goals and instruments adopted in this subsector. The entry of other actors such as First Nations, environmentalists, small business, and forest communities into this subsystem after 1970 resulted in some contestation over these goals and instruments, but these actors have been unable to significantly alter these elements of tenure policy or their corresponding objectives and settings. In the particular circumstances of the times, although these new actors brought new ideas with them into the policy regime, they have only agreed on the need to halt the completion of the tree farm system by the key players in the old policy network. They have been able to block adoption of new policies consistent with the old regime and have obtained minor concessions from governments in areas such as small business, Aboriginal, and community tenures, but they have not been able to substantially alter the status quo.

It bears repeating, therefore, that a central factor affecting stability in this subsector, and accounting for nondecisions, is the institutionalization of past tenure decisions both in law and in terms of the interrelationships found between policy actors. As Margaret Weir and others have argued, these kinds of policy legacies are a very important contributor to policy stability.[116] As Weir asked in her study of American employment policy, when an issue is routinized or institutionalized, "what accounts for the pattern of periodic innovation set within a broader historical trajectory in which arguments favouring a government role in employment steadily lost ground? To answer this question I show how the interaction of ideas and politics over time created a pattern of 'bounded innovation' in which some ideas became increasingly unlikely to influence policy. Central to this narrowing process was the creation of institutions, whose existence channeled the flow of ideas, created incentives for political actors, and helped to determine the political meaning of policy choices."[117]

Weir's notion of bounded innovation calls attention to the path-dependent nature of BC tenure decisions. That is, contemporary tenure decisions occur within the context set by past decisions, and the policy legacies of legal and compensatable tenure rights bias decisions toward the status quo by making changes very costly while institutionalizing the rights of some actors but not those of others.

For most of the first 100 years of provincial history, the closed industry-government policy community portrayed the tenure system as one that maximized government revenues while promoting the economic health of the province and ensuring responsible corporate behaviour. The fact that this policy community broke down after 1970 without substantially altering any of the goals, objectives, instruments, or settings of tenure policy illustrates the relationship that policy communities have to institutions. That is, the policy community was restructured by the entrance of new actors, such as environmentalists and First Nations, and the image of tenure policy was sullied at both the popular and the elite levels as inefficient and inequitable. However, the institutionalization of tenures through legislation and the courts, coupled with the need to compensate companies for their alteration, has continued to prevent change from occurring in this critical subsector.

Although most firms have threatened litigation to resolve these problems, in the recent proposed transfer of 30,000 hectares of land or $84 million to MacMillan Bloedel for compensation for park withdrawals from its licences, the company offered to relinquish its claims as an incentive to the government to enact the company's proposals for major tenure reform.[118] This example illustrates the close links between historical tenure policies and present-day proposals for reform, and how the institutionalization of past decisions continues to shape, direct, and constrain present and future choices.

The BC tenure case, then, confirms some elements of contemporary policy theory and poses additional questions for existing models and concepts. It provides a good example of the manner in which certain change processes, such as policy community expansion, and certain stability processes, such as path dependency, interact to block or limit policy change.[119]

The case also reveals how elements of the policy-making process, such as policy communities and networks, exist and interact at both the sectoral and subsectoral levels. Although most analyses in the past have examined policy making virtually exclusively at the sectoral level, an analysis focusing on the subsectoral level helps to illustrate how issues and sectors are related and how they bring about policy change.[120]

That is, as this case underscores, the constellation of actors in the subsectoral tenure community was heavily influenced by events occurring at the sectoral level and beyond. Actors such as environmentalists and First Nations, presumably, were able to enter tenure policy deliberations as a result of their activities in other venues and sectors. Perhaps more significantly, however, as the other chapters in this book also reveal, events and activities at the sectoral level have been and continue to be blocked by events and occurrences at the subsectoral level. In the case of the tenure subsector, for example, path dependencies have not only blocked change in the subsector, such as the completion of the rollover proposal, but this blockage has prevented the realization of alternative patterns of activity at the sectoral level, such as the significant restructuring of policy goals to create a new system of bioregional or ecosystemic forestry. As such, in the context of BC forest policy, the tenure subsector provides a good example of the creation and operation of a critical subsector, one capable of influencing the structure and shape of the entire provincial forestry sector.

Appendix 4.1

Major tenure types, British Columbia

Tenure	Rights	Responsibilities	Term/size/ replacement provisions
Tree Farm Licence (TFL)	Rights to carry out forest management on a specific area of Crown land and almost exclusive rights to harvest an AAC from the licence area, under cutting permits.	Licence responsible for resource inventories, strategic and operational planning, road building, and reforestation. Licence must maintain a manufacturing facility if required in original licence.	Twenty-five-year term; replaceable every five years. Typically very large-scale operations.
Forest Licence (FL)	Rights to harvest an annual volume of timber within a TSA, under cutting permits.	Licensee responsible for operational planning, road building, and reforestation. Licensee must maintain a manufacturing facility if required in original licence.	Typically fifteen-year term; replaceable every five years. Typically medium- to large-scale operations.
Timber Sale Licence (TSL) (mostly SBFEP)	Rights to harvest timber from a specified area of Crown land within a TSA or TFL area.	Ministry of Forests is responsible for operational planning, road building, and reforestation on TSLs sold under the SBFEP. Licensee must maintain a manufacturing facility if	Typically six months to five to ten years; most not replaceable. Typically small- to large-scale operations.

Woodlot Licence (WL)	Rights to carry out forest management on a specific area every five years (maximum 400 hectare on coast, maximum 600 hectare in interior) of Crown land and exclusive rights to harvest an annual volume of timber from the licence area, under cutting permits.	planning, road building. Licensee must not own or operate a manufacturing facility.	strategic and operational and reforestation. Small-scale (family-focused) operations.
Pulpwood Agreement (PA)	Rights to harvest up to a maximum annual volume within the TSA or TSLs, in the event that its holder cannot meet its fibre requirements privately.	Licence requires management plan. If harvest occurs, responsibilities are similar to FL. Licensee must maintain a manufacturing facility.	Up to twenty-five years; new contracts may or may not be replaceable. Potentially very large-scale operations.
Timber Licence (TL)	Exclusive rights to harvest merchantable timber from a defined area of Crown land, under cutting permits.	Operating plan required. Licensee responsible for operational planning, road building, and reforestation. Once forest is reestablished, area reverts to Crown and becomes part of the TSA or TFL.	Variable; not replaceable. Individual licences are relatively small.
Miscellaneous	Rights described below.	Very limited responsibilities.	Short term; not replaceable. Very small scale.

Source: BC Ministry of Forests. Available at www.for.gov.bc.ca/pab/publctns/timbtenr/table1.htm.

Appendix 4.2

Harvest area by tenure type: Private and public lands 1981-98

Area harvested (ha)[a] in BC from 81-2 to 97-8 by land status

	Crown land				Private land				
Year	TFL[b]	TSA[c]	Other[d]	Subtotal	TFL[e]	TSA[f]	Other[g]	Subtotal	**Totals**
81-2[h]									**147,889**
82-3[h]									**162,172**
83-4	23,212	161,847	3,169	188,228	5,356	4,280	10,388	20,024	**208,252**
84-5	26,368	164,519	7,566	198,453	7,491	3,842	6,631	7,964	**216,417**
85-6	28,560	175,306	6,531	210,397	5,333	4,511	11,489	21,333	**231,730**
86-7	28,142	188,875	9,447	226,464	496	12,917		13,413	**239,877**
87-8	39,020	184,867	15,082	238,969	3,468	17,545		21,013	**259,982**
88-9	29,423	201,109	6,924	237,456	3,040	17,512	12,393	32,945	**270,401**
89-90	20,790	166,932	6,412	194,134	1,235	14,363	8,652	24,250	**218,384**
90-1	19,068	131,207	1,737	152,012	447	27,903	1,168	29,518	**181,530**
91-2	20,141	138,143	8,066	166,350	4,736	18,305	3,598	26,639	**192,989**
92-3	23,457	163,077	10,067	196,601	3,709	21,289	[i]	24,998	**221,599**
93-4	33,271	142,571	6,071	181,913	751	25,084	[i]	25,835	**207,748**
94-5	23,914	128,576	7,516	160,006	644	29,594	[i]	30,238	**190,244**
95-6	23,654	123,737	5,345	164,869	921	23,819	[i]	24,740	**189,608**
96-7	22,448	150,671	5,920	179,039	760	19,230	[i]	19,990	**199,029**
97-8	21,457	128,484	4,504	154,474	682	18,645	[i]	19,327	**173,801**

[a] Includes only blocks where harvesting is completed in the relevant fiscal year.
[b] Includes Crown land within Tree Farm Licences.
[c] Includes Crown land within timber supply areas excluding Tree Farm Licences.
[d] Includes Crown land within woodlot licences, farm leases, licences to cut, federal lands, and First Nation reserves.
[e] Includes private land within Tree Farm Licences.
[f] Includes private land outside Tree Farm Licences and within certified tree farms for 1983-4 to 1985-6. Includes private land outside Tree Farm Licences for 1988-9 to 1991-2.
[g] Includes private land within woodlot licences, farm leases, licences to cut, federal lands, and First Nation reserves.
[h] Includes all Crown and private land within Tree Farm Licences, tree farms, and farm woodlot licences.
[i] Estimate only because private landowners are not required to report area harvested. Includes private land within woodlot licences.

Source: MOF Annual Reports. Available at: www.for.gov.bc.ca/hfp/forsite/jtfacts/hrvlndst.htm.

5
Policy Venues, Policy Spillovers, and Policy Change: The Courts, Aboriginal Rights, and British Columbia Forest Policy

Recent decisions by the Supreme Court of Canada have resulted in significant change to one major element of the provincial forest policy regime as the development of new forms of Aboriginal property rights has added a new set of actors – First Nations – to the forest policy community. This example of regime alteration is an instructive one for students of policy change and development because how the regime has changed is readily apparent. Blocked at the political level in their quest for recognition and cultural affirmation, in the 1970s and 1980s Native groups and their advisors shifted venues and pursued their claims to title to sizable areas of the province through the judicial process. The courts ultimately upheld the principle of unextinguished Aboriginal title in many parts of Canada, including British Columbia, and forced politicians and administrators to include the province's Aboriginal peoples in any consideration of a wide range of land use practices and policies, including forestry.

Judicial developments in the area of Aboriginal rights affect the general operation of the Canadian political system, but they also spill over into many specific policy sectors, including virtually all natural resource and environmental areas.[1] Although Aboriginal property rights rulings are recent and their ultimate effects unknown, this new category of rights presages a greatly enhanced role for Native organizations and actors in many areas of natural resource and environmental management. However, as will be discussed below, the ability of First Nations to bring about substantial change in policy content may well be limited by their need to continue to succeed in the courts in future decisions related to the operationalization of the new notion of Aboriginal title, including its declaration, the nature of acceptable infringements upon it, and the means and mechanisms by which First Nations can be compensated for its violation. It may well be the case that even if First Nations continue to enjoy success in the courts, the use of Aboriginal lands for large-scale resource exploitation activities like forestry will necessitate the land's conversion from Aboriginal to fee simple forms of title.

As such, since many elements of the new Aboriginal property rights regime remain to be developed,[2] only the general outlines of the most likely scenario for resource management in the near future can be suggested here. Examination of the experiences with Native claims in the rest of Canada suggests that the most likely result will be the establishment of a three-tiered system of resource co-management on affected lands through the conclusion of modern treaties and other forms of administrative agreements between First Nations and provincial governments. This system, consisting of exclusive First Nations control over only small areas of land and provisions for co-management on larger areas of shared lands would leave most Crown and privately held lands untouched.[3] Nevertheless, in BC forestry, any changes in the nature of property rights in the forestry sector will directly affect critical subsectors such as tenure. Given the key position occupied by that subsector, those changes can be expected to lead to changes in other subsectors, including employment, land use, and cut allocation.

The Characteristics of the Canadian Aboriginal Policy Regime

Actors

A variety of policy actors are involved in the Canadian Aboriginal policy regime and have been active in the Aboriginal title issue subsector. They include numerous First Nations and their organizations; the federal government, which enjoys constitutional jurisdiction over Indian affairs; and provincial governments in their role as principal land and resource owners. In addition, a variety of third party interests are involved, ranging from small private landowners to major resource companies with leases over resources located on lands claimed by First Nations.

The role played by First Nations in this issue area is, of course, critical. In areas of the country covered by modern treaties, First Nations have negotiated a variety of proprietary and usufructory rights that can impinge on governmental land and resource use decision making. This situation represents a significant change from an earlier era in which only very limited powers over reserve lands were held by First Nations. It developed slowly over fifty to sixty years as Native organizations pursued a multifaceted attack on the image and venue of the existing Aboriginal rights regime.[4]

As Douglas Sanders observed, this process was at times delayed and at others aided by the actions of Canadian governments.[5] It was definitely delayed in 1927 when the federal government moved to prohibit bands from raising funds to press Native land claims without government consent.[6] It was aided beginning in 1950 when a major governmental restructuring transferred jurisdiction over Indian affairs to a new department of citizenship and immigration. The separation, for the first time, of Aboriginal issues from the administration of physical resources and land resulted in increasing attention to their social, political, and cultural conditions.[7] One important consequence was the revocation in 1951 of the ban on land claims activity in a major revision of the federal Indian Act undertaken at that time.[8]

The 1951 revisions allowed First Nations organizations to articulate a separate vision of how Canadian Native policy should develop and marked their reentrance into the formal Aboriginal policy network. Aboriginal organizations continued to grow and began to engage in an increasingly sophisticated range of political activities.[9] These organizations were in a constant state of turmoil as they attempted to achieve consensus among various Aboriginal groups about their visions of the future. Nevertheless, the groups began to develop the political expertise required to wage a long-term campaign in favour of their interpretation of Aboriginal rights as inherent to a people rather than as granted from above by colonizing states.[10]

The role of the federal government in this policy issue area has already been alluded to. Under the terms of Section 91(24) of the Constitution Act, 1867, the federal government is given exclusive jurisdiction over "Indians, and lands reserved for Indians." The extent of federal responsibilities under this clause is set out in the Indian Act, which was first passed in 1876 and later amended on several occasions. The federal Department of Indian Affairs, under various administrative titles, has implemented the Indian Act, providing a range of services to First Nations that most other Canadians receive from provincial governments, such as health care, welfare, education, law enforcement, housing, employment, and agriculture.[11]

The federal government assumed responsibility for the negotiation of treaties with Canada's Aboriginal population following Confederation in 1867 and succeeded in negotiating a variety of agreements covering a large part of Canada's land mass between 1867 and 1921.[12] Claims in many parts of Canada remained unresolved, however, leading to a new round of modern treaty talks in the 1960s and 1970s in areas such as northern Quebec, the Northwest (including Nunavut) and Yukon Territories, and, most recently, Labrador and British Columbia.

The federal government has been faced with numerous types of claims, ranging from very minor ones to ownership and control of large areas of the country, and developed a wide range of policies and procedures for dealing with these issues. In the 1970s, it attempted to divide claims into two types: specific claims affecting the interpretation of existing treaties and comprehensive claims involving the negotiation of entirely new ones. Many talks in both categories are under way between First Nations and the federal government, which has led to the accumulation of experience and precedents on the part of federal negotiators. Both types of treaty talks, however, are often criticized by First Nations as slow, ponderous, and sometimes unfair, given the advantages enjoyed by the federal government in terms of expertise and the availability of financial, legal, and information resources.[13] For its part, however, the federal government is bound by a series of judicial decisions that have set out the duty of the federal government to treat First Nations in a favourable and respectful manner, consistent with the relationship of trust in which First Nations find themselves vis-à-vis the government.[14]

The role of provincial governments toward First Nations is different from that of the federal government. Although provincial governments deliver a wide range of federal services to First Nations through a complex set of intergovernmental agreements, their relationship with Native peoples is distinct from that of the federal "trust" relationship.[15] In many cases, the relationship between the provinces and First Nations is inherently conflictual because, for example, any expansion of First Nations territorial control reduces provincial jurisdiction over lands and resources, directly conflicting with provincial land management and ownership rights set out in Sections 109 and 92(5) of the Constitution Act, 1867.

Ideas

At least three distinct sets of images or ideas about the nature of relations between non-Aboriginals and First Nations have existed in Canada over the postcontact period. Each has been institutionalized in a set of treaties formalizing First Nations-government relations.[16] The first image was of pragmatic utility and was embodied in the various treaties of friendship and military alliance signed by both British and French colonial regimes with hundreds of Native bands between the time of first contact and the mid-1800s, primarily in eastern Canada and Quebec.[17]

The second image was of paternalistic modernization and assimilation and was embodied in a set of treaties signed by both British and Canadian authorities between 1850 and 1921. These treaties generally sought to extinguish special Aboriginal rights and title in exchange for "presents," land, and other guarantees of traditional Native resource and land use practices.[18] They include the Robinson-Superior treaties in Ontario and run through the negotiation of the so-called numbered treaties with various Native nations in the prairie provinces, up to the conclusion of Treaty 11 in the Northwest Territories in 1921.[19]

The third image of limited coexistence lies behind the "modern" treaties negotiated since the signing of the James Bay and Northern Quebec Agreement in 1974.[20] It extends through the seven comprehensive land claims settled since (the Cree Naskapi or Northeastern Quebec, the Inuvialuit, Yukon, Gwich'in, Sahtu, Nunavut, and Nisga'a Final Agreements[21]) to the over fifty such claims currently under negotiation in British Columbia, Quebec, and Labrador.[22]

Institutions

The primary relevant policy institution in this issue area is the legal treaty setting out the nature of the relationship existing between the Crown and First Nations as well as the land and resource rights, if any, a First Nation possesses. For contemporary Aboriginal land and resource jurisdiction, the current Canadian situation is complex due to the overlapping spatio-temporal nature of the three sets of treaties described above and the policy images they contain. The old treaties of military alliance were often thought to have had little impact on resources, though they have recently been given renewed vigour by the courts and now affect a variety of resource rights. These rights range from hunting and trapping wildlife to harvesting in the crab fishery as well as, potentially, oil and gas and other large-scale resource development.[23] In areas covered by early treaties of settlement, First Nations control a mixed variety of different aspects of resource management, depending on the nature of the treaty involved. In most cases, however, Native peoples have control over only land and resources included on Indian reserves, though they are unable to alienate these lands except through an extremely cumbersome political-administrative process. Some bands, such as those located in parts of southern Alberta with substantial oil and gas reserves, do very well from the sale of resource rights. But again, control over the expenditure of funds remains governed by complex arrangements between local bands and federal "Indian" administrators.[24]

For modern treaties signed after 1975, the situation is quite different. Many of these treaties cover vast areas of land (especially in the Northwest and Yukon Territories) and are extremely dense and complex. Based loosely on the precedents set by the Alaskan Native land settlements in the United States, these agreements contain provisions concerning educational, political, social, and cultural life as well as provisions relating to the economy and environment of the area covered under the treaty.[25] Although each agreement is unique, the general tendency has been for the First Nations involved in each land claim to receive a small area of land for which they have exclusive surface and subsurface rights. Each also receives larger areas over which they may exercise only surface and subsurface rights of specified kinds.[26] What has received the most attention in these treaties to date has been mineral, oil, gas, and, especially, hunting and trapping rights, though in Quebec issues related to hydroelectricity generation have also been important.[27] Because most of these agreements have covered areas in Canada's far north, forestry issues have not received much attention, though this situation is changing as agreements are negotiated and signed for heavily forested areas of British Columbia.[28]

For the most part, modern settlements have traded off actual or potential First Nations control over mineral, oil and gas, and hydroelectricity development in large areas of the country in favour of financial compensation, recognition of Aboriginal title over restricted areas of land, and protection of fishing, hunting, and trapping rights.[29] Nevertheless, the fact that hunting and trapping rights intersect with provisions governing resource project approvals – especially concerning habitat and environmental protection measures – this has given Aboriginal groups covered by modern treaties a major voice in most resource policy areas.[30] Thus, for example, Native groups may not actually own or directly control a mineral or forest resource, but they may control access to the resource across lands dedicated to hunting and trapping.[31] Similar controls exist with lands covered by older treaties, all of which contain trust, or "fiduciary," arrangements and responsibilities requiring governments to consult with First Nations regarding alienation or use of reserve lands and resources.[32]

This patchwork of resource ownership and management rights has become more complex and far-reaching in recent years as the Supreme Court of Canada has gradually developed jurisprudence concerning issues of Aboriginal rights and title in the remaining areas of the country not covered by treaties and agreements, including virtually all of British Columbia.

As the discussion below shows, the courts effectively created the modern treaty system by actively promoting the attainment of a policy image of limited coexistence over the continued adherence of federal and provincial governments to the earlier policy paradigm of modernization and assimilation. This kind of overt judicial policy making is rare in Canada, but it represents, in this instance, the results of a case of successful venue change on the part of organizations representing Canada's indigenous population. The institutionalization of this pattern of rights has significant implications for many policy sectors and subsectors because recently created Aboriginal property rights spill over into many areas of social and economic life, not the least of which is forestry.

Historical Background

Shortly after Confederation, the Judicial Committee of the Privy Council had ruled in *R.* v. *St. Catherine's Milling,* a case originating in a federal-provincial dispute over the Ontario-Manitoba border, that Aboriginal title existed in Canada but was a phenomenon that traced its origins to the Royal Proclamation of 1763.[33] In other words, Aboriginal title existed only in areas covered by the proclamation and only because it had been granted to the indigenous population by the British monarch. This ruling confirmed the logic that lay behind the settlement treaties negotiated at this time. That is, Aboriginal land entitlements were not inherent or absolute, and they could be altered and extinguished by governments through a variety of means, including the adherence of Native bands to federal treaties. It also confirmed the belief of some provincial governments, notably British Columbia, that Aboriginal entitlements did not exist in that province because it was not part of the area covered by the original proclamation of 1763.[34]

This legal-political status quo was first altered, accidentally, through the impact of the natural resource transfer agreements signed between Ottawa and the western provinces in 1930. Those agreements aimed at reconciling the anomalous situation that had existed with respect to federal ownership of provincial lands and resources since the creation of Manitoba in 1870 and Alberta and Saskatchewan in 1905. Because these agreements mentioned Native rights, they had the effect of constitutionalizing such rights for the first time. This fact provided Native groups in the west, and others by implication, with an avenue through the courts to address their concerns and challenge the validity of federal government actions.[35] As Douglas Sanders put it: "While the Department of Indian Affairs would not [even] support the hunting rights of prairie tribes, the transfer of natural resources from the federal government to the provinces in 1930 placed policy making in different hands. In 1930, as in 1982, a constitutional opening occurred for reasons unrelated to Indian issues, which resulted in the constitutional recognition of Indian rights."[36]

After the ban on land claims organizing was lifted in 1951, Native organizations throughout the 1950s and 1960s engaged in numerous activities ranging from protests and demonstrations to court challenges and media campaigns aimed at altering the political and administrative status quo. These activities were primarily designed to force government recognition of Aboriginal rights to self-government on Native lands.[37] In these efforts, the issue of the relationship existing between Native and non-Native societies and governments was redefined from removing barriers to, and promoting, the assimilation of First Nations into the cultural and socioeconomic mainstream to being a question of the recognition and restoration of historical Native rights and entitlements. This redefinition allowed Native groups to use a variety of means and new venues, including the courts, appeals to the United Nations, and First Ministers meetings, to challenge their exclusion and marginalization by Canadian federal and provincial policy makers and administrators.[38]

British Columbia First Nations were prominent players in this process of shifting policy venues, and many of the significant court cases were decided on the facts of BC cases.[39] This situation was due not only to the merits of the cases but also to the continued refusal of the provincial government to recognize any form of unextinguished Aboriginal title in the province. Although the colonial government before Confederation recognized Aboriginal title and negotiated several treaties in the southern Vancouver Island area to extinguish it, the provincial government immediately preceding union with Canada in 1871 denied the continued existence of Aboriginal title in the province, downplaying the legal status of the treaties already signed by former governor James Douglas and arguing that unencumbered Crown title existed upon the province's entry into Confederation.[40]

When the federal government assumed control over Native affairs under the Terms of Union with the new province, it initially disputed the provincial position that Aboriginal title had been extinguished.[41] But it soon came to accept the argument put forward by the Judicial Committee of the Privy Council in *St. Catherine's Milling* that Aboriginal title originated with the Royal Proclamation of 1763 and that made by the provincial government that the document did not apply to British Columbia. This doctrine guided provincial government policy for close to 120 years until a series of Supreme Court cases in the 1970s and 1980s made it clear that the argument was no longer tenable.[42]

Although forays into federal-provincial politics yielded only limited results for British Columbia First Nations,[43] the thrust toward the courts in the 1960s produced more substantial results, especially in the area of jurisdiction over, and management of, lands and resources.[44] The *Calder* decision in 1973[45] was the first major victory for Native organizations, serving as the event that forced the land claims issue onto government agendas at both the federal and provincial levels.[46]

In *Calder,* BC First Nations secured a ruling that Aboriginal rights and title had not been extinguished in areas not covered by treaties and that affirmed the Royal Proclamation of 1763 extended throughout the country. Although the case was lost on a technicality, the ruling produced immediate results. It allowed Native groups to challenge through court injunctions resource development on lands not covered by treaties, a tactic that led to the negotiation of the first modern treaties in Quebec and the Northwest Territories as well as the creation of a new federal policy on both specific and comprehensive claims.[47] The ruling also eventually led to the reversal of the century-old BC policy on land claims and the commencement of treaty talks in the province.

The Policy Cycle

The development of provincial policy relating to the new role to be played by First Nations in BC forest policy occurred very much as a spillover from the successful venue-shifting activity undertaken more generally by Aboriginal groups in Canada. Beginning with *Calder,* this strategy on the part of First Nations ultimately culminated in the reversal of the Supreme Court's century-old *St. Catherine's Milling* doctrine on Aboriginal rights with the December 1997 decision in *Delgamuuk'w.*[48] Decisions in other cases related to the nature of government responsibilities toward First Nations, the continued existence of Aboriginal rights in the fishery, and the definition of Aboriginal title all combined to affect the substance and process of provincial government policy making in the forest sector in the 1990s.

Agenda Setting

The issue of Aboriginal participation in provincial forest policy making arrived on the provincial agenda as a result of the legitimization by the Canadian courts of the political militancy and activism of provincial First Nations. Successive Supreme Court decisions following *Calder,* such as *Sparrow, Van der Peet, Gladstone,* and *NTC Smokehouse,*[49] fleshed out the elements of the new relationship between First Nations and Canadian governments in the resource area. They effectively forced the inclusion of First Nations in resource management and land use decisions on lands not covered by treaties to which Aboriginal title apples, the largest contemporary example of which is most of British Columbia.[50]

The decisions created a new category of Aboriginal title as a sui generis category of property rights somewhere between public and private property. The court argued that these inherent rights can be surrendered or extinguished through the treaty process. However, when treaties are not in place, Aboriginal title continues to exist as an interest or encumbrance upon landownership and title that can only be infringed by governments under certain very specific circumstances and conditions.

Acting in accordance with the tenor of these decisions, the Clayoquot and Ahousat First Nations managed to obtain an injunction preventing logging on Meares Island in 1985, given the possibility of forestry operations interfering with the land claim of the Nuu-chah-nulth people. After a series of Aboriginal blockades of various projects throughout the province in 1985-90, the Vander Zalm government established the Ministry of Native Affairs in 1988 and ended the province's historical refusal to recognize Aboriginal title. This began the process of creating new provincial resource policies compatible with Supreme Court jurisprudence and the new property regime established in the post-*Calder* era.

Policy Formulation

The effect of the court decisions on Aboriginal rights and title was to provide provincial First Nations with a set of property rights enforceable by the courts as constituting an "interest" in the land area covered by

Aboriginal claims. The actions of provincial First Nations in pursuing this interest through the courts resulted in the issue being placed on the provincial policy agenda. However, despite the long buildup to the late 1980s round of judicial decisions, the province remained unprepared and unsure of the steps required to resolve this issue and of the impact modern notions of Aboriginal title would have on resource development and management.

The provincial response to this uncertainty was twofold. First, it established in 1990 a BC Claims Task Force with First Nations and federal government representatives. The task force report urged the permanent resolution of the uncertainty surrounding land claims through the negotiation of modern treaties with BC First Nations. It recommended the establishment of a treaty commission to coordinate the land claims and treaty negotiations process, and it also set out a timetable and detailed description of the procedures to be followed in so doing.[51]

After 1991, the new NDP provincial government moved quickly under Premier Mike Harcourt to adopt the proposals of the Treaty Commission, and both Harcourt and his successor Glen Clark vigorously supported and defended the negotiating and passage of the landmark Nisga'a final settlement through the provincial legislature. However, when the NDP initially gained office in 1991, the content of any future treaty provisions relating to forest issues was unclear, as was the question of what arrangements should prevail in the period before the successful conclusion of treaty talks.

In the forest sector specifically, since the early 1970s provincial First Nations had been involved in only a very minor way. At that time, the first provincial NDP government, in an effort to promote economic development on reserves, had amended the Forest Act to provide woodlot licences of up to 400 hectares of Crown land for terms of up to fifteen years to persons or "a band as defined by the Indian Act (Canada)." At least thirteen First Nations in BC had taken advantage of this provision, combining the forested portion of their reserves with leased Crown land to create opportunities to participate in the forest industry.[52] At least one band, the Tl'azt'en Nation in the Fort St. James region, operated a Tree Farm Licence. To obtain the licence, forest management on some 2,500 hectares of the Tl'azt'en reserve was combined in 1981 with that on 49,000 hectares of nearby Crown forest land.[53]

One initiative of the new provincial government was to try to expand through the treaty process similar types of limited forest sector involvement on the part of First Nations. Between the commencement of talks in 1991 and 1999, forty-seven First Nations representing about 70 percent of the Aboriginal population entered the BC Treaty Commission process to settle their grievances. Total settlement cash costs were expected in 1996 to range between $5.7 and $6.2 billion.[54] Canada was expected to contribute the bulk of these costs, with BC providing between $0.8 and $1.4 billion. BC's contribution would include credit for revenues on land that the province would contribute, reducing the province's share of the cash costs of treaties. It was estimated that First Nations would receive outright ownership of between 24,000 and 29,000 square kilometres of mainly rural land, which is roughly 3 percent of total provincial Crown land. Overall, the province suggested that a total of 5 percent of the productive provincial forest base could be transferred to Aboriginal groups as a result of the entire treaty process. This transfer could amount to 1.15 million hectares containing as much as 3.45 million cubic metres of timber. Existing licence holders would have to negotiate new arrangements with Aboriginal landlords, though they would be entitled for compensation for any losses in cuts which might result.[55]

The only major claim to be finalized with a substantial forest component is the Nisga'a claim, for which a final agreement was signed in April 1998 and royal assent granted in April 2000.[56] The agreement provides some details about the kind of forest policy the province expects to develop on lands covered by Aboriginal title. The Nisga'a Final Agreement contains provision for a Nisga'a TFL, transfers approximately 45,000 hectares of productive forest land to the Nisga'a, accounting for an allowable annual cut (AAC) of about 150,000 cubic metres of timber, or about 7 percent of the total AAC in the local Kalum timber supply area.[57] The agreement provides for substantial Nisga'a control over the forest resource on treaty lands:

> 5.3. On the effective date, the Nisga'a Nation owns all forest resources on Nisga'a Lands.
> 5.4. Nisga'a Lisims Government has the exclusive authority to determine, collect, and administer any fees, rents, royalties or other charges (in respect of timber resources on Nisga'a Lands).
> 5.6. Nisga'a Lisims Government will make laws in respect of the management of timber resources on Nisga'a Lands.[58]

Some clauses in the agreement, however, provide only for consultation and notification on issues related to timber lands near Nisga'a lands. It also specifies that Nisga'a foresters must maintain standards of forest management equal to or exceeding provincial government standards. Other clauses call for cooperation in areas such as wildlife management and fire protection.

Hence, this case illustrates that, under pressure from the courts, the provincial government has moved to formulate a system of First Nation-provincial forest resource co-management on Native lands. This is in keeping with the general desire, enunciated by the National Aboriginal Forestry Association (NAFA), that a system of joint management of forest resources by Aboriginal communities and provincial governments be established throughout Canada.[59] As NAFA argued in documents submitted to the Royal Commission on Aboriginal Peoples:

> There is a need to gain access to harvest resources such as timber. Second, there is a need to gain access to resource management decision making so that resources will be managed on an integrated basis taking Aboriginal cultural and traditional uses into account along with the interests of Canada's industrial society.[60]

NAFA contended that:

> First Nations feel entitled to partnership roles with the provinces in forest management in their traditional areas. Provinces must be encouraged to improve access to timber resources for Aboriginal firms, either through direct licences or third party agreements with industrial holders of large forest management agreement areas.[61]

In their view:

> The ultimate in cooperation between a province and an Aboriginal community is the truly "joint management" venture agreement in which Aboriginal organizations and the province sit as true partners in decision making. This type of partnership is usually sanctioned by the province through a formal agreement and even legislation.[62]

Decision Making

While entering into the treaty process to develop a long-term solution for Aboriginal title issues and their resource management implications, the provincial government also moved in the short term to enhance the presence of First Nations in existing land use decision-making processes, including the provincial forest policy-making process. Under continuing pressure from the courts,[63] in the early 1990s the provincial forest ministry formally recognized that Aboriginal rights are protected by the Constitution and cannot be infringed upon by activities undertaken or authorized by provincial law.[64] It mandated that Forest Development Plans must identify the location of areas of Aboriginal sustenance and of cultural, social, and religious activities associated with traditional Aboriginal life. Aboriginal representatives were to be involved in forestry activity planning through consultation and negotiation, and a First Nations Forestry Council was formed to increase their involvement in all areas of forestry.[65]

Under this system, various nontreaty memoranda of understanding were negotiated with provincial First Nations covering forestry issues. The Xaxli'p First Nation, located in the Lillooet area, for example, entered a joint stewardship agreement and memorandum of understanding with British Columbia in 1992. The agreement applies to any disposition and use of land, water, and resources within the Xaxli'p traditional territory. It provided for increased involvement for the First Nation in land and resources disposition; integration of Xaxli'p traditional knowledge in decision making; recognition of the Xaxli'p decision-making process; notification of and information about any proposed land and resource disposition by the government; consideration of Xaxli'p positions on proposed dispositions; structures and processes to seek consensus between the parties; and a dispute settlement mechanism. A memorandum of understanding under the joint stewardship agreement provides for the joint undertaking of an integrated resource management plan for the territory and for some employment on Ministry of Forests projects.[66]

The Claims Task Force recommended that this system of administrative agreements be expanded into a series of Interim Measures Agreements (IMAs) that would serve to establish the rights and responsibilities of provincial and First Nations governments while treaty negotiations were in progress. By October 1997, seventy-five IMAs had been signed, and sixty-one others were being negotiated. Of these, thirty-six covered forestry-related issues, from jobs and employment concerns to the mechanisms of consultation and timber allocations.[67]

Probably the most well-known IMA was negotiated in the Clayoquot Sound area with the Nuu-chah-nulth First Nation. Signed in March 1994, the IMA established a framework for the joint management of resource extraction in the area. The agreement created a central region board to administer the agreement, with equal numbers of provincial government and Nuu-chah-nulth members, a joint-chair system, and the requirement of a double-majority vote for major decisions.[68] The agreement smoothed the relations among the province, First Nations, and the forest industry;[69] however, environmentalists remained opposed to aspects of the agreement they felt did not protect nontimber values in the area. In June 1999, a further agreement was negotiated for MacMillan Bloedel, Greenpeace, Western Canada Wilderness Committee, Sierra Club of BC, the Natural Resources Defense Council, and the Nuu-chah-nulth First Nation, in which the company and the First Nation agreed to use ecoforestry principles and promote nontimber activities such as ecotourism in exchange for the environmentalists' help in marketing timber products from the area.[70] Finally, in October 1999, MacMillan Bloedel turned over its logging rights in the area to Iisaak Forest Resources, a new joint-venture in which five Clayoquot bands have a 51 percent interest, with MacMillan Bloedel owning the remaining 49 percent share.[71]

Like the treaty process, these developments moved the province in the direction of co-management of Aboriginal forest lands. However, while the Supreme Court expressed a clear desire in the *Delgamuuk'w* decision for such negotiated solutions to the title issue, some First Nations preferred to continue to use the courts to resolve claims issues. The first major effort to halt a logging operation under the *Delgamuuk'w* precedent failed in the province's Supreme Court and Court of Appeal, given the existence of overlapping claims and the inability of the First Nation involved to establish the sacred character of the lands in question. However, there was little doubt that such appeals would continue to be made.[72] This expectation was borne out in September 1999 when the Siska resorted to roadblocks to stop logging in the band's claim area near Lytton,[73] and the courts refused to grant a provincial government injunction to prevent logging undertaken by the Westbank First Nation on Crown lands it claimed as its traditional territory.[74] As a result, several other bands quickly emulated the Siska roadblock.[75]

The potential for disruption of resource exploitation caused by these cases continued to worry the provincial government. Those concerns led to the nomination in 1998 of high-profile former Yukon NDP premier Tony Pennikett as lead provincial negotiator on the BC Treaty Commission[76] and to a variety of proposals to "fast-track" or "prebuild" land settlements into claims negotiations to reduce any remaining uncertainty over property rights in large areas of the province covered by Native claims.[77] This effort bore some fruit in the year 2000 when a "political accord" was signed between the federal and provincial governments and the Wet'suwet'en First Nation, committing all three parties to work together to increase Wet'suwet'en involvement in the forest sector.[78]

Implementation

The essence of the three-tiered landownership and management regime toward which the province and First Nations are moving is very similar to the system put in place in the successfully concluded modern treaties located in Quebec and the Arctic. The general model they contain, and the one followed in the Nisga'a Final Agreement, was described by the Royal Commission on Aboriginal Peoples:

> On lands in the first category (which would include those lands now called Indian reserves), full rights of beneficial ownership and primary, if not exclusive, jurisdiction in relation to lands and resources would belong to the Aboriginal party in accordance with the traditions of land tenure and governance of the people in question. Aboriginal understandings of their title with respect to such lands could be recognized more or less in their entirety, leaving the people free to structure their relationship with the lands in accordance with their own world view.

> On lands in the second category, which would comprise a portion of the Aboriginal party's traditional lands, a number of Aboriginal and Crown rights with respect to land would be recognized by the agreement, and rights of governance and jurisdiction would be shared among the parties. Cojurisdiction or co-management bodies, which could be based on the principle of parity of representation among parties to the treaty, could be empowered to manage the lands and direct and control development and land use.
>
> On lands in the third category, a complete set of Crown rights with respect to land and governance would be recognized by agreement. Even on lands in this category, however, some Aboriginal rights could be recognized, to acknowledge that Aboriginal peoples enjoy historical and spiritual relationships with such lands. For example, Aboriginal people, as a matter of protocol, could serve as diplomatic hosts at significant events of a civic, national or international nature that take place on their territory.[79]

Most of these elements remain to be negotiated with most of the province's First Nations, let alone implemented in the framework of a treaty or administrative agreement. Although implementation is likely, the process at the rate of signings anticipated by the provincial government will take well over two decades to complete.[80]

The biggest threat to successful implementation of this new co-management system involves the unwillingness of some First Nations to shift venues from the courts back to the political system. Some First Nations, for example, argued that *Delgamuuk'w* strengthened their hand and their ability to develop and manage forest resources in their traditional territories.[81] They refuse to participate in treaty negotiations, preferring to remain with the court system, looking for a court-imposed solution to the questions of operationalizing Aboriginal rights.[82] The First Nations Summit organized shortly after *Delgamuuk'w* was released, for example, issued a statement to then federal Indian affairs minister Jane Stewart and provincial Aboriginal affairs minister John Cashore, in which they argued:

> All provincial and federal government alienation of lands and resources must be suspended until arrangements are made with First Nations and our informed consent is obtained. Without this it is our position that all actions (licences, leases, permits and so on) by the provincial and federal governments constitute unlawful infringements ... Where First Nations consent to infringement, fair compensation must be provided ... This may include such things as revenue sharing (i.e. stumpage, royalties). As well, First Nations need to be involved in the policy and legislative agenda of the governments. Existing policies and legislation must be reviewed and revised.[83]

Although some First Nations did pull out of treaty talks,[84] the federal and provincial governments ignored calls for a freeze on development and agreed only to minor changes in the mandate given to the provincial Treaty Commission.[85] The majority of First Nations – including the Nisga'a[86] – continued to work within the existing treaty and administrative processes toward the creation of co-management regimes covering their territories. They did so because, although *Delgamuuk'w* clearly established the need for consultation with First Nations over resource management issues on lands covered by Aboriginal title,[87] many other issues remained unresolved by the ruling and will require lengthy and expensive court cases for their resolution, with no guarantee that the outstanding issues will be resolved in favour of First Nations.

For example, the extent and nature of the required consultation is not specified in the ruling. First, it must be recognized that the requirement for consultation only applies to land and resources over which Aboriginal title exists. First Nations may assert title, but only a court decision actually establishing title would be of weight and effect. And obtaining recognition of a title claim is not a simple matter. According to the criteria set out in *Delgamuuk'w,* a First Nation must prove that:

(i) the land must have been occupied prior to sovereignty,
(ii) if present occupation is relied on as proof of occupation presovereignty, there must be a continuity between present and presovereignty occupation, and
(iii) at sovereignty, that occupation must have been exclusive.[88]

Second, even if title can be established, the court has also ruled that Aboriginal title is not absolute and that a wide variety of infringements on that title can be made.[89] The chief justice argued that the range of legislative objectives justifying the infringement of Aboriginal title is very broad. He stated:

> In my opinion, the development of agriculture, forestry, mining, and hydroelectric power, the general economic development of the Interior of British Columbia, protection of the environment or endangered species, the building of infrastructure and the settlement of foreign populations to support those aims, are the kinds of objectives that are consistent with this purpose and, in principle, can justify the infringement of aboriginal title. Whether a particular measure or government act can be explained by reference to one of those objectives, however, is ultimately a question of fact that will have to be examined on a case-by-case basis.[90]

Third, there is the question of exactly what constitutes a "consultation" along with whether lack of consultations or failure to obtain Aboriginal consent amounts to a veto over an infringement. As the court put it, this point too will vary case by case:

> The nature and scope of the duty of consultation will vary with the circumstances. In occasional cases, when the breach is less serious or relatively minor, it will be no more than a duty to discuss important decisions that will be taken with respect to lands held pursuant to Aboriginal title. Of course, even in these rare cases when the minimum acceptable standard is consultation, this consultation must be in good faith, and with the intention of substantially addressing the concerns of the aboriginal peoples whose lands are at issue. In most cases, it will be significantly deeper than mere consultation. Some cases may even require the full consent of an aboriginal nation, particularly when provinces enact hunting and fishing regulations in relation to aboriginal lands.[91]

Finally, there is the question of exactly what kind of compensation is required after title has been recognized, and consultation on an infringement has been found to be inadequate or an infringement ruled as unjustified. Here, the court refused to rule, except to say that these matters would be very difficult to resolve:

> The amount of compensation payable will vary with the nature of the particular aboriginal title affected and with the nature and severity of the infringement and the extent to which aboriginal interests were accommodated ... Since the issue of damages was severed from the principal action, we received no submissions on the appropriate legal principles that would be relevant to determining the appropriate level of compensation of infringements of aboriginal title. In the circumstances, it is best that we leave those difficult questions to another day.[92]

The court did say, though, that:

> No doubt, there will be difficulties in determining the precise value of the aboriginal interest in the land and any grants, leases or licences given for its exploitation. These difficult economic considerations obviously cannot be solved here.[93]

Hence, the results of any future court decisions on these questions, even if they are favourable to First Nations, may not have the impact that some First Nations expect. That is, the judiciary may not necessarily develop a process that greatly enhances First Nations powers to promote resource exploitation as opposed to their power to veto such development. First Nations, for example, are required to continue to use lands covered by

Aboriginal title for purposes consistent with their claim of traditional use.[94] Hence, the chief justice argued that surrender of Aboriginal title in exchange for various "considerations" would be *required* if Aboriginal peoples wished to use the lands for purposes inconsistent with their title. He stated very clearly: "If aboriginal peoples wish to use their lands in a way that aboriginal title does not permit, then they must surrender those lands and convert them into non-title lands to do so."[95]

Thus, although affirming Aboriginal rights and the continued existence of Aboriginal title in lands not covered by treaties, the *Delgamuuk'w* decision in most respects reinforces the principal components of the existing treaty process. Although a court-imposed settlement of implementation problems related to Aboriginal title is not out of the question, the difficulties, dangers, and costs associated with continuing a lengthy court process were not lost on the majority of provincial Aboriginal leaders. Nevertheless, by June 2000 several prominent First Nations, including the Sechelt Band, which had gone so far as to negotiate an agreement-in-principle in their own treaty talks, continued to express dissatisfaction with the treaty process and began, or threatened to begin, court actions.[96]

Conclusion: Policy Venues, Policy Spillovers, and Policy Change

In their work on policy formation in the United States, Baumgartner and Jones noted several strategies employed by excluded members of policy communities to gain access to policy deliberations and affect policy outcomes. One of these strategies involved the redefinition of a policy issue to alter the location where policy formulation occurred. Such manipulations of policy images and policy venues were intricately linked. As they noted: "Policy venues tend to become involved sequentially in an issue because of their differing perspectives and responsibilities ... Image change and venue access proceed simultaneously and in an interactive fashion. Changes in policy images facilitate changes in venue assignment. Changes in venue reinforce changes in image, leading to an interactive process characterized by positive feedback."[97]

Although not all policy issues are susceptible to manipulation, and not all political systems contain any, or as many, alternative policy venues, Baumgartner and Jones argued that actors outside of formal and informal policy processes have an incentive to attempt to alter an image or venue when they are excluded from an existing policy community.[98]

In their study of American agenda setting processes, Baumgartner and Jones outlined one typical pattern of image/venue manipulation that involved a shift in the venue for policy discussion from the executive to the legislative political arena. If the image of a regulatory issue, for example, could be redefined as an issue involving budgetary or legislative activities, this redefinition could transfer the issue from the executive to the legislative sphere, resulting in previously excluded parties attaining a position in policy deliberations denied to them by the reigning administration. Or as Baumgartner and Jones also noted, "some types of images may be well accepted in one venue but considered inappropriate when raised in another institutional arena."[99] Hence, "losers always have the option of trying to change the policy venue from, say, the national government to subnational units, or from so-called iron triangles to election politics, and such efforts are a constant part of the policy process."[100]

The evolution of Aboriginal title in British Columbia and its impact on the forest sector illustrates how this latter type of venue manipulation succeeded in altering the status quo and led to changes in the sector of policy regime. Just such a shift in venue occurred with respect to the pursuit of Aboriginal title by First Nations in Canada. The shift, however, was from the executive branch to the courts rather than to the legislature. As a result of this change in venue, a series of decisions by the Supreme Court of Canada provided working definitions of Aboriginal rights and title that had the effect of altering the nature of property rights throughout the country.

As the discussion presented here suggests, the mechanism of image manipulation and venue change can be added to those already well-established sources of policy change, such as external crises and perturbations and policy learning.[101] However, the case also illustrates another aspect of the processes and dynamics of policy change: how changes in one policy sector – here, federal Aboriginal policy – can spill over and affect activities in another, such as provincial forest policy. As this discussion has shown, although only loosely related throughout the course of Canadian history, events and activities in the Aboriginal sector recently have had a major impact on events and activities in many other sectors, notably those related to resource use, even though they involve a different level of government in the Canadian federal system. Thus, this case not only illustrates the impact and importance of venue change on a policy change process but also provides a good example of the impact and importance of cross-sectoral policy spillovers as a fundamental source of policy dynamics.[102]

A third lesson for the analysis of policy change can be derived from the case study. The case not only highlights the roles of venue shifting and policy spillovers, but also helps to clarify the relationship existing between sectors and subsectors in the process of policy change.[103] Here, the actual changes in Aboriginal policy that ultimately affected most areas of provincial forest policy were related to court-imposed alterations in property rights. These changes undermined the basis of the critical provincial forest policy tenure subsector by placing limits on the ability of the provincial government to dispose of Crown lands without first taking into account Aboriginal interests. The new definition of Aboriginal title emerging from the court system forced changes in aspects of the tenure subsector that reverberated from one subsector to another, as they immediately affected closely linked issue areas such as cut allocations and land use as well as other more distant subsectors such as employment and forest practices. This finding helps to clarify the patterns and mechanisms of subsectoral and sector policy change, underlining the critical nature of some subsectors and the linkages between critical and noncritical ones.

6
Fine-Tuning the Settings: The Timber Supply Review

Timber supply is a central issue of BC forest policy in large part because of the close links between timber supply and tenure. Government control over the location and volume of timber harvesting is an integral part of the system of renewable leases described in Chapter 4. The long-term goal of postwar forest policy, the orderly liquidation of the province's natural forests and their replacement by managed stands of commercially valuable timber, was in large part to be achieved by regulating timber supply. Instead of the forest industry cutting according to its own schedule, licence holders are assigned an allowable annual cut (AAC) determined by the chief forester. This system of yield regulation, as the instrument is generally known, has proved immensely flexible. As this chapter will argue, the objectives that yield regulation is supposed to promote under the general concept of "sustained yield" are many and varied, including a sustainable flow of wood fibre, a sustainable flow of revenue, a sustainable industry, and sustainable communities. However, the addition of a vaguely specified objective of ecological sustainability has put increasing strain on yield regulation.

With so many different objectives in play, and so many different interests organized around them, yield regulation will inevitably generate disputes about whether AAC is set too high (or too low) to reach the desired objectives. Within ten years of adopting yield regulation, a dispute between the forest industry and government foresters over whether AACs were too conservative was an important part of a royal commission on forest policy. By the late 1980s, however, more generalized concerns were being expressed about the sustainability of the province's timber supply. These concerns ran the gamut from the familiar ones about "overcutting" to worries that timber supply calculations were so flawed that nobody really knew whether present cut levels were sustainable or not. At the very least, there seemed to be a general lack of confidence in the process by which AACs were calculated, focusing on lack of transparency and accountability. The first Timber Supply Review (TSR1), initiated in 1992, was designed to address these concerns, above all by improving the quality of the data on which timber supply calculations depend, by taking a more sophisticated approach toward the uncertainties of forecasting and by providing a much clearer rationale for the ultimate AAC decision.

As we have noted in Chapter 1, the new government found itself in a propitious environment for major forest policy innovation, with key background conditions in place. Yet, although there appeared to be little in the TSR proposal that the forest industry could quarrel with, many in the environmental community believed that a thoroughgoing review of the provincial timber supply must result in fairly dramatic reductions in cut levels across the province. And beyond the narrow issues of data quality and process, environmentalists had larger game in view; namely, the much wider and potentially more contentious theme of the broader goals and objectives of timber supply policy. Like the other issues explored in this book, timber supply policy became entangled in a debate about the meaning of a "sustainable" timber supply, with the introduction of new ideas about a rate of cut that could meet ecological rather than purely social and economic objectives.

With the benefit of hindsight, these wider expectations were clearly doomed to disappointment. Like many of the other policies discussed in this book, TSR1, though associated with the NDP's peace package, has deep roots in much older forest policy controversies. Decisions taken many decades ago continue to constrain contemporary policy makers. Indeed, partially obscured in the blizzard of forest policy initiatives launched by the new government is a continuous theme of incremental policy adjustments disguised as novelties, a theme that can be traced to the tenacious persistence of some very basic assumptions about the connections between timber supply, security of tenure, the larger system of licences and tenures in place in BC, and the jobs and revenues generated by the forest industry. While a significant part of the story is told in other chapters, this one will focus on the central place of timber supply planning in BC forest policy. It will argue that, in spite of all the pressure for fundamental change, with alternatives visible on the wider policy agenda, the traditional instruments of timber supply policy (namely, yield regulation and cut controls) survived. After TSR1, the first round of timber supply review, the result is clearly a change in the settings of the original instrument designed to realize a modified ordering of the multiple objectives of yield regulation, not a change in the instrument itself or a major reevaluation of the goals of timber supply policy. Early indications from the second round, scheduled for completion by 31 December 2001, confirm this conclusion.

Regime Characteristics

As a result of the postwar decisions described in Chapter 4, state direction of timber harvesting by its licensees through the setting of AACs became a central feature of BC forest policy. Not surprisingly, a distinctive timber supply subsectoral regime, one dominated by a tight-knit and relatively closed policy community made up of government and industry actors sharing a common vision of timber supply policy, emerged to manage both the overall direction of yield regulation and its finer-grained features. The structure and membership of the timber supply community was, in part, determined by the character of the forest policy sector as a whole, notably the paramount provincial jurisdiction over provincial Crown land and the concentration of undivided administrative authority in the MOF. The legislation is unusually clear, requiring an AAC to be set for each TFL and TSA, and identifying the chief forester (in recent years, always a career bureaucrat and professional forester with the rank of assistant deputy minister and, thus far, always a man) as the official with sole statutory responsibility. The office of the chief forester was created in 1912, part of the broader North American movement known as Progressive Conservation that sought to base resource policy on sound scientific and technical grounds, insulated from political "interference." The chief forester thus finds himself in an unusual position in a Westminster political system, where executive authority is traditionally vested in the minister and delegated to administrative subordinates.

In fact, his position is all the more unusual given that the legislation explicitly requires him to make what is, in effect, a political decision by balancing multiple statutory objectives. In setting the cut, Section 8(3) of the Forest Act states that the chief forester shall consider:

> (a) the rate of timber production that may be sustained in the area ...
> (b) the short and long term implications to the Province of alternative rates of timber harvesting from the area;
> (c) the nature, production capabilities and timber requirements of established and proposed timber processing facilities;
> (d) the economic and social objectives of the Crown, as expressed by the minister, for the area, for the general region and for the Province; and
> (e) abnormal infestations in and devastations of, and major salvage programs planned for, timber on the area.

The existence of these multiple objectives in the legislation together with the permissive "consider" gives the chief forester substantial discretion in how to balance the objectives in any particular AAC determination. In practice, of course, it would be a very rash chief forester who failed to consider the government's overall forest policy direction in making

AAC determinations. As we shall see, one effect of the TSR has been to formalize this "direction" in the shape of public letters from the minister to the chief forester stating the "economic and social objectives of the Crown." The effect of the minister's letters is nicely illustrated in the AAC rationale documents. In the pro forma sections that describe the policy framework for making determinations, the language is permissive: the minister "suggests" and "asks." When a decision is actually taken and justified with reference to the letters, the language changes: "I must consider the direction given by the minister of forests to minimize AAC reductions unless they are necessary to avoid compromising sustainable forest management."[1]

On the industry side, the number of actors had been reduced and the monetary interest of any particular actor in timber supply decisions increased by the concentration of cutting rights in the hands of a relatively small group of large forest products corporations brought into being by the cooperative sustained yield policy. The concentration of cutting rights, in the form of TFLs and "quota" Forest Licences on the rest of the TSAs, meant that both the individual companies and their peak interest organization, the Council of Forests Industries (COFI), had a great deal at stake in setting the AAC and were carefully organized to influence AAC decisions. The special position of TFL holders in this subsector was given legal recognition by according a licence holder (and no one else) a statutory right of appeal against an AAC determination.

Although COFI is no longer the unchallenged peak organization representing the industry, a development that reflects wider changes in the forest products sector,[2] industry actors remain important in this subsector because technical planning functions have been devolved from the MOF to licence holders. Thus, the relatively closed nature of the subsectoral regime is reinforced by the nature of the timber supply issue. First, the technical nature of AAC planning means that only those who are capable of contributing to the debate at the appropriate technical level are included in the community, reinforcing its character as an especially tightly knit and closed group. Timber supply analysis requires the capacity to generate and analyze substantial amounts of data about the present state of the forest and to project into the future on the basis of knowledge about the consequences of past and present decisions for that future state. Second, by comparison with many other policy subsectors, the very long planning horizon of timber supply policy creates acute path dependency. That is, we are now living with the consequences of timber supply decisions taken decades ago that force us along a particular path chosen then. To retrace our steps and reopen possibilities that were rejected by those early decision makers would be expensive or even impossible. Path dependency reinforces the technical character of timber supply analysis by foreclosing or, at least, putting serious obstacles in the way of a radical change in direction. Timber supply planning was not a subsector likely to attract outsiders brimming with excitement about the possibility of new policy directions.

By the late 1980s, this policy community had achieved institutional form in the shape of "TSA steering committees" composed of regional representatives of relevant government ministries dominated by the MOF and representatives of the major licensees operating in each TSA (even though the AACs for their TFLs were determined separately). These steering committees effectively set management goals at a regional level throughout the province.[3] From the beginning, this group of senior forestry professionals in the MOF and senior forest industry executives, often foresters themselves, were bound together ideologically by the Sloan vision of convergent interests. That is, liquidation-conversion is the best of all possible worlds: good for the industry, good for the provincial treasury, and good for the people of BC. Add to this belief the very long planning horizon of sustained yield timber management, in which conditions on the ground today are the outcome of decisions taken decades previously, and timber supply policy begins to look rather like a special case of the supertanker going full speed at sea. Changing course is going to be an agonizingly slow process, even in the unlikely event of attracting the notice of the small group of officers on the bridge by addressing them in a language that commands their respect and attention.

On the other side, environmental opponents of the traditional policy community faced an uphill battle just to organize around the issues. The determination process is highly technical in nature and thus resists the simple "save the *x*" campaigns available in land use conflicts. Although activists argue that the public does understand the rate of cut issue and, they claim, also feels that the rate is too high, they also note that addressing the issue is connected in the public mind first and foremost with increasing protected areas.[4] Explaining that designating more protected areas may simply intensify the cut over the rest of the landscape if the AAC is not reduced and organizing to prevent this outcome has proved difficult, resulting in an elite rather than a populist approach. However, all efforts to develop a legal strategy of the kind that had worked so well in the US Pacific Northwest floundered in the face of the discretionary language of the Forest Act. With the partial exception of the TFL 44 case, discussed below, courts have consistently deferred to the chief forester's professional judgment and declined to apply a stringent standard of review that would allow "second-guessing" of AAC determinations.[5]

History

In his well-known book on agenda setting in public policy, John Kingdon argues vehemently against the idea that we could ever find the "origins" of a policy decision, however deeply we delve into the history of a particular policy area.[6] In the approach to policy analysis that Kingdon promotes, policies are responses to the perception of problems that are themselves very often the unintended consequences of previous policy initiatives and so on, if not ad infinitum, at least back into some misty past of limited relevance to present decisions. If they are not careful, students of contemporary Canadian welfare policy find themselves investigating the actions of the Speenhamland magistrates in the 1790s and thence back to Queen Elizabeth I's Poor Law, whereas students of BC forest policy are confronted with the timber supply problems of medieval shipwrights and Pliny the Younger's concerns about the disappearing forests of Ancient Rome. It is true, of course, that any event or decision has a context that is potentially infinite in scope, but the task of the analyst is precisely to close the context in a way that illuminates the problem at hand. In this instance, we need to understand how and why BC adopted the policy that Jeremy Wilson has aptly called "liquidation-conversion" of its extensive publicly owned forests[7] to appreciate both the broad and the narrow agendas to which the TSR was attempting to respond. This effort takes us back at least a half a century.[8]

Reflecting on the future of the forest industry after the Second World War, the British Columbia government was faced with the legacy of decisions taken in 1912 and described more fully in Chapter 4. In brief, apart from some small private forests concentrated on the east coast of Vancouver Island, the province retained ownership of nearly all forested land. The bulk of this land lay in the Forest Reserve managed by the provincial Forest Service, "management" that amounted largely to inventory and fire protection, with the balance in a variety of old leases and licences under which the land reverted to the province after the trees were removed. The government wished to encourage the development of a modern forest industry in a province where forest operations were hampered by rugged terrain and distance from markets. But the government also wanted to avoid a repetition of the speculative frenzy that had been summarily ended in 1912 by the freeze on new leases while, at the same time, looking to sustain the resource into the future. In the latter respect, the older tenures offered a model to avoid because reversion to the province provided no incentive to leave the land in a fit condition to grow more trees. Surveys in the 1930s showed that as little as 10 percent of this land was regenerating to a second-growth forest of commercially desirable species.[9]

The solution favoured by the province's chief forester, C.D. Orchard, and duly canvassed by a royal commission conducted by the province's chief justice, Gordon Sloan, was to adopt a then current idea known as cooperative sustained yield management. In the Forest Act of 1948, the government created a new Forest Management Licence (FML) that (in its original conception) combined private land, Crown land held under old licences, and new Crown land released from the Forest Reserve to create management units large enough to support a major processing facility in perpetuity if managed according to sustained yield principles.[10] "Sustainable" in the context of this policy meant a sustainable flow of fibre. To achieve this goal, all the land in the management unit, including private land, would be subject to yield regulation, an AAC being set by the province's chief forester for each unit on the basis of estimated present and future timber supply. Smaller operators without large permanent mills would be supplied from Public Sustained Yield Units (PSYUs) carved out of what remained of the Forest Reserve and managed on sustained yield principles by the Forest Service.

Somewhat to the surprise of Sloan, the FMLs proved extremely popular. In spite of opposition from established operators like H.R. MacMillan, several new consortia came forward with proposals to build production facilities to be financed from loans secured against the perpetual timber supply promised by an FML. Critics had to fall in line or risk being overtaken by rivals, and the BC forest industry entered a period of sustained expansion.[11] As it turned out, of course, things were not so simple. Cooperative sustained yield was a policy that, temporarily, presented itself as a satisfactory solution to at least three rather different problems. First, from the BC government's point of view, it served to spur an industry vital to regional economic development in the province without surrendering any of the province's almost unchallenged constitutional jurisdiction over forest resources. It also promised to stabilize and render predictable the revenues generated by the forest industry because the latter would now be required to cut within a narrow band around an AAC set by the Department of Forestry instead of matching its cutting to the vagaries of the business cycle. Second, from industry's point of view, and particularly that of the new entrants to the business, cooperative units solved the problem of how to acquire a secure timber supply with minimal capital investment upfront, an achievement for which it seemed well worth sacrificing some control over cutting decisions, especially as the government soon began to award FMLs without requiring companies to pool their private land. Finally, professional foresters, concerned about the status of their profession, saw cooperative sustained yield providing a critical boost to their importance, marking the beginning of a new era of active forest management in BC in which they would inevitably be called on to take a leading role.

Over the ensuing fifty years, all actors in the forest policy community have struggled to make this hybrid of private and public ownership work. It did not take very long before tensions arose over the statutory responsibility of the chief forester to set the AAC for each FML and PSYU. The question of the appropriate rate of cut for a sustainable timber supply was a central feature of two further inquiries, Sloan's second royal commission in 1955-6[12] and Peter Pearse's twenty years later.[13] As the debate progressed, it became clear that there were really four issues: how quickly to liquidate the existing forest; how to plan for the disappearance of the characteristic large, old trees the existing forest provides; how to deal with the one-time "falldown" in volumes from older to younger stands when the second rotation begins; and, finally, how best to determine the present state of the forest and to predict its future characteristics.

The first issue – how quickly to liquidate the existing natural forests – would become a central focus of later dispute once the environmental movement had become concerned with the fate of old-growth forests, but there was little disagreement among the witnesses who appeared before Sloan or Pearse. Tree growth tends to follow a well-defined curve in which growth is slow in the early years as seedlings establish themselves, moving into a period of rapid development as the tree canopy closes and suppresses competition, finally slowing down as trees reach maturity. If the AAC is calculated to maximize the amount of fibre available, trees should be cut when their growth rate reaches a maximum (the "culmination age") so that the trees whose growth rate is starting to decline can be replaced with fast-growing seedlings that contribute as much increment as possible to the total growing stock. In areas like much of the BC Interior, where fire is common, nature may interrupt the growth curve too soon, and the management problem may be how to protect trees until they reach their culmination age. In coastal BC, by contrast, where conditions may be very wet, or where trees like Douglas fir have evolved very thick bark to protect against all but the most intense wildfire, trees may live on long after their growth rate has slowed. The result is the classic old-growth forest composed of many large, very old trees contributing little or no increment, dying occasionally from disease or windthrow and opening up gaps for a few new seedlings to survive. From a timber production point of view, this mess of "overmature" timber needs to be cleared up, and the only question is how quickly to do it. In BC, Sloan had decided to follow the practice of private landowners in Washington and Oregon and attempt to liquidate all the old growth in a single rotation, thereby converting the forest to managed second growth as quickly as possible. If we are looking for an objective that was generally agreed on by the policy community at the time, "liquidation-conversion" would be that objective.

The next two issues around rate of cut are a consequence of the adoption of liquidation-conversion as the objective of forest management. One concerns the future characteristics of the forest. Although, as its advocates ceaselessly remind us, the point of liquidation-conversion is to ensure that forests regenerate after logging, the kind of forest that regenerates is intended to be very different from the original one. Sloan saw this change as an industrial problem. How would an industry that had grown up on a diet of big old trees be able to adapt to their planned disappearance? It is not only a question of mills processing smaller logs but also of how to adapt various categories of wood products to trees grown faster and lacking the characteristic strength and fine grain of slower-growing old trees. Although perfectly aware of these issues, Sloan was confident that the future disappearance of large old trees was a problem that could safely be left to "the men of that time" to solve, an explicit rejection of what later came to be called the precautionary principle.[14] A self-conscious policy of liquidation-conversion at least introduces an element of predictability into the liquidation of the old forests: we not only know from the beginning that large old trees are going to become increasingly scarce, but we know how soon that scarcity will make itself felt and can make our plans accordingly. Pearse, as we shall see below, was obliged to treat the desired future forest as a more complex question than one of log processing technology.

Equally serious, however, was the third issue arising from the liquidation-conversion project, an issue not much discussed by Sloan. Though participants at his hearings clearly understood the concept, they lacked the word that would later describe it so well: "falldown." Trees that are "overmature" from an industrial forester's point of view don't stop growing; they just grow more slowly. Thus, a hectare of 240-year-old trees will, other things being equal, produce a greater volume of fibre than a hectare of eighty-year-old trees. However, if eighty years is the culmination age of the trees in this example, then three rotations can be expected to produce substantially more fibre than one rotation that cuts only at 240 years; this is the productivity gain promised by sustained yield management. Falldown occurs because, between the initial cut of the old trees and the first cut of the managed trees, there will often be a significant, one-time difference in volume. Again, the defenders of BC's version of sustained yield management could point to the fact that falldown is a well-known effect that is a planned-for outcome of the liquidation-conversion project and can be dealt with by a gradual reduction of the AACs over the first rotation toward the sustainable managed cut level, originally known as the Long Run Sustained Yield (LRSY). Moreover, the difference in volume can itself be reduced by silvicultural techniques that make the most of the site for growing only well-developed, commercially valuable tree species instead of the apparently haphazard arrangement of species, ages, shapes, and sizes preferred by nature and by practices such as the commercial thinning of managed stands. Liquidation-conversion, it was confidently asserted, minimized the time when valuable timber idled away its life on the stump, prey to fire and disease, and maximized the benefits to industry, government, and, ultimately, the public as owners of Crown forests. Just how quickly we should proceed to the LRSY is a policy decision in which we must balance the considerable economic cost of forgoing present production against the potential for very large and disruptive downward revisions in the AAC at some time in the future.

The fourth and final issue is how to calculate the AAC and the nature of the assumptions used in the calculation itself. Although the questions of the appropriate level for AAC and the methods used to arrive at it may appear to be drily technical in nature, they actually go to the objectives of yield regulation and the problem of agreeing on those objectives when a policy addresses multiple problem definitions. The question of "overcutting" can only arise in the context of some agreed-upon objective: the claim is simply that too many trees are being cut for that objective to be realized. Perhaps the simplest objective on offer was the original one put forward by the forestry profession: the objective of yield regulation is to balance the total loss of wood fibre through fire, disease, and logging against the additional fibre provided by tree growth over the same period. A successful policy maintains this balance, neither taking out more fibre than can be replaced, nor, and this point is equally important, wasting fibre by taking out less than is grown, allowing nature to rob the landowner of potentially valuable timber. Given this definition of sustained yield, the calculation of timber supply becomes a matter of determining the productive land base – excluding areas that will never be logged because they are physically inaccessible, required for other uses, or economically not worth accessing – working out the volume of standing timber on the productive area and the annual growth that can be expected from it (expressed as a mean annual increment [MAI] to allow for differences in site conditions); subtracting the expected losses for fire and disease; and, finally, dividing the total growth by the number of years in the rotation. Even this greatly oversimplified account of how to calculate an AAC reveals critical uncertainties and imponderables. What should be counted as the area of productive forest? How accurate is the inventory of standing timber, the classification of stands into different species and site classes (done from aerial photographs checked by sample cruises when resources allowed), and how accurate are the tables predicting growth for each species and class? What should be the rotation age in the divisor? The biological MAI or the age at which trees become commercially useful? Should there be a cushion of theoretically available timber supply that is set aside against the possibility that the calculations are wrong?

On almost all of these questions, both Sloan and Pearse sided with industry critics of the provincial Forest Service, who argued that the AACs had been set too conservatively and underestimated future timber. Their decisions subtly transformed the rank ordering of the multiple goals being pursued by the sustained yield policy. Because there is a considerable degree of uncertainty in the estimates of future wood supply on which present cut levels are based, conservative estimates protect the goal of timber sustainability from inadvertent overcutting. However, there is a cost attached to being conservative: the loss of present production that might, in fact, be perfectly sustainable. Pearse's report, in particular, led to the elimination of the "cushion" that the Forest Service had maintained to guard against the possibility of errors and uncertainties in the timber supply analysis. This decision to be aggressive rather than conservative was, in effect, a policy decision to favour short-term production, employment, and revenue goals over longer-term biological ones with potentially large costs if, in fact, the analysis proves to overestimate timber supply.

A similar bias in favour of production goals can be detected in the debate over cut controls and other elements of yield regulation such as utilization standards. The volume of timber that can be recovered from a given area of forest depends on how that area is cut. Upward pressure on the AACs was sustained by the adoption of a policy of clearcutting as the dominant approved silvicultural system throughout BC, and by ever closer regulation of how much of the cut was required to be removed for processing and of how much could be left behind as waste. Closer utilization standards were used to justify AAC increases, and, once more, those decisions had serious consequences for future policy options. For example, any change to the policy of clearcutting would change the assumptions on which future wood supply and present AACs are calculated, almost always in a downward direction. At the same time, licensees were given some leeway in the area of cut controls; that is, how close to the AAC they had to cut in any one year during the life of a five-year management plan, though never enough leeway to silence criticism that the requirement to cut roughly equal volumes every year whatever market conditions may be makes no economic sense. Worse, it exacerbates the already cyclical character of the forest products industry, creating artificial shortages in booms and forcing more wood onto already glutted markets in slumps.

Thus, from the very beginning, yield regulation in BC could not be a simple technical calculation. The rate at which "overmature" timber would be liquidated ("rationed" was the preferred term), the amount of time that industry and timber-dependent communities would be given to cope with falldown, and the degree of confidence with which policy makers approached uncertainties in forest science and in the data available, all had to be decided on the basis of what the policy was trying to sustain: commercially valuable trees, forests of a particular kind, the forest products industry, timber-dependent communities, or the provincial treasury, to name but a few of the options. To Sloan himself, the different objectives of yield regulation, where they were distinguished at all, were presented as sitting happily one with another. It would be possible to spur the development of a modern forest products industry through guaranteed supplies; to improve and stabilize provincial revenue from Crown forests by reducing the impact of cyclical movements in the forest products markets on the amount of wood cut; and to bring modern forest management to BC, improving the state of the growing stock to maximize production. To Sloan, there was no need to choose among these objectives.

Revisiting all these issues at the next royal commission in 1975, Peter Pearse at least understood that hard choices would have to be made. His solution was to revise the Forest Act, making it absolutely clear that setting the AAC is a policy decision. Policy makers should have the widest possible administrative discretion to make a determination in accordance with the larger goals of provincial policies on employment and regional development. In this respect, one great virtue of Pearse's approach to AAC determination as a policy problem lay in his understanding of the linkages between yield regulation and other policy issues. In response to the problems he identified, he offered two solutions, policies that were in his mind clearly complementary but that subsequently attracted rather different coalitions of interests. On the one hand, he believed that many of the looming problems of timber supply, already visible in the 1970s, could be mitigated if not entirely solved by doing a better job of forest management. Pearse, a resource economist, helped popularize the discourse around investment and security of tenure. Private companies operating on public land are reluctant to make the silvicultural investments necessary to expand future timber supply, it is said, because they fear losing these investments if their tenures are altered or abolished by government policies. If outright privatization is not politically feasible, then at least the terms and conditions of TFLs (as FMLs had been renamed in the 1950s) could offer more long-term security. To boost future timber supply, a larger area of the provincial forest should be managed under this kind of licence.

Both industry and the forestry profession are strongly supportive of this approach, and as Chapter 4 has detailed, the terms of TFLs were changed in the 1978 Forest Act to offer increased security. A decade later, Forest Minister Parker's proposal to roll over existing volume-based licences into new TFLs was, in his mind and that of the industry and professional coalition that backed him, simply the belated implementation of Pearse's most important recommendation. Far from perpetuating "overcutting," the rollover, by demanding more management responsibilities from licensees in return for greater security of tenure, was the key that would unlock the considerably underused productive potential of BC's forests. And, it should be said, the connection between greater security of tenure and enhanced silvicultural investment was strongly reasserted by the Forest Resources Commission and by Clark Binkley during the latter's appointment as dean of the UBC Forestry School.[15]

On the other hand, both the fiery blast of public discontent that greeted the rollover proposal and Parker's bemused intolerance toward his opponents were related to the relative neglect that Pearse's other important line of recommendations had received. As we have seen, Pearse was clear that AAC determination is not a purely technical matter dictated by the results of the timber supply analysis but, rather, that both the assumptions made in the analysis itself and the interpretation of the results are driven by political decisions about desired future states. Although acknowledging that the Westminster tradition leaves the political executive ample scope to make such decisions unfettered by legal or administrative restraints, Pearse thought that any sensible government would certainly wish to be in touch with public opinion on so important an issue. He therefore recommended a greatly expanded system of public involvement. As Pearse was well aware, in addition to concerns about the effect of falldown on timber-dependent communities, the 1970s saw the first stirrings of the aesthetic revulsion against the successful completion of the liquidation-conversion project itself, a growing public uneasiness with a landscape in which juvenile Douglas fir march in serried ranks to a distant horizon.

Although the arguments about the ecological uniqueness of "ancient forests" lay some time in the future, Pearse was concerned that the political decisions about the AAC be taken with full knowledge of public concerns about the desired future characteristics of forested landscapes.[16] Influenced, perhaps, by the planning literature popular in the early 1970s, he was particularly insistent that serious public consultations should take place at the neglected regional level, the level, in effect, where AAC decisions for TSAs and TFLs were being made and where their consequences were most felt. Successive governments attempted to use the requirement of many TFLs that the licence be tied to the operation of a particular mill as an engine of regional development, but companies still cut and moved logs according to an older logic of development. A calculation in the early 1990s, for example, showed that around two million cubic metres of logs were sent annually from the north coast to the south coast for processing, whereas 1.2 million cubic metres travelled from Vancouver Island to the southern mainland.[17] Serious public involvement in determining regional AACs would complicate this freedom of movement on which companies had come to depend, and, not surprisingly, the implementation of this strand in Pearse's report lagged a long way behind that of the strand stressing security of tenure and silvicultural investment.

Efforts to improve public involvement were largely symbolic. Instead of finding a significant place in the forest policy community, opponents of the political direction of provincial forest policy found themselves writing the footnotes to logging plans. In consequence, as Chapter 2 details, a pattern of debilitating watershed-by-watershed conflict emerged in which diminished portions of key watersheds were eventually given protected status but only after substantial logging – and conflict – had taken place. More important for this chapter, the pattern of regional differences in timber supply, fibre flow, and mill requirements was not seriously addressed, leaving the looming possibility of critical local wood supply shortages.

We have looked at the historical development of timber supply policy in BC in some detail because the 1992 decision to embark on the TSR was in large part a response to the appearance of overcutting as an issue on the policy agenda. This section has attempted to demonstrate that the question of overcutting hinges on the goals of timber supply policy, which, for most of this period, were the goals of sustained yield timber management modified by the decision to liquidate the province's natural forests in a single rotation. However, different conceptions of the problem to which this variant of cooperative sustained yield is supposed to be the solution result in each interest having a rather different conception and ordering of these goals and hence a different understanding of the objective in view. A managed forest can only be overcut in relation to a management objective. If there is general agreement about management objectives, it is a relatively simple matter to determine whether too much (or too little) wood is being cut to achieve these objectives. However, the presence of different goals and goal rankings in the policy community means the potential for irreconcilable disputes on this issue.

Very early on, as royal commissions reveal, such disputes arose between the foresters' goal of maximizing biologically sustainable fibre production, the government's goal of maximizing a sustained stream of general revenue, and the forest companies' goal of short-term profitability and freedom to respond to business opportunities. These disputes were resolved through a process of accommodation within the policy community without disturbing the subsectoral regime. However, the appearance of a more complex goal of ecological sustainability was altogether more serious. Not only did it threaten to bring new actors and new ideas into the subsector and to put more downward pressure on the AACs, but it was not at all clear that the basic instrument of yield regulation around which the old actors were organized was suitable to achieve such ecological objectives as biodiversity conservation.

In a belated response to Pearse's recommendations, members of the public had been given the right to review the highly technical draft five-year management plans for TFLs and TSAs, and an adequate response to such public comments was required from licensees before the plans were approved and an AAC set. In practice, these consultation provisions, essentially passive and reactive, were not an especially effective means of public involvement. On TFLs, the MOF acted in its regulatory role to determine whether the licensees had satisfactorily responded to the issues raised by the public at open houses. On TSAs, when planning took place at all, the MOF was both regulator and manager at once. Moreover, much of what concerned the members of the public who bothered to take part in these exercises was considered to be government policy to which licensees were unable to respond except by restating the policy. The obvious example was clearcutting, a major concern judging by the number of comments received but dealt with by a form letter stating that the MOF had determined clearcutting to be the appropriate silvicultural system in the management area. Nonetheless, we can here detect the first signs of movement that led to the Timber Supply Review.

Policy Cycle

Agenda

In *Agendas, Alternatives and Public Policy,* Kingdon makes the case for a clear distinction between the two elements of policy change identified in his title: agendas and alternatives. In his view, although the process of agenda setting often involves both politics in the widest sense and the impact of events in the world, the generation and choice of alternative policies is largely the work of relatively closed and specialized policy communities. Moreover, the vital transition of an issue from the broader policy agenda to a firm position on the government's decision agenda is linked by Kingdon to the presence of a viable alternative ready to go.[18] Thus, we would expect that getting the issue of overcutting as defined by the environmental community onto the policy agenda is one thing, but formulating and implementing alternative policies is another, and so it proved with timber supply. There is no doubt that the failure of the rollover policy (which meant the departure of Parker from the forests portfolio), the intensification of public protests around logging in candidate protected areas, and the ability of the environmental movement to exploit the growing salience of environmental issues in the late 1980s and early 1990s all helped put overcutting onto the policy agenda. How the issue played out after it reached the agenda illustrates the special significance of narrower policy communities in formulation and decision making.

The unlikely lever that began to loosen the timber supply community's sclerotic grip on the AAC was a new issue, biodiversity conservation. Starting with the very high profile TFLs in the Clayoquot Sound area of Vancouver Island, TFLs 44 and 46 (the latter later divided into 46 and 54), environmental organizations employed their forestry consultants to highlight concerns about the impact on native biodiversity of what they considered an excessively high rate of cut combined with reliance on clearcutting.[19] As noted, liquidation conversion as practised in BC means that AACs will typically be substantially greater than the level of cut sustained in the long term (the LTHL) for several decades as "overmature timber" is removed, usually by clearcutting and replanting. To be sustainable on its own terms, forest management must then see AACs fall to the LTHL, either gradually in a series of controlled stages or steeply if AACs have not been reduced to accommodate the falldown.

By the early 1990s, two related problems had emerged here. First, licensees appeared to have taken advantage of their position in the policy community to keep the AACs as high as possible. In effect, they were putting off the gradual reductions that the conversion project requires for a "soft landing" at the LTHL in favour of maintaining present levels of fibre throughput for as long as possible. Second, the conversion project itself, even if adequately managed for a soft landing, presupposes that overmature timber, with its higher than average volumes of timber per hectare, will continue to be cut until the new, managed forests come on stream. Any change in this assumption based on claims about the need to maintain habitat within the working forest will dramatically alter the equation, requiring immediate reductions in the AAC to compensate. In pursuit of their goals of preserving more old forests from conversion, either by outright preservation or by substantially reducing clearcutting in old forests, environmentalists correctly saw the AAC as a key lever of change. Without AAC reductions, even preservation of a part of the old forests of BC could only result in greatly intensified pressure on the rest. Here, the biodiversity argument was critical in focusing attention on the provision of suitable habitat to maintain native biodiversity across the entire forested landscape. At this scale, the landscape is arguably being modified in the direction of younger seral structures at a rate and over an area greatly in excess of historical levels, which must necessarily put stress on species that depend on older seral structures for habitat, especially if they lack mobility.[20]

The biodiversity preservation argument provided a fourth, ecological, problem definition of what yield regulation is trying to sustain, adding to the already overrich mix of revenue, fibre, and industrial objectives in the subsector. Over the protests of many in the industry and the forestry profession in BC, the licensees in question were made to accept "biodiversity guidelines" that promoted new forest practices. The most significant for yield regulation were new rules for the spatial distribution of cutblocks ("adjacency constraints") that prevented new blocks being scheduled for cutting before adjacent ones had reached an age where they could provide suitable habitat for species deemed at risk. In effect, these guidelines meant that the AAC would have to be reduced unless the licensees could find new areas to locate cutblocks free from the adjacency constraints, a special problem in TFLs that had already been heavily logged. Although the licensees felt they could do so, mainly by cutting in high-elevation or steeply sloping areas previously considered inoperable, the chief forester apparently did not agree. Indeed, he was especially concerned with what he perceived as a growing trend on the part of licensees to schedule such difficult areas for harvest to keep up the AAC and then not to follow through and cut them, intensifying the cut on more accessible areas to reach their AAC instead. He determined to "send a message" to all licensees by reducing the AAC for these west coast TFLs to take account of the impact of the biodiversity guidelines on the assumption that the licensees' plans to log previously inoperable areas were bogus.[21]

TFL 44 was held by MacMillan Bloedel (MB, now Weyerhaeuser), then the province's largest licensee by area, and the company found the chief forester's message unacceptable. MB exercised its statutory right of appeal to a three-person board, which not only ruled in favour of the company but substituted the AAC in MB's original draft plan for that in the determination. The chief forester appealed in turn to the Supreme Court of BC, which upheld the board's original decision, casting an unaccustomed and most unwelcome glare of publicity on the AAC-determination process itself. It became clear from cross-examination of witnesses that the MOF lacked the capacity to carry out a thorough analysis of the impact of the biodiversity guidelines and that the chief forester had arrived at the new determination for TFL 44 by splitting the difference between MB's original request and a self-described quick-and-dirty impact analysis performed by ministry staff. In the circumstances, both the court and the appeal board felt that the chief forester failed the legal test requiring him to have a "tenable evidentiary basis" for his determination.[22] Whatever the truth of the matter and the legal issues at stake, to a sceptical public it seemed that the MOF, the custodian of the public interest, had so far devolved the technical side of forest management planning to licensees that it now lacked the elementary ability to operate on the same technical level and to act as an effective regulator.

While events in the politics stream were forcing overcutting onto the larger policy agenda, the validity and effectiveness of yield regulation in BC had been the subject of considerable debate within the policy community, including two studies of the Timber Supply Analysis process undertaken in 1990-1, one for the Forest Resources Commission and the other by the MOF.[23] These technical studies indicated a broad array of problems, though neither study had a great deal to say about the critical issue of problem definition and goals in determining whether overcutting was taking place. The FRC's report, which might have been expected to range more broadly, took a narrow, populist view of the overcutting thesis as a question of "political influence." Unsurprisingly, it found "no evidence of political influence on the determination made by the chief forester,"[24] focusing instead on the reliability of the AAC figure in relation to the ostensible management goals of the provincial forest estate. As the report points out, from this perspective the fact that the AAC is above the LTHL at any given time is not the problem. However, so long as the province remains committed to the general management goal of keeping cutting in balance with the productive capacity of the forest, the AAC and the LTHL must eventually meet. At the present point in the conversion project, the AAC must be reduced or the LTHL increased, or there must be some combination of the two. The size of these reductions and increases is determined by the yield analysis data. In this respect, the report noted, among other things, the weakness of the provincial inventory, delays in carrying out the determinations, the problems around defining the area of productive forest land, the MOF's failure to use modern techniques of "sensitivity analysis" that would show the result of variation in the key assumption of the yield analysis, and the lack of a monitoring program to follow up on the operational validity of the assumptions in the analysis. These flaws led the report's authors to conclude that the reliability of the provincial AAC is "questionable." In other words, setting aside the wider issue of determining management goals, the MOF lacked the technical capacity to evaluate whether the avowed objective of balanced liquidation-conversion was being achieved within acceptable limits of uncertainty.

Ironically, the MOF's internal report, while producing a similar list of technical shortcomings, did manage a somewhat larger view. At one point, the unresolved broader issue was actually identified: how to integrate timber supply planning, which determines the location and rate of cut on managed forest land, with the government's new interest in more extensive and participatory land use planning, which determines the total area of land available to forest management. In the MOF action plan drawn up in response to its internal study, timber supply planning was to be demoted from its preeminent position, in which the AAC acts as a constraint on land use planning, to a supportive role, providing timely information on the impacts of alternative land use strategies.[25] A hostile critic, the editor of *Forest Planning Canada,* put his finger on the difficulties with this approach. It would require a revolution in BC forest policy, starting with a reexamination of Sloan's basic assumption that the goal of the policy is to maximize timber production and proceeding to implement Pearse's call for meaningful regional-level land use planning that would actually set timber production goals in the larger context of changing social, environmental, and economic objectives.[26] It was not merely that, even fifteen years after Pearse's report, no such planning had been undertaken, nor even that, as soon as the CORE regional planning process described in Chapter 2 began, it at once uncovered the deep divisions about the goals of provincial forest policy that had been effectively covered over for decades. Altogether more important was that fact that none of the major existing policy actors in the timber supply subsector stood to gain anything from such a change. More than that, they stood to lose a great deal: timber rights in the case of the companies and the undisputed lead role in forest policy formulation in the case of the MOF. So, although the MOF's study had reemphasized the difference between the technical yield analysis and the much wider grounds for an AAC determination set out in the Forest Act, the community proceeded to ignore the disputes about the larger context of yield regulation, plunging with suspicious enthusiasm into addressing the host of technical issues brought up in the reports. Inside the community, it was in everybody's interest to tinker with the policy in place rather than to undertake the broad examination of policy goals demanded by the proponents of the overcutting thesis. Fine-tuning the gradual convergence of the AAC and the LTHL was exactly what the community proceeded to do.

Formulation and Decision Making

As noted in Chapter 1, the policy cycle model has often been criticized as an oversimplification because, in practice, the stages of the cycle are often difficult to distinguish. Issues often proceed from one stage in the cycle to the next without being addressed. In the case of the TSR, formulation and decision making are difficult to separate because the "review" is an artificial label attached to a number of related initiatives in the timber supply subsector. In one sense, the "decision" can be identified with the announcement that Section 7 (now Section 8) of the Forest Act would be amended to require by law that an AAC be set for every TFL and TSA at least once every five years. However, in making this change to the legislation, the government also served notice that the next round of AAC determinations was designed to address the technical criticisms of the determination process, the criticisms that had come to dominate the agenda-setting and problem definition stages of the cycle. In addition to the statutory requirement to keep AAC determinations up-to-date, changes would be made in the process aimed at improving the information available to the chief forester; identifying the effect of current forest management practices on future timber supply and modelling the impact of alternative practices; and generally making the chief forester's exercise of his discretionary authority in setting an AAC more transparent by requiring a more extensive rationale for each decision.[27] The overall thrust of this policy package was identified by referring to the new set of determinations as the Timber Supply Review, a phrase whose rhetorical force suggests that each new AAC determination would be based on a review of the adequacy of the existing determination rather than on an incremental adjustment based on long-standing assumptions. In short, it was to be a new start.

The review did nothing to resolve the rather different expectations about the goals of the TSR from within the timber supply policy community and from the larger issue networks affected by AAC determinations. Within the policy community, it was understood that the TSR would respond to the technical criticisms about delayed determinations and poor analysis. It would take into account the impact on timber supply of the government's other forest policy initiatives, notably the regional land use planning exercises undertaken by CORE and the LRMPs, but also the revised regulatory environment that would be created by the Forest Practices Code then under discussion. It was understood, grudgingly perhaps, that the TSR might mean substantial reductions in the AAC in some coastal areas that combine a history of aggressive cutting, low levels of silvicultural investment, and significant withdrawals from the forest land base. Certainly, few in the policy community wished to rerun the damaging public controversy of MB's appeal. The provincial government needed to be seen to assert its control of timber supply management to restore public confidence that the liquidation conversion project would, on its own terms, provide a stable and predictable flow of wood fibre to the mills rather than a crash-landing at some greatly reduced level of timber supply.

In pursuit of this understanding of the policy objectives of the TSR, the government marshalled its considerable statutory resources. The government had already amended the Forest Act in the wake of the MB decision in order to remove a licensee's statutory right to appeal an AAC determination. As we have seen, it also made the determination of a new AAC within five years of the old one a statutory requirement. In addition, a concerted campaign was launched to draw attention to the wording of the Forest Act, which, as we have noted, makes the clearest possible distinction between the timber supply analysis as one component of an AAC determination and the chief forester's discretionary authority to make the determination itself based on a broad range of factors, of which the analysis is only one element. Of these factors, the "social and economic objectives of the Crown" were given special emphasis. The MOF commissioned a report on the legal status of these objectives from the dean of the University of Victoria's law school, resulting in a "for the record" letter from the minister to the chief forester being appended to every AAC determination.[28] In making determinations, the chief forester is reminded of the importance that the government attaches to "the continued availability of good forest jobs and to the long-term stability of communities that rely on forests" and instructed that "any decreases in allowable cut at this time should be no larger than are necessary to avoid compromising long-run sustainability."[29] In the context of the letter, there could be little doubt that "sustainability" here means precisely the sustainability of industrial wood fibre. Just in case any doubts remain, the new determination process includes a socioeconomic impact analysis of the crudest possible kind, in which the impacts of the various timber supply options on employment and revenue are calculated using multipliers provided by the government.[30]

However, TSR1 also had to confront the fact that, outside the timber supply community, there was now an alternative problem definition of what the critical phrase "long-run sustainability" means – one that focuses on biodiversity and habitat conservation and the provision of ecological services. The alternative had been propelled onto the larger policy agenda by events in the politics stream, yet TSR1 as just described notably fails to address any of its proponents' concerns. At the time, this omission was explained by noting that ecological objectives were driving other key initiatives, such as the 12 percent target for protected areas and the habitat protection provisions of the Forest Practices Code and that the outcome of these other initiatives would be promptly reflected in revised AACs. However, the "whole landscape" approach of the alternative problem definition was not amenable to this kind of compartmentalization. As we have observed, proponents of the alternative approach believe that expanded protected areas, though welcome, would not achieve the goal of protecting native biodiversity unless they were set in a matrix of working forests managed to maintain significant habitat for species that need to move from one protected area to another or that are incapable of moving at all. Much of what is meant by overcutting in this view refers precisely to what is seen as a rate of cut in the working forest that is too excessive to achieve habitat conservation objectives.

More seriously for those who were fighting for a modified status quo, the ecological critics argued that the timber supply management process is currently being done backward. At present, the "sustainable" rate of cut is determined by the traditional timber supply analysis and then distributed across the landscape in ways that are supposed to minimize loss of habitat. This rate of cut then constrains all other planning processes. The ecological critique demands that adequate habitat requirements, including both reserved areas and practices that maintain habitat in the matrix, be determined first and that the AAC come out as a residual figure: the amount of timber that can be taken once all such requirements have been met with an adequate margin of safety. Although proponents of this view acknowledge that determining these minimum requirements is fraught with uncertainty and largely untested, they point to the American experience in the Pacific Northwest, in which a court-ordered plan that took biodiversity conservation as its primary objective resulted in Allowable Sale Quantity (roughly the equivalent of the AAC) on national forests in the region being reduced by roughly a factor of four.[31] In their view, such a substantial reduction is convincing evidence that the traditional way of calculating the AAC is a recipe for overcutting in relation to conservation goals. As we shall see, they were confirmed in their belief when the recommendations of a scientific panel struck to devise a conservation-oriented approach to forest management in the highly contested Clayoquot Sound area of Vancouver Island resulted in local AAC and harvest reductions of a similar magnitude to those under the American plan rather than those found in TSR1.[32]

Given the approach to TSR1 set out in the minister's letter on social and economic objectives, any such expectations of a radical departure from traditional timber supply management would clearly be doomed to disappointment, and so it proved. Although failing to keep the ecological critique off the wider policy agenda, the timber supply community clearly saw the danger of allowing AAC to become the residual of calculations designed to achieve forest policy goals other than their own. In response, they reduced the scope of the alternatives by emphasizing their novel and untested character and by a heroic public relations campaign stressing their potential impact on the jobs and revenue generated by the industry. On the government side of the policy community, in addition to the emphasis on job creation and community stability, the high symbolic value created by calling the new round of determinations the Timber Supply Review, the extra resources committed to completing the project (albeit a year later than originally scheduled), and the value to the MOF of reasserting the ministry's lead role in this subsector all worked to give the traditional interpretation of sustainability a considerable boost. TSR1 sets out to reassure the policy community that the liquidation-conversion project remains on track. Far from downgrading the importance of the AAC calculation as a regulatory instrument, the effect of TSR1 was to further entrench traditional yield regulation while allowing for some recalibration of the instrument. Indeed, in a counterattack on the ecological critique, more "realistic" yield regulation was now put forward as the only tested and feasible way of realizing the goals of habitat protection and biodiversity conservation.

Implementation

Formulating TSR1 in this way had two main effects on implementation. The predictable effect of the emphasis on short-term continuity has been a renewed emphasis on finding ways in which timber supply may have been underestimated, mitigating the downward pressure from other initiatives. Certainly, the crude old claims about transforming falldown into an endlessly expanding wood supply through massive silvicultural investment no longer appear. Moreover, the published rationales for TSR1 show many examples of the chief forester holding the line on licensees' unrealistic promises to expand operations into previously inoperable areas by unconventional and high-cost methods, such as helicopter and balloon logging. On the other hand, there is little doubt that the major thrust of timber supply planning is to maintain the AAC at its present level of around 65 to 70 million cubic metres and to avoid the falldown to a level currently predicted as little more than 50 million cubic metres. In the short term, much is made of the possibility of relying on the regulatory impacts of other elements of the Forest Practices Code to relax visual quality objectives (VQOs), identified in an impact study as particularly likely to cause timber supply reductions. So important has the trade-off between VQOs and other elements in the code proved that the relaxation was specifically authorized in a supplementary letter on the economic and social objectives of the Crown.[33] And early results from the Old Growth Site Index (OGSI) study of the actual difference between old-growth and managed volumes suggesting that the second-growth volumes will be larger than predicted are also cited as an unexpected and welcome mitigation of falldown effects.[34]

Above all, in an entirely new departure for AAC determination, the TSR1 rationales adopt the highly fashionable rhetoric of risk management, stressing the role of transparent AAC determinations in transforming uncertainty into quantifiable risk. On the surface, the risk management rhetoric has made the whole determination process more open. Determinations are supported by an extensive rationale in which the chief forester details how he handled each of the factors in the Section 8 list in arriving at his conclusion. All sources of information are listed. Much of this material is now available on the Web, and the rest, being clearly identified, could be obtained by any interested party through BC's Freedom of Information and Protection of Privacy Act. Compared with the days when determinations were announced in a quick press release, a considerable increase in transparency has been achieved.

However, the risk management discourse has also introduced upward pressure on AACs into the timber supply equation. Obviously struggling with a potentially very large reduction in the Kingcome TSA, for example, Pederson argued that the major uncertainties in the yield analysis simply cancelled each other out. On the one hand, the amount of sensitive soils may have been underestimated and the remaining stand volumes overestimated, leading to a potential 17 percent AAC reduction. On the other hand, regeneration delay may have been overestimated and site productivity in the new forest underestimated, resulting in an 18 percent increase. Of course, these figures only cancel each other out if all the estimates are equally uncertain, and this point is simply asserted in the rationale. Pederson moves on with evident relief to the quantified effects of the Code, where he assumes that reductions attributable to riparian and biodiversity constraints can be more than compensated for by relaxing visual quality requirements, making the net impact of the FPC on timber supply positive![35] By these and other shifts, the base case analysis calling for an immediate 33 percent reduction in AAC is worked down to 22 percent. Many other examples could be given here.

In general, the determinations in TSR1 (and those that have been made in TSR2) show a marked reluctance to depart from the status quo unless not doing so will clearly result in unacceptably large declines in AAC in the future. The decision rule here is stated in the TSA discussion papers: "If the predicted long-term sustainable timber supply [is] lower than the initial harvest level, reduce the timber supply within a defined range of eight to 12 percent per decade, until the long-term level [is] reached."[36] Thus, it is one thing to note that the evidence and inferences on which a determination is based are now transparently available for public inspection. Given the great weight of statutory authority behind the chief forester's unfettered discretion to draw any conclusion that is not "patently unreasonable" from the information, it is another to imagine that this new transparency creates a system that is more accountable to constituencies outside the policy community.

In the event, the overall impact of TSR1 on the provincial AAC was a reduction of 0.5 percent, a figure that has been met with derision by most environmental groups.[37] Interestingly enough, the reduction is also well below the 6 percent prediction for the impact of TSR1 in the industry-generated Price Waterhouse report that was used to influence the political climate of TSR1.[38] In response to such criticism, the chief forester stressed that this result had been achieved only by almost doubling the lower-value and unconventional component of the AAC – marginal timber, salvage, and deciduous stands – to nearly 11 percent of the total provincial AAC. TSR1 resulted in a 5.2 percent provincewide reduction in the traditional coniferous sawlog component of the AAC. Moreover, the decision to proceed on information currently available meant that less than half of the AAC determinations in TSR1 fully account for the impact of the Forest Practices Code, and many did not include reductions for new protected areas not yet finally designated when the determinations were made.[39]

More important, the aggregate figures disguised considerable regional and local variations, including variations between different tenures. At the regional level, Figure 6.1 clearly shows the difference in conventional volume reductions between the regions where a longer history of logging meant that falldown from old-growth volumes could no longer be

Figure 6.1

Impact of TSRI on AAC by region

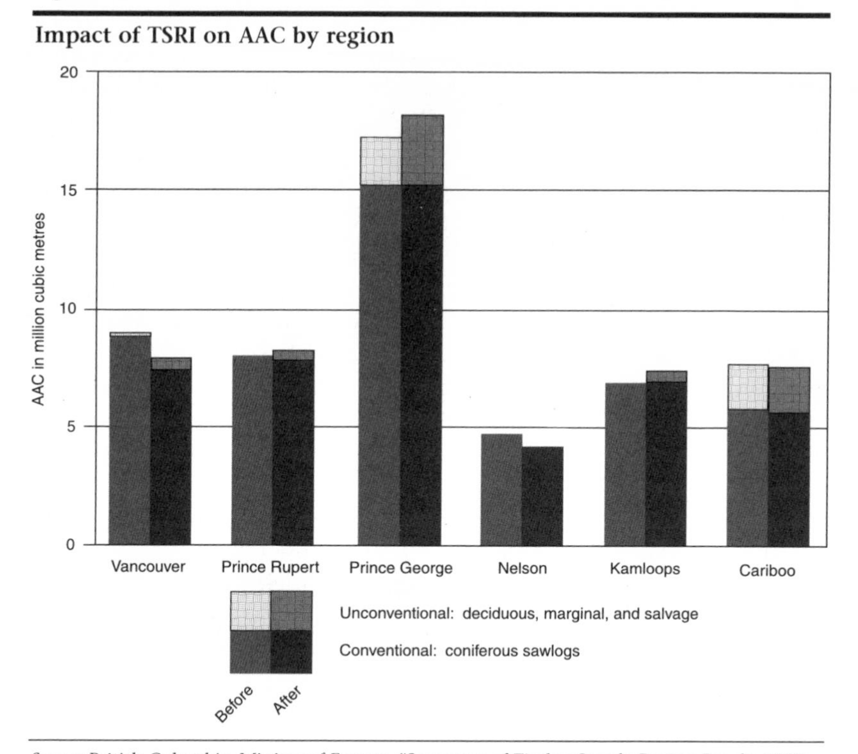

Source: British Columbia, Minitry of Forests, "Summary of Timber Supply Review Results 1992-1996," http://www.for.gov.bc/tsb/other/review/review.htm.

avoided – Vancouver (-9.3 percent) and Nelson (-11.7 percent) – and the rest of the province, where reductions in conventional volumes averaged around 1 percent and were usually more than compensated for by increases in the unconventional AAC. For Vancouver Island, an area with a long history of logging, the figures for the total AAC (including unconventional volumes, which make up less than 5 percent of the total) are summarized in Table 6.1. These figures, which only estimate the impact of the Clayoquot Land Use decision and Scientific Panel recommendations and do not include the potential impact of the Vancouver Island Land Use Plan at all, show a roughly 6 percent decline in aggregate AAC. However, the striking difference between the magnitude of the TFL and TSA AAC reductions warrants closer study, with important implications both for small processors whose total cut comes from TSAs and for those larger operators whose Forest Licences supplement their TFL volumes.

Apart from focusing on the small aggregate reduction, the main response of environmentalists was to point to the continuing overcut, the difference between the current AAC and the LTHL. Provincially, the AAC

Table 6.1

AAC changes on Vancouver Island (excluding mainland portions of tenures)

TFLs (excluding mainland portions)	Old AAC	TSR1 AAC	Change in percentage	Date set
6	1,300,000	1,288,000	-0.9	1995
19	978,000	978,000	0.0	1996
25	487,809	485,317	-0.5	1996
37	1,085,000	1,068,000	-1.6	1994
39	1,786,824	1,750,320	-2.0	1996
44	2,680,000	2,450,000	-8.6	1993
46	558,860	535,000	-4.3	1996
47	628,524	764,660	+21.7	1996
54	138,000	75,750	-45.1	1996
Total TFL	**9,643,017**	**9,395,047**	**-2.6**	
TSAs (excluding mainland portions)	**Old AAC**	**TSR1 AAC**	**Change in percentage**	**Date set**
Arrowsmith	468,500	400,000	-14.6	1996
Kingcome	444,173	345,553	-22.2	1996
Strathcona	1,644,626	1,378,820	-16.2	1996
Total TSA	**2,557,299**	**2,124,373**	**-16.9**	
Total Vancouver Island (cum)	**12,200,316**	**11,519,420**	**-5.6%**	

Source: Compiled from BC Ministry of Forests Timber Supply Reviews rationales, available at www.for.gov.bc.ca/tsb/tsr1/ration/rstate.htm.

is 21 percent higher than the total LTHL, and this finding is exactly the same difference as existed before TSR1.[40] Table 6.2 shows how the aggregate figure once again disguises the fairly dramatic differences between regions and tenures, with reductions in most tenures being balanced by an increased emphasis on cutting up to the LTHL in the remote, northern TSAs where harvest levels have traditionally been small in relation to the timber theoretically available. More important, on the traditional view of sustainability, cutting above the LTHL does not mean overcutting. Liquidation of old growth in a single rotation requires temporary AACs above the long-run unmanaged timber yield of the forest. On the surface, the AAC reductions found in the Vancouver and Nelson regions in TSR1 are exactly the kind that the liquidation-conversion project has always assumed will come and provide evidence of the government's renewed determination to absorb some of the costs of falldown and not to keep passing them on to future generations. It is far from clear what is happening in the other regions.

Table 6.2

Excess of AAC over LTHL before and after TSR1, by region and tenure (in percentages)

	Falldown before TSR1	Falldown after TSR1
Vancouver	124	116
Prince Rupert	128	130
Prince George	100	106
Nelson	154	140
Kamloops	130	137
Cariboo	138	137
Coast	124	117
Interior	119	122
TFL	110	107
TSA	125	126
BC	121 (137)*	121 (137)*

* Using an estimated impact on the LTHL of protected area decisions not considered in TSR1.
Source: Compiled from BC Ministry of Forests Timber Supply Reviews rationales, available at www.for.gov.bc.ca/tsb/tsr1/ration/rstate.htm.

Altogether more significant, though rather more difficult to explain, is the evidence of "overshooting." Planned falldown implies a series of small, incremental reductions over many decades with a soft landing at the LTHL. Where a tenure has been overcut according to the traditional definition, harvest levels must at some point fall below the LTHL, possibly for several decades, while an overabundance of immature trees caused by the past overcut reach commercial size. Of course, this outcome could be avoided by more intense rationing of old-growth forest now, but that in turn would mean even more immediate and dramatic reductions in AAC. The sense in which there is an overcut on the traditional biological definition of sustainability is the sense in which the chief forester continues to maintain present cut levels by planning for the AAC to fall *below* the LTHL at some time in the future on many tenures. In TSR2, some tenures have already gone into the trough, notably TFL 5 and the Fraser TSA.

The second, an unforeseen outcome of TSR1, was the appearance of AAC impact "caps" on other initiatives. In retrospect, a number of factors promoted the caps. The first cause was simply the adoption of a deadline for the completion of TSR1 at the same time that the government embarked on all its other forest policy initiatives. It soon became clear that TSR1 would never be completed at all, let alone completed within the five-year statutory time frame, if the chief forester had to wait until the AAC impacts of land use decisions, Forest Practices Code changes, or one-off events like the Clayoquot Scientific Panel recommendations had been fully evaluated. As it was, the completion date for TSR1 had to be postponed for a year. In consequence, consideration of some impacts, notably the CORE land use decisions and the Scientific Panel recommendations, was put off until TSR2, the next round of determinations. In other cases, notably the Code, it was decided to estimate the impacts of new initiatives as a guide to making determinations. One way of reducing the uncertainty around these estimates is simply to turn them into caps, directions to MOF regional and district managers that Code planning requirements must be implemented so as to limit their impact on timber supply to a predetermined percentage, the AAC cap. In this respect, the caps were also a response to the TFL 44 debacle. If that had resulted from sloppy impact assessment of the original biodiversity guidelines, one way of ensuring that there would be no similar embarrassment is simply to state the timber supply impact in advance.

In part, however, the caps were also a response to developments in the political stream, including a coordinated public campaign by the industry promoting the idea that the cumulative effect of the forest policy initiatives would be a dramatic downward shift in timber supply with consequent loss of jobs and revenue. The report by Price Waterhouse, commissioned by the industry-union Forest Alliance, predicted a 17 percent decline in provincial AAC by 2004 as a result of all initiatives, costing 70,000 jobs and 5 percent of provincial GDP. At the same time it was casting doubt on both the technical and political feasibility of initiatives like the TSR, industry was working behind the scenes on the idea of incorporating timber supply targets into regional planning.[41] After all, industry reasoned, if the land use planning process was being driven by a 12 percent protected area target, then one way of fending off the most dangerous alternative problem definition, in which timber supply becomes a residual quantity after allowing for other forest objectives, would be to demand targets of its own. Better still, timber targets could be linked to the new rhetoric of zoning by arguing that, provided land was carefully chosen for timber management emphasis (in other words, with a regime of relaxed practices in place), more timber could be found on a smaller land base.[42] Seen in this light, land use planning is no longer a zero-sum game, and the industry was offering the classic win-win beloved of the alternative dispute-resolution types running the CORE tables and LRMP process. In the event, the government would not publicly commit to positive timber targets for the CORE regional processes in the face of strong opposition from environmental organizations. However, as Jeremy Wilson notes in this book, such targets have surfaced again in the LRMP process, and more important, the ground had been prepared for "negative targets" in the form of caps on AAC impact.

The basic cap is 6 percent for the total impact of the Code over and above any reductions required by TSR1 for other reasons, such as falldown or protected areas. This figure, the 6 percent solution as environmentalists were quick to dub it, was drawn from a study of potential Code impacts by a joint MOF/MOELP working group. The study, though, cannot in any sense be read as recommending that impacts be capped at this figure. It concludes that "operational interpretation of the guidebooks will affect to a large degree the impacts that FPC requirements have on provincial timber supplies."[43] In one of the most curious examples of a self-fulfilling prophecy, more common in the social than the natural sciences, the original study was published as *The Forest Practices Code Timber Supply Analysis* and widely referred to as the authority for the 6 percent cap. "Operational interpretation" is now guided by the conclusions of the study so that the study's impact predictions will be true. Even more remarkable, the estimated impacts of the different elements of the Code that were aggregated to produce the 6 percent figure also found their way into caps on specific initiatives, roughly 2 percent for riparian protection, 4 percent for biodiversity conservation, 1 percent for watershed assessments, and, most controversially of all, 1 percent for BC's nearest approach to an endangered species conservation policy, the Management of Identified Wildlife Species (MIWS). The relaxation of visual quality requirements noted earlier contributes a net gain of 2 percent, for an aggregate negative impact of 6 percent. Dissent from within the MOELP about the possibility of achieving meaningful habitat protection within these limits briefly surfaced, but official evidence of continuing conflict has been limited to MOELP insisting on the original report's exact figures for biodiversity conservation, 4.1 percent in the short term and 4.3 percent in the long term, a distinction that represents about 650,000 cubic metres or a medium-size TFL.[44]

The early results from TSR2 show no evidence of any change of pattern. As of late June 2000, twenty-one new AACs had been decided. The AAC in nine of the units has been maintained, and it was increased in three and decreased in nine. Almost all of the significant reductions continue to come on the coast, including an 18 percent decline for the Fraser TSA on the mainland, 10 percent for the Strathcona TSA on Vancouver Island, and a further 15 percent cut for TFL 44. The overall impact on the provincial AAC thus far has been modest – a further reduction of almost 750,000 cubic metres, which is only 1 percent lower than the final TSR1 total. With many determinations in the Interior and north still to come, the final impact may be even smaller.[45]

Evaluation

As in so many other subsectors described in this book, the evaluation of the TSR has been different inside and outside the relevant policy community. Outside, the response has been almost uniformly negative. As noted, environmentalists seized on the minuscule reduction in overall provincial

AAC to argue that TSR1 had been a failure, and this interpretation was carried over into the "failing grades" given to BC forest management in the various report cards issued by environmental organizations for international consumption. A report on TSR1 coauthored by a prominent academic forest commentator continued to promote the idea that any cutting above the LTHL constitutes overcutting, without making any notable impact on the terms of the debate.[46] More significantly, Drushka has noted that, as long ago as Pearse's royal commission, concerns were being expressed that yield regulation is an instrument that is failing to achieve its stated objectives even as the continuing relevance of those objectives is being questioned.[47] TSR1, an exercise in fine-tuning the settings of the instrument, has done nothing to address the larger issues. And picking up on this critique of the effectiveness of yield regulation, an effort is now being made to revive the strategy that proved so successful in TFL 44: taking the expanded public involvement opportunities and the increased transparency of the determination process seriously. The BC Environmental Network (BCEN), with support from an American foundation, has published a citizens' guide to AAC determinations that attempts to demystify the process and identify areas in which awkward questions could be asked.[48] This strategy, essentially an internal critique that attempts to force the policy community to live up to its own commitments under threat of loss of legitimacy if it fails to do so, has potential. By combining "citizen scientists" with the strategic use of professional consulting where funds permit, campaigners could at least narrow the chief forester's room for manoeuvre and reduce the fudge factor in future determinations. What they cannot do this way is change the overall policy direction.

Within the community, reaction has been much more muted once the 6 percent battle was won. Although one licensee did launch a court challenge to a TSR1 AAC determination (and lost), licensees have settled down to play under the new rules without protest. Technical issues continue to dominate the discussion among insiders, notably the problems of spatial constraints on timber supply planning. In the old days, once an AAC had been determined, it was up to the licensee to decide how to find that volume on the licence area. As we have seen, this approach led to concerns that licensees were including areas within the TFL land base that contributed timber to the AAC calculations but would never, in fact, be logged. TSR1 and TSR2 determinations continue to squeeze this flexibility out of the system, sometimes by stern warnings about the distribution of cutblocks and sometimes by actually "partitioning" AACs, requiring so much to come from marginal stands, deciduous trees, or even from specific areas. More seriously, even in these days of expanded computing power and more sophisticated software, it remains a challenge to model spatially such things as adjacency constraints and wildlife corridors in a way that translates smoothly into volumes of timber gained or lost on different types of terrain.[49] At present, this modelling is usually done by extrapolating from smaller-scale pilot studies, and there is room for disagreement in the future. Finally, larger licensees, especially those operating in high-profile areas like the west coast of Vancouver Island, have long dedicated resources to public consultation as a cost of doing business. If expanded public consultation as envisaged by BCEN does become a reality, this development may put pressure on smaller licensees unused to the demands that can now be made for information and consultation, with unpredictable consequences.

The chief forester himself has stressed the achievement of the TSR in drawing attention to the changing composition of the province's timber supply, a change that can easily be overlooked if we focus only on aggregate figures. Faced with a diminishing stock of high quality, high value timber, Pedersen confirms that he has "looked aggressively at mitigation opportunities" by bringing stands into the AAC that would formerly have been considered economically inoperable: "because we have historically projected developing only about one-half of the total forest area of the province, there is a large marginal area that has come under consideration for future development given changes in technology, pricing, innovation and markets."[50] If licensees have indeed got the message and are becoming more concerned with problems of the appropriate end uses for different kinds of timber, then the TSR can certainly claim to have contributed to the improvement of BC forest policy. We would at last be explicitly addressing the issue of timber quality in the transition from old-growth to second-growth that Sloan raised in his second Royal Commission and left for the "men of [our] time" to resolve. But there are at least two significant drawbacks to the approach stressing the economic opportunities to be found in marginal stands taken in the TSR. First, world markets may not cooperate. "Technology, pricing, innovation and markets" may not change in the way required by the policy, with marginal BC timber remaining too expensive in the face of competition from plantation forests. Second, and equally important, the fact that only one-half of the provincial forest was originally slated for development does not mean that the undeveloped half has served no purpose at all. It has been used for recreation, community watersheds, viewscapes, habitat, and a host of other benefits. The policy of keeping up the AAC by more extensive use of the whole land base, rather than by more intensive use of a smaller working forest, threatens to reopen debates about saving favourite places and about biodiversity conservation that the other initiatives described in this book were supposed to have laid to rest. This would be a heavy price to pay.

Significantly from the point of view of evaluation, the review was overtly presented as an exercise in policy learning. From the original technical reviews of the determination process to the published rationales and their accompanying justificatory rhetoric, the story line was one of learning from the mistakes of the past.[51] Mistakes have been made, we are told, and in some parts of the province, it's time to pay for those mistakes. But overall, the situation is under control and new opportunities lie just over the horizon. It is clear that what counts as "learning" – who is learning what and with what effect – is heavily influenced by regime characteristics, especially the tight policy community and the technical nature of the issue.

The kind of policy learning in the timber supply subsector has been well described by Knoepfel and Kissling-Naef as the "redefinition" of an instrument by existing actors, "when actors positively take on board external changes and innovations in order to consolidate their own position."[52] They show how Swiss farmers have been able to maintain direct agricultural subsidies in the face of international, regional, and domestic pressures by claiming to use the payments to promote more ecologically acceptable farming practices. As they note, "many of the recipients of these direct payments have yet to undergo real conversion to ecologically based practices: the main aim is to ensure the maintenance of the status quo ... existing agricultural practices are being offered a new legitimacy ex post facto."[53] And in another parallel with timber supply policy in BC, Knoepfel and Kissling-Naef link this phenomenon with the benefits to be gained from transforming a one-off decision, which is often, as in the BC timber supply case, a zero-sum game, into a sequential game with the opportunity to continually debate and provide retrospective justification for each decision in the sequence. In fact, of course, this sequential element is built into the TSR process and made more visible by the statutory requirement for five-year determinations. The TSR emphasizes that each individual determination is not a once-for-all decision but part of a regularly scheduled "review," with the promise that the decision can always be revisited. The critical sense in which each decision further locks forest policy into a particular path with ever increasing costs, if we wish to retrace our steps, is obscured by the opportunity to employ all kinds of new arguments to justify that path at each successive review. When the salience of forestry issues is low, then visual quality standards can be relaxed without public resistance. When there is more informed concern about falldown, then the OGSI project can "discover" more fibre and so on.

Equally significantly, Knoepfel and Kissling-Naef link this learning outcome to the existence of highly structured and organized policy networks, policy linkages to other subsectors, and a common knowledge base built up in part through standardization and centralization. Each of these features is evident in the timber supply subsector. In addition to the well-organized policy community that had formed around the legal obligation of licensees to carry out timber supply planning and the requirement of reasonably standardized procedures and assumptions for doing so, AAC calculations involve just those complex exchanges "with features of different policy fields being played off against each other" that Knoepfel and Kissling-Naef identify.[54] As we have noted, the lesson that the Forest Alliance successfully persuaded the government to draw was the need for caution in adopting untested approaches to timber supply management with potentially far-reaching consequences on revenue and employment when so many other initiatives were on the table. As a result, traditional AAC calculations were not only maintained in place but became a central coordinating instrument limiting the impact of the other initiatives, reducing uncertainty for industry and government alike.[55] The MOF's public information guide on timber supply analysis continues to stress that analysis can be used to support land use planning and decisions about silvicultural investment, and it cautions against confusing the Timber Supply Review with land use planning processes. However, we have seen how the existence of AAC impact caps based on underlying decisions about timber targets transform "support" into "direction."[56]

There is a price to be paid for this kind of victory, of course, a price most clearly evident in the latest government-sponsored review of BC forest policy. At yet more public hearings, the Timber and Jobs commissioner heard the familiar litany of complaints and concerns about timber supply. He recommended a "further review" of the sustainability of BC's timber supply in the future.[57] In this respect, the TSR has clearly failed in its objective of reassuring a sceptical public that timber supply planning can meet sustainability goals. It is the kind of failure that results from dealing with contested problem definitions in the formulation stage by choosing to address one definition and ignore the others. Whatever its other virtues, TSR must be added to the long list of forest policy initiatives that attempted to transform broad questions of social, economic, and environmental sustainability into narrow technical ones, with predictably disappointing results.

Conclusion: TSR1, the Lion That Squeaked

As the historical section makes clear, yield regulation and cut controls have long been key policy instruments in the BC forest sector. Cutting within a narrow band around an allowable annual cut for the life of a five-year management plan remains a condition of holding long-term forest tenures on public land. As we would expect, in the real world of public policy just why this condition is prescribed and what policy objectives are being sought by demanding and enforcing an AAC are much less clear. Conflicting problem definitions and policy goals were incorporated into the policy from the very beginning, conflicts that were only partly suppressed by the convenient ambiguity surrounding the original idea of sustained yield. The appearance of an ecological goal, encompassing complex and poorly specified objectives like forest health, biodiversity protection, or the maintenance of ecological services, and the general transformation of sustained yield into the even more ambiguous "sustainability" only added to an already overrich mix. It was never likely that the Timber Supply Review could satisfy all the constituencies that were promoting these ideas. As in other cases discussed in this book, conflict over problem definitions persisted right through the policy cycle and continues to colour evaluation of the success or failure of the TSR.

While the internal objectives of the TSR process – more timely AAC determinations, better information and more transparency about the rationales – have been largely achieved, both the technical nature of the issue and the path dependency of policy in this subsector worked to prevent the classic fragmentation of the timber supply policy community into a more open issue network with new actors, new ideas, and new directions. Much of the literature on policy change has linked significant policy change to venue shifting, where the new venue creates a very different decision context characterized by a loose issue network rather than a closed policy community. Where there is no history of negotiation between the new players and the old, and no presumption of the virtues of incremental adjustments over radical change, so much more is possible. In a sense, the TSR case provides a negative confirmation of this thesis. Because it was difficult, if not impossible, to make a media issue out of the technical minutiae of AAC determination, the opportunities described by Baumgartner and Jones never arose: "Technical expertise, inside contacts, and legal skills may prove to be of no value where an emotional public media campaign is waged. So, if the challenging group is able to choose an arena where its special skills are reinforced and where the skills and resources of opponents are rendered useless, then it may win."[58]

Unable to shift venues, those outside the technical community faced an uphill battle to break into it, lacking both the legal levers for change and the ability to make a media issue out of the AAC determination. In the absence of tighter limits to administrative discretion, the fact that determinations have become more transparent is of very little consequence in promoting alternative conceptions of sustainability.

New ideas were certainly not lacking. Conservation issues had appeared on that larger agenda, for which traditional yield analysis methodologies seemed irrelevant. Even more threatening from the subsectoral policy community's point of view, alternative models were also available in which timber supply was treated as a residual after habitat protection and biodiversity conservation. But what we learn from the TSR1 case is the impotence of new ideas if the actors promoting them are successfully excluded from policy formulation. With help from some noisy interventions in the wider politics stream, the traditional policy community was able to maintain a focus on sustaining short-term employment and revenue as the key policy problems in this subsector, problems for which improved yield regulation seemed far and away the most plausible solution. The alternative models, stigmatized as experimental and untested, were kept off the restricted selection of policy alternatives that made up the actual decision space for the government. As this chapter has emphasized, the decision space was already bounded by the nature of the issue itself, characterized by acute path dependency as a result of the very long planning horizons of forest management and the consequent need to take account of timber supply decisions made many decades before. In the end, the choice lay between a tarnished status quo not without sentimental support from licensees still nostalgic for the good old days and a careful recalibration of yield regulation dressed up in the newly fashionable language of risk management. As the minister's letter on socioeconomic objectives and the special emphasis on crude socioeconomic impact analysis make clear, TSR1 was formulated in response to this problem definition and this set of alternatives.

In addition, the TSR provides further evidence that a sector in which policy making has been disaggregated into subsectors is not necessarily a sector in which policy will have no direction at all. Timber supply joins tenure as another critical subsector, one in which decisions taken within the subsector are able to constrain decisions taken in other subsectors. Maintaining timber supply to support good forest jobs and timber-dependent communities, as the minister's letter on the socioeconomic objectives of the Crown directs, will necessarily involve an intensification of cutting over the diminishing area of working forest when new protected areas are designated. This situation is an example of the resilience of a critical subsector in the face of contradictory policy development in other subsectors. However, in the TSR case the use of the AAC cap to limit the impact of other initiatives on timber supply is a particularly clear example of a critical subsector actively constraining other subsectors. The MOF's public information guide on timber supply analysis continues to stress that improved estimates of timber supply are used to support land use planning and decisions about silvicultural investment. As we have noted, the guide cautions against confusing the TSR with land use planning.[59] But the confusion is understandable. Although the cap ostensibly applies to the impact of Code provisions on timber supply, the fact that differences in management emphasis across the working forest will be handled by a flexible application of the Code means that the cap has also significantly constrained landscape unit designation and the species conservation strategy (MIWS). Critical subsectors like timber supply are not merely resilient in the face of contradictory policy proposals in other subsectors, they actively provide direction across the sector as a whole.

Rumblings of discontent continue to be heard. Perhaps even more significant than the ongoing efforts of the local environmental movement to persuade a largely indifferent and uncomprehending public that cutting above the LTHL constitutes an overcut, licensees continue to chafe against the loss of control over timber cutting that yield regulation entails. More frequent use of alternative silvicultural systems packaged as phasing out clearcutting have important implications for the AAC, effectively allowing licensees rather than the government to determine how much public forest is going to be cut. The continued effort to zone public lands with weaker regulatory control for high productivity sites has similar implications. Either of these proposals has the potential to dethrone yield regulation by making the AAC the residual outcome of decisions taken elsewhere in the forest policy sector, this time really in corporate boardrooms. Taken together, they offer a plausible strategy for an industry still concerned about the possibility that Clayoquot-style practices might eventually spread right across the landscape and ever eager to recover that flexibility to match timber cutting with market conditions lost in the Faustian bargain that created sustained yield tenures in BC.

7
Timber Pricing in British Columbia: Change as a Function of Stability[1]

Childhood lessons notwithstanding, the history of BC forestry reveals that money does indeed grow on trees. The wealth generated from extracting timber from publicly owned provincial forests has created thousands of well-paid jobs, corporate profits, and governmental revenues. But the question as to which groups get to pick money off the tree, and how much they get to pick, has been a controversial issue in BC forest policy history. At the heart of this issue is what economists refer to as resource rents, or the difference between the cost of converting standing trees into forest products and the price the forest product can command in the marketplace.[2] How this "difference" (or surplus) should be allocated has been a central concern to government, industry, labour, environmental groups, and even the BC timber industry's American competitors. This chapter focuses primarily on stumpage policy (the price the government sets for the right to harvest standing timber) because this policy has emerged as the dominant method of resource rent collection in the province.[3]

Timber pricing[4] is a fundamental part of forest policy and intersects with every other subsector reviewed in this book. It directly affects corporate profitability, general government revenues, forest employment levels, whether companies choose to undercut their permitted allowable cut, and trade relations with the United States. Timber pricing is affected by, and inexorably linked to, the existing tenure system, but, as we will see, there is some room to manoeuvre over instrument choices and a considerable degree of policy choice at the level of settings, even when tenure does not change.

Timber pricing is also an indication of how governments value the timber resource, which to date, environmental groups argue, has shown a bias toward timber production and away from environmental issues.[5] This policy area reinforces the importance of subsector comparative analysis because, while criticized by environmental groups in the 1990s,[6] no corresponding change to subsector membership or policy occurred as a result. This finding contrasts with our review of what transpired in land use and forest practices.

An enduring tension exists between the idea that stumpage rates should be low enough to encourage economic growth and forest employment, and the idea that rates should be high enough to provide a revenue stream to the provincial treasury to help defray health, social, and other governmental services costs.[7] This chapter reviews the historical manifestation of this tension, which enables a better understanding of timber pricing change and stability in the 1990s policy cycle.[8]

The 1990s timber pricing policy cycle presents a case of both policy change and stability. On the one hand, the long-standing goal of encouraging a healthy forest industry remained durable, and new goals such as ecosystem protection failed to find their way into the existing problem definition. Similarly, the use of administered pricing (where government, rather than the market, sets the price) as a policy instrument remained constant.[9] Proposals for using new innovative policy instruments either resulted in negative decisions (see Chapter 4), blocked from progressing in the policy cycle, or resulted in only minor changes.[10]

On the other hand, the policy cycle produced significant changes to timber pricing policy settings, the actual stumpage rate. The provincial government substantially increased the cost of harvesting timber in 1994 and then reduced costs in 1998 (see Figure 7.1). Why were there significant changes in the setting, whereas the goal and instrument remained unchanged?

We argue that the explanation lies in understanding the implicit historical compromise that has emerged among members of this policy subsector. The seed of the compromise can be traced back to the beginnings of BC forestry when timber pricing policies were limited to the single goal of promoting a healthy forest industry.[11] The compromise accepts this outcome as the only legitimate goal and further narrows objectives within this goal to four, with three of these championed by different members of the development coalition.[12] Industry has focused on the objective of facilitating industry profitability, labour has promoted a high-wages objective,[13] and government actors have bolstered the revenue objective. A lesser employment objective has been asserted from time to time, most notably by government and labour members. But the reason the compromise explains policy change is that members recognized a hierarchy of objectives. The profitability objective is paramount, followed by the high-wages objective, then the revenue objective. When the employment objective is asserted, it is usually at the expense of the revenue objective. Objectives lower on the hierarchy enter the problem definition only when those higher up are not considered a serious policy problem.

This compromise means that the profitability objective will be a dominant policy problem when the decision-making stage of the policy cycle

Figure 7.1

BC target stumpage rates, per cubic metre, 1987-98 (2nd and 4th quarters)

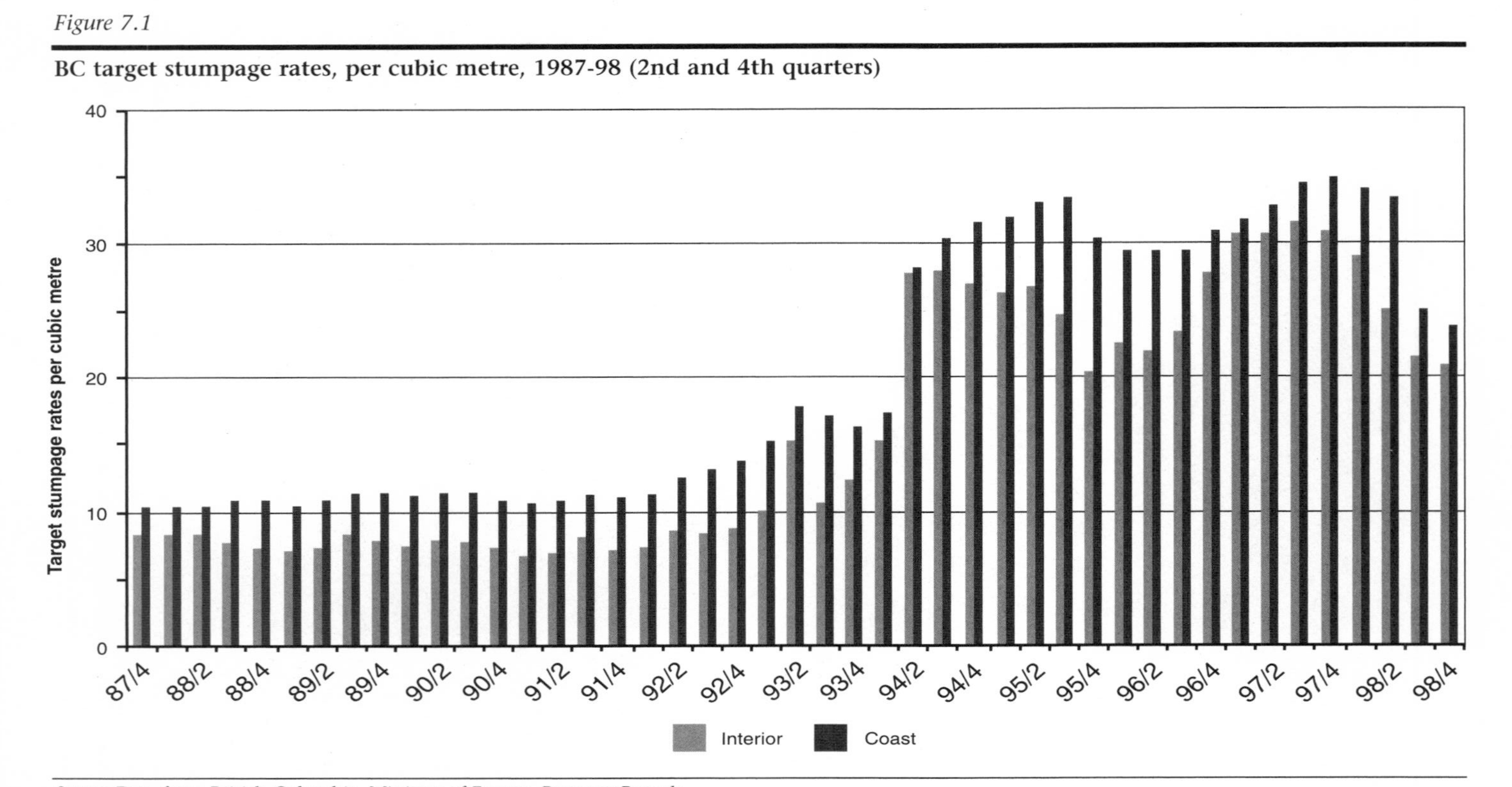

Source: Data from British Columbia, Ministry of Forests, Revenue Branch.

intersects with the downside of the business cycle. That is, when corporate profits are down, the profitability objective becomes the most important concern in the policy cycle, whereas the revenue objective is at the bottom. However, the reverse is also true: when the forest sector is at the top of the business cycle and companies are making high profits, the profitability objective drops to the bottom of the objectives pecking order because it is not considered a policy problem. As a result, high wages and governmental revenues become more important objectives in the short run. In other words, the revenue objective will be given increased priority when governments face a revenue crunch that coincides with an upswing in the business cycle and when the high-wages objective is relatively satisfied.

This long-standing compromise explains why the revenue and high-wages objectives took precedence over the profitability objective in 1994. It was not because these objectives were suddenly more important in the timber pricing subsector (they were not) but because companies were experiencing high profits (see Figure 7.2).[14] Indeed, when a serious downturn in the forest industry began in 1996, the profitability objective quickly reasserted itself in the policy subsector as the dominant objective, the wages objectives took second place, and the revenue objective was displaced by a last ditch effort on the part of the government to reassert the employment objective (see Chapter 8).

The compromise explains the dominance of administered pricing because this policy instrument gives government the ability to make swift changes to the settings when objectives fluctuate. Quick change to settings would be much more difficult under a different policy instrument such as competitive bidding, where the market would more directly dictate changes in prices.

This argument supports the hypothesis raised in Chapter 1 that business interests have more political influence during difficult economic times. It also adds a twist to Baumgartner and Jones's concept of a policy equilibrium as an explanation of policy stability because, in this case, the equilibrium is changing in the short run in order to provide long-term subsector stability.

Recognition of this compromise reveals the importance of distinguishing why policy settings change. Changes in settings that are the result of stable goals, a closed subsector, and fluctuating but exclusive objectives are arguably different from changes in settings that result when objectives are altered (for example, they no longer address the wages objective) or when objectives and/or goals expand (for example, to promote environmentally responsible logging through timber pricing). Accordingly, we must be careful to analyze the underlying causes of setting changes before we assess the degree of change that has occurred. The act of describing policy change does not, by itself, capture its significance.

Figure 7.2

Percent return on capital employed, 1989-98

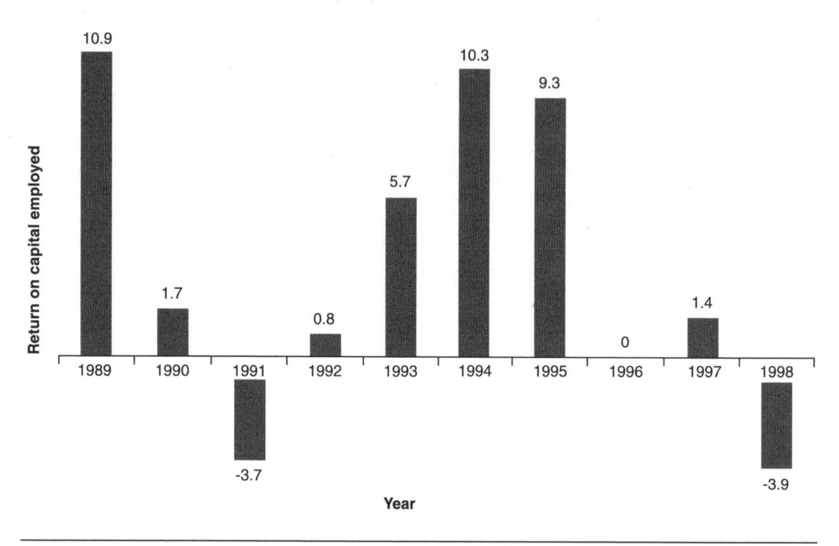

Source: PricewaterhouseCoopers prepared for Council of Forest Industries.

This chapter proceeds in four parts. First, it outlines key characteristics of the policy regime. Second, it provides a history of timber pricing policy before 1990. This section focuses on policy choices made in previous policy cycles and their legacies. Third, the chapter reviews the policy cycle in the 1990s by looking at how policy change and stability occurred. It explores how the policy regime and background conditions combined to bring changes in policy settings along with stability in goals, objectives, and, for the most part, instruments.

Finally, the chapter concludes by examining how policy changes would represent a break from the historical compromise. We focus on how external perturbations from the American timber lobby and the economic crisis in Asia have placed significant fatigue on the timber subsector, and for the first time since the 1940s, innovative ideas are being offered by business interests that would represent a change in the existing compromise.[15] The section examines what types of changes might take place if an environmental goal was added to this sector. It also allows us to better explore how goals and objectives might be altered fundamentally within this subsector and what their impact might be on instrument and setting choices.

Subsector Regime Characteristics

Timber pricing subsector regime characteristics diverge and converge with the broader forest sector regime. Like the broader forest sector, provincial government landownership, provincial authority over natural resources, and the Westminster model of government all combine to produce a considerable degree of state capacity to implement timber pricing policy choices. Unlike other more politicized subsectors such as land use, the historical compromise described above, the technical nature of timber pricing, and the relative lack of attention on the part of environmental groups have resulted in a closed subsector that discourages new ideas and new interests.

Actors

The three development coalition subsector actors are distinguished by their core interests in stumpage policy. Government sees this policy as a source of revenue, so it is treated seriously by the highest levels of government. Because of this direct relation to government revenues, the government has a more direct interest in timber pricing than it does in the other subsectors described in this book. The government understands that a healthy forest sector is also a source of revenue, through indirect means such as corporate and income taxes flowing from the sector, so it must be mindful that using stumpage as a revenue grab does not injure the sector. The industry is interested in receiving profits and an adequate return on investment. Labour is interested in maintaining high wages. At a certain level of prices for forest products, these three interests compete directly for how much of the available rent it obtains. (The outcome of this struggle was the historical compromise outlined above.)

Two key outside interests also try to influence this subsector. The first outside interest comes from American forest companies that argue BC stumpage prices are so low they amount to a subsidy to BC forest companies, giving them an unfair advantage over American companies. Since 1982, the American companies have used American trade law in an effort to impose import taxes in Canadian lumber or to force up the price of stumpage. Their original complaint (though now expanded to other policies such as raw log export restrictions) was directed at BC's use of administered pricing as a policy instrument. However, because administered pricing is a key part of the historical compromise, policy responses to the American pressure have not resolved the core of this dispute. At the same time, American pressure does not challenge the goal of promoting a healthy forest industry.

The second outside interest is from environmental groups, which argue that low stumpage rates lead to overharvesting by making economically marginal areas (areas that would have produced important ecological benefits if left unharvested) profitable and by encouraging an unsustainable level of cut (see Chapter 6). Because environmental groups have been unable to gain membership in this subsector, they act through advocacy efforts and by supporting American forest companies' efforts to increase harvesting costs in BC.[16] Whereas American timber industry pressures may lead to changes in objectives, environmental group pressures are aimed at adding ecological goals to this subsector, and they thus represent the most fundamental source of proposed subsector change. However, given the limited access to this subsector, environmental groups have focused their attention on land use and forest practices policy, where their influence was more direct and where their actions appeared to have more impact.

Different government, labour, and industry organizations are involved at different stages of the policy cycle. Choices over goals and instruments are made (or not made) by the provincial cabinet in consultation with industry associations (for example, the Council of Forest Industries) and labour (primarily the Industrial, Wood and Allied Workers Union). Choices over objectives are made by the provincial cabinet within the constraints (noted above) of the existing compromise. Technical choices about just how the settings will be raised or lowered are also determined by the provincial cabinet after closed and usually formal consultations with industry.

Detailed technical work on how to implement setting changes is left to the Ministry of Forests, which consults with "individual licensees, forest industry associations and other interested parties through the Interior Appraisal Advisory Committee (IAAC) and the Coast Appraisal Advisory Committee (CAAC)."[17] Forest timber companies and their associations dominate membership in these committees. Indeed, at this stage, technical issues limit participation mostly to affected forest companies and Forest Service Valuation Branch officials, who have some discretion in how pricing will affect individual companies.

Institutions

Like the broader subsector, provincial government landownership, provincial authority over natural resources, and the Westminster model of government affect this sector by giving ultimate policy authority to the provincial government. However, instrument choices must be consistent with the existing historical compromise and tenure system, limiting the scope of policy choices. These constraints are evident in the job and mill (appurtenancy) requirements of most forest harvesting licences, in effect institutionalizing employment objectives through the current tenure system. These appurtenancy requirements would severely hinder efforts to move toward adopting a pure competitive bidding system as a policy instrument because such rules would drive down the market price.[18]

In this sense, the institutions combine to allow the provincial government to take decisions that may be at odds with the forest industry officials, but these decisions are limited in scope. The institutions in the sector set the stage for how new ideas are developed and how a closed and limited set of actors influence policy choices.

Ideas

The subsector remains lodged in a frontier mode of development.[19] New ideas regarding ecosystem management and biodiversity protection have been largely blocked. The range of ideas in the timber pricing policy subsector is narrowly focused on how to best achieve the three related objectives of facilitating corporate profitability, creating employment, and raising government revenues. As a result, new ideas are usually technical in nature and focus on ways to alter instrument settings. Ideas that focus directly on conservation have difficulty being seriously entertained. Similarly, broader philosophical ideas promoting free market principles have had difficulty permeating this sector, despite the ideological predisposition of many members of the Social Credit party, which dominated BC partisan politics from the 1950s to the early 1990s.

The use of government-controlled administered pricing fits within the ideational capitalist framework that has developed in BC, where government's role as a facilitator of economic development on a grand scale has long been considered appropriate, indeed necessary, for the ongoing development of its frontier mode of economic development. Thus when competitive bidding withered away, there were few voices of disapproval. This bias toward statism is limited to government as facilitator of capitalist enterprise rather than as hostile toward it.

We briefly review below the interaction of regime and background characteristics before 1990 and then give a more detailed treatment of what transpired in the 1990s policy cycle.

Policy Choices before 1990

The history of resource rent as a policy problem goes back to the mid-1800s when the timber industry began to replace the fur trade as the key economic development staple for the province.[20] The late 1850s marked a transition in which the timber industry began its ascendancy as the province's key economic engine.[21] During this period, resource rent issues first became a key policy problem. The goal was to encourage a healthy forest industry that would promote the settlement and economic development of the province. (This goal remains relatively unchanged 120 years later.) The key objective was to create a climate for corporate profitability so that investment would be encouraged. The objectives of high wages and revenues were secondary, treated as the results of creating a hospitable climate and not, at this stage, as objectives that would affect resource rent policy choices.

Three competing ideas were prevalent among the policy regime regarding just how to encourage economic development: resource rents should be collected on a one-time sale (alienation) of Crown land to the private sector; provincial landownership should be maintained and forest harvesting rights given to forest companies at no initial or subsequent costs; and provincial landownership should be maintained to provide a long-term revenue stream in the form of payments to the provincial treasury. None of these ideas challenged the notion of the staple state,[22] in which the role of state actors was to facilitate a vibrant and productive domestic economy through the development of natural resources.

At first, the sale of Crown land to private interest was the chosen instrument to achieve economic development goals and mirrored earlier policy choices affecting forestry in eastern Canada and the United States.[23] Under this mode, few resource rents were collected because the undeveloped land was deemed of little or no value. At this time, virtually anyone expressing an interest in owning the land could obtain title.[24] Through policy evaluation, provincial government officials eventually learned that this policy was often ineffective in achieving a productive timber industry because the sale of Crown land often resulted in its purchase by absentee landlords who failed to manage the land for industrial development.

Government then turned to the second policy instrument in which the provincial government would maintain ownership of the land but would grant licences (called timber leases) for the rights to harvest the timber. The government would require no immediate or future resource rents (no stumpage or royalty fees would be collected), but ownership of the leases was "restricted to those actively engaged in commercial exploitation of the [forest] resource."[25] As Sloan has noted, "There was a definite policy of discouraging investment in timber land without operating it, indicated by limiting the lease to those who were actually engaged in industry."[26] What the provincial government did require was industrial development commitments (known as appurtenancy requirements), which for the first time explicitly raised the objective of employment considerations in the policy instrument choices.[27] Aside from these industrial development commitments, this form of tenure was "almost as secure as outright ownership."[28] Resource rents were not collected and there were no export restrictions, few company size restrictions, and no limit to the duration of the licences.

Once provincial officials felt that the timber lease policy instrument had secured a burgeoning timber industry, the revenue objective began to emerge in the policy subsector, spurred by a two-decade-long governmental revenue crunch beginning in the late 1870s.[29] Timber leases were replaced with a new form of tenure titled timber licences, which were created "to derive revenue from the cutting of timber on Crown lands"[30] primarily through the imposition of annual timber licence fees.[31]

By the early 1900s, the provincial government was experiencing another revenue crisis so severe that annual timber licence fees were deemed insufficient to meet revenue objectives.[32] As a result, special regulations were created that temporarily altered timber licence requirements in the hopes of raising immediate cash through increased sales of special temporary licences. With these plans, resource rents were captured primarily through one-time licence sales, with less attention on future annual fees. Licences bought during this temporary period contained relaxed appurtenancy requirements and, most important, would be renewable for up to twenty-one years.

So attractive were these conditions to investors that between 1905 and 1907, 9.6 million acres came under these special tenures, ten times the amount of land under the old rules. By 1907, the government had eliminated its budgetary crisis and stopped issuing these temporary timber licences.[33]

Instead of simply returning to the rules on the original timber licences, the province established the Fulton Royal Commission in 1909 to advise on the use of future tenure and resource rent policy instruments. The Fulton Commission recommended a new policy instrument to collect resource rents – a competitive bidding instrument in which "anyone could initiate a sale of timber on a defined tract of Crown land."[34] A new 1912 Forest Act created the authority for such a policy instrument and a new Forest Service was established to administer the competitive bidding process and develop a method for establishing a floor bid.[35] In addition, all previous systems of licences, leases, or temporary tenures became reorganized under one category, now called Old Temporary Tenures (OTT), and would occupy a special status in future timber pricing development.[36] Because the granting of Crown and Old Temporary Tenures had stopped, the competitive bidding process for the right to harvest timber on a designated area of Crown land was the only way for a company to gain new sources of fibre from 1912 until the 1940s.[37] The decision to move to a more technical and constantly changing upset price – the minimum price for a competitive bid sale – from the use of yearly fees marks a key policy legacy in BC timber pricing history. This practice remains the dominant form of resource rent collection eighty-eight years later.[38]

The timber pricing policy subsector was relatively stable until 1945, when changes in domestic economic development and the international political economy caused new concerns to arise on the policy agenda. At the domestic level, phenomenal growth in the forest industry and the rise of the pulp and paper sector raised the concern that inadequate resource management could lead to significant timber shortfalls in the future. Pearse has described this period as a time when "apprehensions had arisen over the unbalanced pattern of timber harvesting in the province, the lack of secure timber supplies for new industrial ventures, and the inadequate provisions for future forest crops."[39]

At the international level, there was a fear that British Columbia might lose its place as a key exporter of forest products to emerging European competitors, especially in Scandinavia.[40] The idea of managing for timber sustainability held out a promise of addressing both policy problems. The concept had permeated the ideas of the policy regime actors, largely a spillover from ideas circulating south of the border.

As Chapter 6 develops, the immediate response was the establishment of a royal commission under Chief Justice Sloan to study what to do about this policy problem that had pushed forest tenures, timber pricing, and sustainable harvest levels on the forest policy agenda. Sloan's major recommendation was to established a sustained yield program[41] that would give forest companies long-term harvesting rights over large tracts of publicly owned forest land in return for a commitment that they practise sustained yield management.[42] Sloan also argued that this new system would give large BC forest companies a competitive edge over their European competitors. Sloan's sustained yield recommendations were adopted by the provincial government, resulting in an important spillover from the AAC subsector to the timber pricing subsector.

The new sustained yield policy largely spelled the death knell of competitive bidding as a timber pricing policy instrument in the province. A decision had been made to give companies the same proportion of harvest they had before this program. Each established company was given a quota in which they had effective cutting rights over new tenure arrangements. Huge, nonrefundable bidding fees were set for firms that wished to outbid a company with a quota in a certain area. As Schwindt notes, "Within twenty years of implementation of sustained yield in British Columbia, competitive markets for the timber resource were closed."[43]

The result was that the process for setting the minimum price for a competitive bid sale became a much more crucial process as this price now established the actual selling price.[44] After this major shift toward administered pricing as a policy instrument, this subsector remained relatively unchanged, whereas broader issues concerning the security of forest firms' licences continued to dominate the policy agenda.[45]

The Modified Rothery, or Residual, System

Before the 1960s, the experience in timber pricing reveals spillovers from the timber supply (AAC) and tenure subsectors because policy choices in those areas eliminated competitive bidding as a policy instrument, resulting in the use of administered pricing. Up until 1987 timber prices were calculated as a residual fee, known as the modified Rothery, or residual, system. Under the residual system, the provincial government was to collect from forest companies through stumpage fees whatever revenues they had "left over" after allowing for operating costs and profit.

The logic of the residual system was that stumpage revenues would fluctuate dramatically as a function of industry profitability because the greater the revenue, the greater the stumpage prices. As long as this approach to stumpage pricing was in place, government had no need, at least in theory, to explicitly modify stumpage prices when objectives fluctuated because the residual approach did this automatically. When profits were up, the residual system collected more stumpage. When profits were down, the residual system collected very little stumpage. (Figure 7.3 reveals fluctuations in average stumpage rates, whereas Figure 7.4 reveals resulting differences in stumpage revenues collected in various years from timber harvesting.[46]) Likewise some have argued that this approach removed some of the incentive companies have to create an efficient workforce and to pay relatively high wages because any increased labour costs would, in theory, be deducted from what they paid in stumpage.[47] It also was relatively easy for companies to conform to their tenure appurtenancy commitments because these costs, too, would simply be deducted from operating costs, resulting in lower stumpage fees.

In short, until 1987 the fluctuating objectives approach within the current historical compromise was addressed by the way timber prices were calculated – calculations that placed priority on corporate profits, followed by wages, and lastly by governmental revenues.

The 1970s: Criticism of the Residual System

The residual system that was supposed to capture resource rents in theory had less success in practice, and as early as the 1960s, American lumber companies and provincial forestry critics were arguing that stumpage rates

Figure 7.3

BC select average stumpage rates 1963-84 in cubic metres

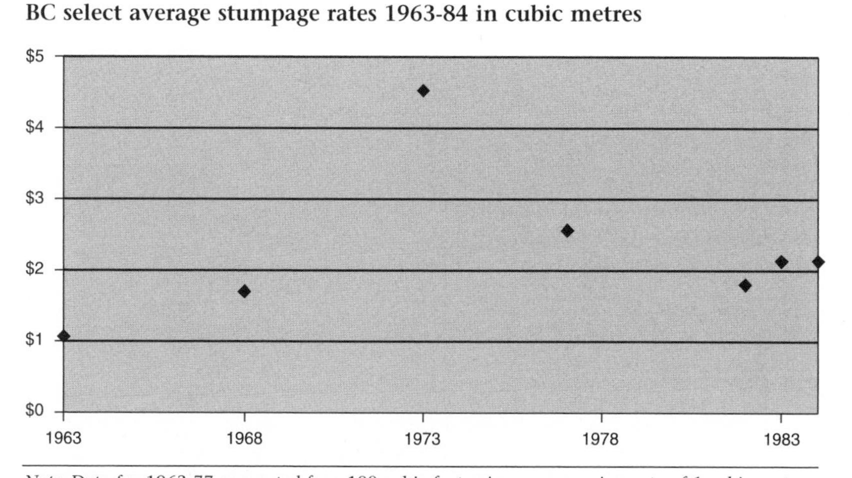

Note: Data for 1963-77 converted from100 cubic feet using a conversion rate of 1 cubic metre = 35.3 cubic feet. See *Southern Wood Conversion Factors and Rules of Thumb,* Wood and Log Volume Conversion Factors at http://members.aol.com/jostnix/convert.htm.
Source: Selected reports of the British Columbia Ministry of Forests.

Figure 7.4

Government stumpage revenues, 1967-98

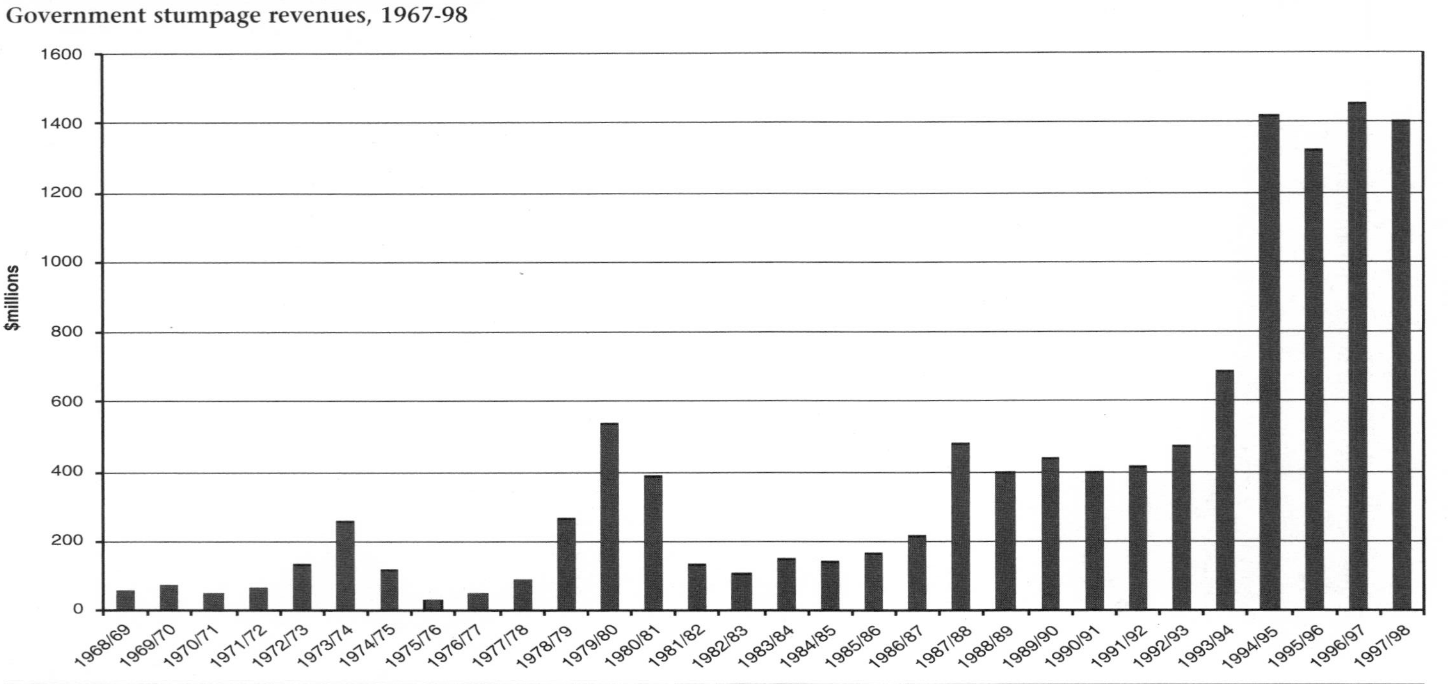

Notes: The stumpage rates from 1980/81 through 1987/88 are net after deducing for stumpage credits given to companies during this time. From 1980/81 through 1987/88 credits were paid to companies for work they conducted on reforestation and roads. After 1987/88 companies were still required to conduct this work and bear these costs, but the credits were eliminated.
Source: Select Annual Reports, British Columbia Ministry of Forests.

were failing to capture full economic rent, resulting in an unfair return to the provincial treasury. The criticism of the residual system heightened in the early 1970s following the brief tenure of Dave Barrett's New Democratic Party government (1972-5). Barrett appointed Bob Williams, who was a long-time critic of BC's stumpage system, as Minister of Forests, Lands and Water Resources. Williams treated the stumpage issues as an equity one, asserting that citizens of British Columbia deserved to get more from their publicly owned forest resource.[48] Williams immediately appointed Peter Pearse to first examine the specific issue of stumpage rates/resource rents and the tenure system[49] and then, more generally, the state of forestry in British Columbia.[50] Williams proposed innovative methods of collecting higher resource rents, including increasing royalty fees paid on OTT lands as well as establishing a government agency that would virtually become a log broker in the province, driving up the cost of lumber to market prices.[51] However, even before the defeat of the NDP government in 1975, an economic slump forced Williams to back off implementing these policy innovations. The experience provides additional evidence that the profitability objective becomes paramount when decision making in the policy cycle intersects with the downside of the business cycle.

The 1980s: External Perturbations and Free Market Ideology

Timber pricing policy again came under criticism in the 1980s from domestic and international sources. Within BC, criticism came from a faction within the governing Social Credit Party who were strongly free enterprise in orientation, ideologically predisposed to favour small business, and viewed vertically integrated industrial forestry as hindering a competitive climate. This group saw the current stumpage system as discouraging access to the "little guy" as well as giving away the public resource at bargain prices. One of the key voices of this faction was maverick Social Credit MLA Jack Kempf, whose ideas resonated with those of Bill Vander Zalm, who became premier shortly thereafter in 1986.

The international scrutiny came from American forest companies, which first began official trade dispute proceedings in 1982, claiming that BC's stumpage rates amounted to an unfair subsidy, putting them at a competitive disadvantage. This exogenous scrutiny from American timber companies came as BC and other Canadian provinces increased their share of the American market from 7 percent in the early 1970s to over 30 percent just a decade later.[52] However, in 1983, the US Commerce Department ruled that BC's timber pricing policy did not amount to a subsidy.[53] Following intensive congressional lobbying and changes in American case law, the US Commerce Department made a ruling three years later that stumpage prices in BC and other Canadian provinces did amount to a subsidy. Kempf and Vander Zalm used this pressure to help justify their efforts to increase stumpage prices and their reform agenda.[54] The timing of these efforts illustrates our argument about the conditions under which the revenue objective is paramount: industry began to report high profits in 1986 while government was experiencing a revenue shortfall.[55] As Kempf argued, "This province has a deficit of over $1 billion ... Maybe we should be channeling that money into other areas."[56]

Far from fighting this pressure, Kempf and Vander Zalm sent out clear signals to the American coalition that they supported their position. Kempf contacted and exchanged information with the Coalition for Fair Lumber Imports, the American industry group.[57] As he later noted: "It was quite clear to me that the US Commerce Department and Coalition for Fair Lumber Imports had a very good case and so it behooved us to try then and get the best deal that we could, and I think we did."[58]

Before a final ruling could be made, Canada and the United States agreed to a compromise that would see Canada impose a voluntary 15 percent export tax on softwood lumber shipments to the United States. This deal was strongly supported by Kempf and Vander Zalm, who clearly used the American pressure to further the governmental revenue objective. The Vander Zalm government moved quickly to take advantage of a provision in a memorandum of understanding (MOU) that permitted the export tax to be reduced if equivalent replacement measures were initiated. The replacement measures constituted an expansion of the Small Business Forest Enterprise Program,[59] a requirement that companies perform silvicultural work without compensation, and, most significantly, a change in how stumpage prices were collected. So significant were these changes that even the American government agreed that they more than offset the 15 percent export tax and therefore agreed to its elimination.

These policy changes marked a schism between government and industry members of the BC timber pricing subsector.[60] Industry officials steadfastly maintained support for the Rothery formula, arguing that "stumpage fees are not low if the anomalies are corrected."[61] British Columbia forest companies were very much opposed to the idea of replacing the export tax because replacement measures would apply to lumber products not intended for the American market. As one official said, "Our position is retain the tariff as a tariff so we can point to it."[62]

Disentangling pressure from American forest companies and the Kempf/Vander Zalm efforts to increase stumpage is fraught with difficulties. Nonetheless, there is evidence that, although the softwood lumber dispute helped government carry forward with revenue objectives by shifting the balance of forces within the subsector, American pressure was not a direct cause of these changes. Indeed, the American lumber companies were pressuring BC to adopt an American-style competitive bidding system, but, instead, BC maintained administered pricing as a policy instrument, temporarily moving settings upward. Even the expansion of the Small Business Enterprise Program, which used a form of competitive bidding, was limited to a small percentage of the AAC, and in many cases it was not a pure competitive bidding system because the program required companies to achieve employment targets. Indeed, the story of how settings were increased reveals much about the durability of the historical compromise.

The 1987 Comparative Value Pricing System

Convinced that the residual system failed to capture the proper economic rent, the Vander Zalm/Kempf solution was to introduce a comparative value pricing system (CVP) in which a target revenue would be established and specific stumpage rates would then be calculated to achieve the revenue objective (see Figures 7.5 and 7.6). Unlike the residual system, government now had the means to explicitly choose how much revenue it would receive. Based on world timber market prices and other measures, the provincial government then derived an average stumpage fee as a derivative of its revenue objectives.[63] Stumpage rates increased sharply after the CVP was introduced, as Figure 7.5 shows.

Often overlooked in analyses of the 1987 stumpage policy changes was that the CVP also included a formula for automatically adjusting stumpage prices up or down. The formula was not based on industry profits but on the Statistics Canada softwood lumber price index.[64] If the market price for timber went up 10 percent, the policy now was that stumpage rates would also go up 10 percent. Although this approach might seem logical, it meant that future increases in stumpage rates would no longer be tied to industry profits but to the price timber commands in the marketplace. Because stumpage rates only increased at the same rate as market prices increase, industry over time would most likely gain an increasingly larger share of the economic rent. That is, once business pays for its operating costs, any small increase in market price could lead to a large increase in profits. However, this new system had no way of adjusting stumpage rates beyond the lumber price index.

Put another way, over time the relatively high stumpage fees in 1987 would gradually collect less and less of the economic rent if prices in world markets for lumber rose. Under this system, government actors would at some point in the future have to explicitly intervene to increase target rates to restore the equilibrium of profitability, wages, and revenue objectives. As Peter Pearse noted, "Though there were some improvements in the proposal, the new system would be even more arbitrary than the current one."[65]

It was therefore not a surprise that, by 1994, rising market prices for lumber had once again resulted in record profits for the forest industry, with comparatively little rent going to provincial coffers. The stage was set for the 1990s policy cycle to restore the historical compromise.[66]

The Policy Cycle in the 1990s

The preceding story illustrates how the AAC and tenure subsectors spilled over in the 1940s to create a timber pricing subsector in which an administered pricing policy instrument carried out the core principles of the development coalition compromise. The 1970s experience revealed the difficulty in challenging this compromise, whereas the 1980s experience revealed how external perturbations could be harnessed by the government to further the revenue objective, though still be constrained by the subsector's core principles. The key legacy of the 1980s policy cycle was that the government now had to act explicitly to raise and lower settings to carry out the principles of the historical compromise. We must therefore be careful to examine governmental timber pricing policy changes to see whether they are developed to maintain the core principles of the subsector or whether they represent a fundamental departure from these principles.

The 1990s experience is remarkable for how the myriad of 1990s forest policy initiatives discussed in this book spilled over to the timber pricing subsector. The Forest Practices Code (FPC) directly increased operational costs, whereas the FPC, the protected area strategy, AAC determinations, and Aboriginal treaty processes reduced the commercial land base. All other things being equal, these spillovers shift the supply curve (to the left), increasing the market price for timber. Tenure reform would have provided the greatest influence on this subsector, but the negative decisions discussed in Chapter 4 show that it had little effect on stumpage policy.

Agenda

Timber pricing came on the policy agenda for three reasons. First, forest companies were experiencing record profits, whereas the provincial government was not only running a deficit, it was looking for sources of funding to offset some of the effects of its environmental initiatives on the forest sector and forest-dependent communities. Recall that the 1987 CVP system meant that, without government action, there was no way to adjust the base timber pricing targets beyond the incremental Statistics Canada index.[67] The minister of forests purposely used this information to let it be known that the issue was squarely on the governmental policy agenda. He publicly announced that recent increases in world market prices for lumber meant that industry profits were out of kilter compared to what was accruing to the Crown.[68] A group of environmental groups also commissioned a study that revealed poor returns to provincial treasury in the late 1980s and early 1990s during a period of high corporate profits.[69] This effort revealed a new-found effort on the part of environmental groups to focus at least some of their attention on stumpage policy despite their historical difficulties in influencing this sector. The equilibrium was ripe for an adjustment of revenue objectives to the top of the priority list.

Second, American forest companies were once again pressuring BC and other provinces to increase stumpage rates. After the Canadian government exercised its right to terminate the 1986 Memorandum of Understanding, the US Commerce Department again ruled in 1991 that BC and three other Canadian provinces subsidized their softwood lumber exports. This time BC's raw log export restrictions were deemed a larger subsidy than its stumpage prices. Though NAFTA binational panels would eventually reverse the ruling, the Americans quickly moved to change the law that had led to the FTA binational panels ruling in favour of Canada. By 1994, American companies were once again threatening to launch countervail action. Forest Minister Andrew Petter used the American pressures to argue that stumpage prices were low and that this inequity would be addressed.[70] BC environmental groups publicly pointed to the ongoing American pressure as proof stumpage prices should be increased. Indeed, BC-based groups such as the Western Canada Wilderness Committee shared information with the American timber company coalition, assisting in their congressional lobbying efforts.[71]

Formulation

The issue soon moved to a consideration of the options for increasing settings. A move away from administered pricing was quickly rejected as being far too encompassing and problematic, because any move away from administered pricing would involve the very difficult task of altering the durable tenure system (see Chapter 4), including its appurtenancy requirements. Instead, choices were limited to ideas regarding how to raise the settings and where to earmark the funds.

In cabinet, the key debate was whether the increases should go to general revenues or to the forest sector to offset the effects of recent initiatives by creating a forest investment fund. Premier Harcourt, who had campaigned on ending the "war in the woods," was strongly in favour of using the increased revenues to invest back in the forest industry, to minimize and even offset the economic effects of his environmental forestry initiatives. Others, such as then minister of employment and investment Glen Clark, supported funds going back to general revenues, fearing that any permanent arrangement on funding would limit future spending policy choices.

In the end, cabinet decided to use most of the stumpage increases to create an investment fund in part designed to offset the effects of the

Forest Practices Code and land use initiatives.[72] However, it was clear that choices over the use of stumpage funds would not change the fact that stumpage rates were to be increased. The forest industry was told by the premier's office that it could support the increases and have the funds go back to the forest sector or oppose the increases and have the funds go to general revenues. With little room to manoeuvre, industry focused deliberations on how to best reinvest these funds. The results of these deliberations focused on the creation of Forest Renewal BC, a new Crown corporation designed to invest in the forest industry and forest communities, funded through the increase in revenues (see Chapter 8).

Decision Making

Administered pricing as a policy instrument did not change, and the actual calculation did not deviate from, the principles of the CVP, which continued to use target prices to guide overall stumpage prices.[73] Instead, the choice was made to increase prices by creating a "super stumpage" fee on softwood lumber. The super stumpage would kick in when Statistics Canada softwood lumber prices hit $250 per cubic metre and would steadily increase as lumber prices rose. When lumber prices went above $400 per cubic metre, sharper stumpage increases would also take effect. (Figures 7.5 and 7.6 show what stumpage would look like at various levels on the Statistics Canada Softwood lumber price index, before and after 1994.) The immediate effects of these increases on target stumpage rates were an increase from $15.17 per cubic metre to $27.47 per cubic metre on the coast and from $17.20 per cubic metre to $28.03 per cubic metre in the Interior.

Although the Forest Sector Strategy Committee facilitated the broad policy decisions, the IAAC, CAAC, and MOF Valuation Branch discussed the "mechanics" and technical issues surrounding an increase in stumpage.[74] Stumpage increases were justified on the grounds that the new world-market prices for timber represented a "structural," long-term change.[75] However, in reality, the new system had built-in adjustments if predictions about "structural adjustment" did not take place: if prices went below $250 per cubic metre, no super stumpage would apply.[76]

By taking an approach that disproportionately increased stumpage the higher market prices rose, the 1994 changes corrected the declining revenue problem of the 1987 CVP system[77] noted above. However, the modifications to timber pricing policy made in 1994 focused only on changing lumber prices and did not build in any long-term formula to adjust stumpage based on increasing operational costs. The provincial government did address the effects of its policies on operational costs in developing its 1994 target stumpage rate, but any future changes to operational costs would require specific intervention to alter the target

Figure 7.5

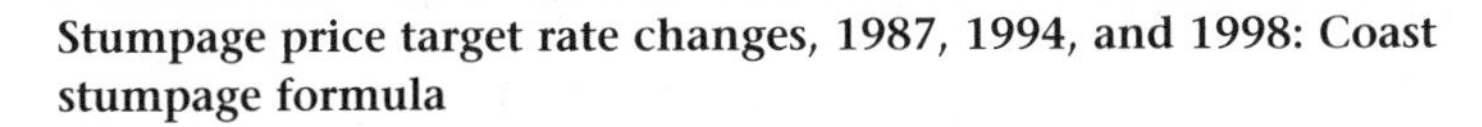

Stumpage price target rate changes, 1987, 1994, and 1998: Coast stumpage formula

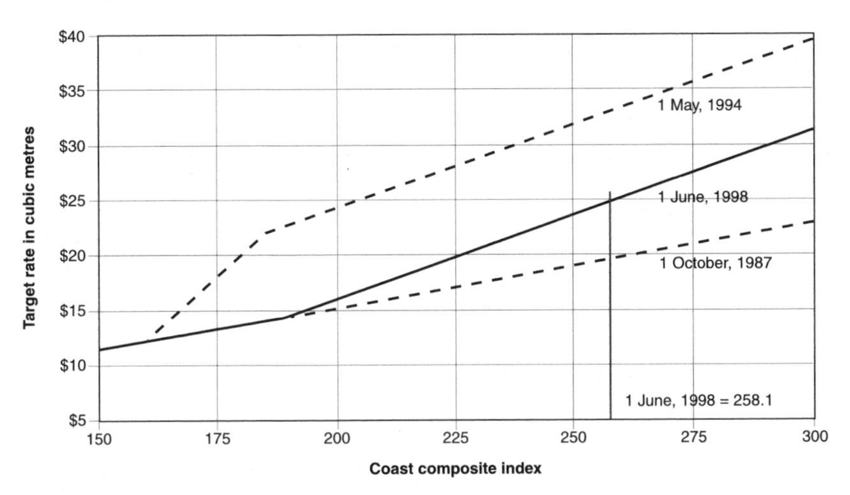

Source: British Columbia, Ministry of Forests, Revenue Branch, "Stumpage changes in British Columbia on June 1, 1998," available at www.for.gov.bc.ca/revenue/timberp/stumpage_June_98/stumpage_June_98.htm.

Figure 7.6

Stumpage price target rate changes, 1987, 1994, and 1998: Interior stumpage formula

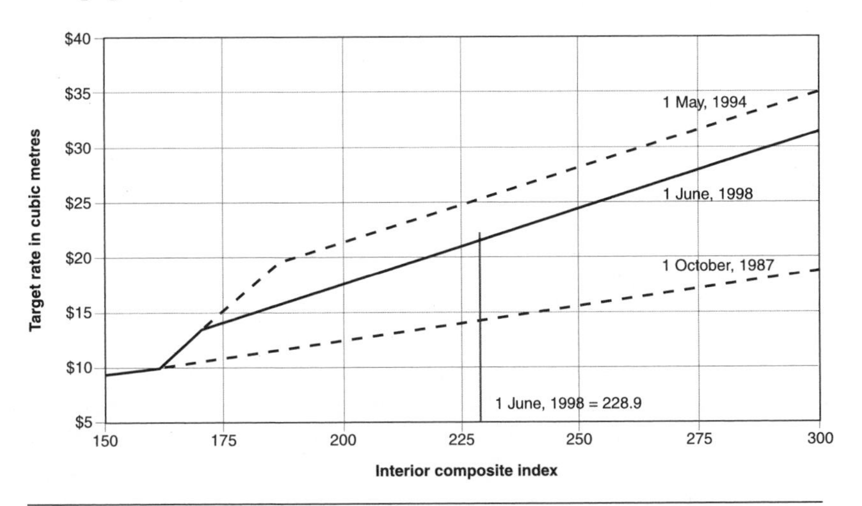

Source: British Columbia, Ministry of Forests, Revenue Branch, "Stumpage changes in British Columbia on June 1, 1998," available at www.for.gov.bc.ca/revenue/timberp/stumpage_June_98/stumpage_June_98.htm.

stumpage rate. In this regard, the 1994 stumpage changes set the stage for further reductions because the provincial government relied on inaccurate estimates regarding the consequences of its environmental and forest policy initiatives on industry operational costs. Estimates were difficult to make because many initiatives were still being developed and implemented.[78] Indeed, even industry's original estimates of the impact of forest policy reforms on operational costs turned out to be below actual costs (see Chapter 3).

Once the choices over the specifics of stumpage rates were agreed on, the announcement of Forest Renewal BC was remarkable for the support it received from within the timber pricing policy regime and from those outside this subsector, such as environmental groups (see Chapter 8). That is, funds were earmarked for environmental clean-up projects and for forest industry workers through the creation of well-paid silviculture jobs and retraining programs. Even industry, which was seeing dramatic stumpage increases,[79] supported the move largely because most of the funds went back to the industry. Industry knew that, in many cases, it would be the beneficiary of Forest Renewal investments.[80] Reinforcing the importance of when the policy cycle intersects with the business cycle, these increases were made at a time of high industry profits when government was in need of revenue as a way of creating new employment and silviculture programs aimed at fostering "peace in the woods."

The choice to place 1994 stumpage revenue increases in Forest Renewal, rather than in the provincial treasury, had important implications for future stumpage reductions at the evaluation stage. Any future reductions in stumpage would have no discernible effect on the provincial treasury. If subsector members implicitly agreed that a stumpage decrease was necessary, it would be much less painful for the government to reduce rates because it would not have to reduce budgetary initiatives that fell outside of FRBC investments.

Implementation

Implementation of the 1994 increases was relatively straightforward because administered pricing remained, and the only requirement was to input the new formula into stumpage price calculations. Technical aspects regarding the complicated issues about how to apply stumpage across different regions and species were addressed by Ministry of Forests Valuation Branch officials and their industry counterparts in the Coastal and Interior Appraisal Committees. Technical issues included administration of the "water bed" rule, which required any reductions in stumpage in one area or region to be increased in another to maintain overall revenue targets. Stumpage increases did initially have their intended effect, with $462 million raised in its first year (see Chapter 8).

Evaluation and Monitoring

Evaluation and monitoring began almost immediately. Governmental officials moved from using American pressure as further justification for a price increase, to using the price increases to show that the American lumber companies could no longer argue that BC's prices were too low. Despite these changes, American pressure persisted. After ongoing consultations among the American, Canadian federal and provincial governments, and forest company officials on both sides of the border, a Canada-US Softwood Lumber Agreement (SLA) emerged as a short-term solution to the ongoing dispute over stumpage policy. Under the SLA, American timber companies agreed not to launch another countervail dispute, whereas British Columbia, Alberta, Ontario, and Quebec agreed to an export quota system. This agreement would see an export tax of $50 per thousand board feet imposed on softwood lumber exports to the United States that exceeded 14.7 billion board feet (BBF) and $100 tax per thousand board feet on exports that exceeded 15.35 BBF.

The result was a cumbersome administrative system in which each province and company fought for its share of the duty-free 14.7 BBF. Although at first seen as an astute move that would force up prices for American consumers, BC coastal forest companies three years later would be decrying the quota, and a coalition of Canadian forest industry officials would be lobbying to find a long-term solution to end the dispute and the quota. The dispute is important for the implementation and evaluation stage of the policy cycle: subsector members learned that increasing stumpage prices was no guarantee for stopping American pressure.

The ink was barely dry on the SLA when the BC forest industry entered the downside of the business cycle (see Figure 7.2), which was exacerbated by the Asian economic crisis and the collapse of demand in that market. The Asian collapse particularly hurt BC coastal forest companies, which export the vast majority of BC forest products destined for Asia. The existing softwood lumber agreement provided a double hit: in addition to making BC companies unable to redirect exports to the American market, any relief they sought in a reduction in stumpage prices would now be scrutinized for whether they violated the nonderogation provisions of the SLA.

In the meantime, the governing New Democrats had changed leaders from the environmentally inclined Harcourt to Glen Clark, who came from a working-class, union background. Clark's first response to this economic crisis was an attempt to find a way to focus on profitability, wage, and employment objectives through the development of the Jobs and Timber Accord, while at least temporarily avoiding stumpage rate decreases. Clark also focused on reducing the effects of the Forest Practices Code on industry operations.[81] Taxes and other incentives were held out for those companies that created jobs, and fines were threatened for companies that did not meet their obligations. Raw log export restrictions were also temporarily relaxed.[82] By focusing on well-paid sustainable forest jobs, the accord can be seen as an effort to keep the high-paid jobs objective on par with the profit objectives by redirecting revenues from the 1994 timber pricing increases. Although the accord was not directly focused on a pricing issue, it is relevant to timber pricing: it used the 1994 stumpage increases to address the profitability and high-paid jobs objectives, and it gave the first indication that the revenue objective had fallen to the bottom of the objective hierarchy.

The political cost in redirecting revenues from 1994 stumpage increases was losing considerable support from the environmental community. And even this ambitious plan could not forestall for long the necessary timber pricing reductions required by the historical compromise. Faced with an ongoing Asian slump and increased operating costs associated with the environmental forest policy initiatives, forest industry executives pushed the provincial government to reduce stumpage. Reducing stumpage would explicitly push to the bottom the revenue objective and place highest priority on the corporate profit objective. Thus, the Jobs and Timber Accord aimed in part to maintain the high-wages objective with the profitability objective, but this approach could not ultimately be sustained.

The 1998 Timber Pricing Reductions

By 1998, the Jobs and Timber Accord was unable to reverse a trend in declining forest sector employment amid increasing corporate losses.[83] The NDP government eventually supported a reduction in stumpage, a decision that reflects the priority profitability objectives have during an economic downturn and that reveals the government's limited room to manoeuvre. Just two years earlier, the government had promised penalties if companies failed to create jobs. It was now saying in official briefing notes that punishing companies would be irresponsible.

Just how a stumpage reduction would be accomplished was fraught with difficulties. The Asian collapse and the American quota system meant that coastal forest companies were more greatly affected by the latest economic crisis than their Interior counterparts, but the 1994 changes had nonetheless taken into account falling prices. In theory, there should be no reason for additional stumpage reductions. Moreover, the Softwood Lumber Agreement (SLA) meant that the government had to strategize over what system of reductions best shielded it from certain American claims that stumpage decreases abrogated the SLA. The BC government saw the SLA as constraining its choices about how to manage the historical compromise rather than as a reason for changing the compromise. As Forest Minister Zirnhelt noted, "Nothing, in dealing with the United

States, is risk-free," but "If our cost structure is out of line, we have to bring it into line."[84]

In the end, a stumpage reduction took place that the government hoped would shelter it from complaints that it abrogated the SLA, arguing that reductions were simply taking place to account for increased operating costs.[85] Working with selected forest company CEOs, the government commissioned a new study to examine the effects of the Forest Practices Code (and other governmental initiatives such as the Protected Areas Strategy) on operational costs (see Chapter 3).[86] Company officials asserted that the current pricing regime inability to account for changing operational costs meant that they were paying more than was economically warranted.[87] The government had hoped focusing on operational costs would show that it was simply bringing the forest industry up to a level playing field with its international competitors. Using the results of this new study (in combination with other estimates), the government concluded that the current stumpage prices underestimated the impact of the Code, the Protected Areas Strategy, and other policies on operational costs by $4.89 per cubic metre.[88] The government multiplied the annual harvest for 1998 by $4.89 to arrive at the amount of stumpage reduction that would take place (see Figures 7.5 and 7.6 for its effect on target prices). The government could therefore say it was changing the target price simply because operational costs had increased.[89]

However, the way in which stumpage reductions were decided revealed that stumpage price decreases were greater than increased operational costs warranted. As described in Chapter 3, the government reduced some of the procedural and substantive requirements under the Forest Practices Code in 1997 and 1998, but the cost estimates used by the government were from cost estimates from studies through 1996, before the measures designed to reduce costs were put into place. Thus, the government's stumpage reductions more than offset the estimated increase in operational costs.

Another implementation decision regarding coastal versus Interior stumpage rates further demonstrated that increased operational costs were as much a rationalization as a reason for stumpage reduction. Instead of simply applying the same one-time reduction to the entire province, months of negotiations between coastal and Interior companies revealed a different compromise – in which the entire amount of forgone stumpage revenues would be split equally between the two regions, and then the target prices would result from this split. Because the Interior produces over two-thirds of the provincial harvest, this approach meant that the coastal companies were getting a much larger reduction on stumpage ($8.10 per cubic metre) than their Interior counterparts ($3.50 per cubic metre).[90]

The evaluation stage of the 1990s timber pricing policy cycle revealed how fluctuating objectives can alter choices made at the decision-making stage as the policy cycle intersects with the business cycle to produce fluctuating objectives consistent with the long-standing compromise among development coalition members.

The 1990s policy cycle illustrates the durability of the historical compromise that has maintained a single goal supported by three fluctuating objectives. The goal of a healthy industry had not changed, the administered pricing instrument was left intact, and profitability objectives remained paramount in the face of an economic downturn. The story provides support for the hypothesis raised in Chapter 1 that business is often most powerful politically when it is economically weak. As timber pricing moves into a new cycle in the year 2000, and timber prices show signs of increasing, we most certainly will see new stories of fluctuating objectives that will lead to changes in stumpage rate policy setting, changes that are necessary to provide stability in this subsector.

The Canada-US softwood lumber dispute revealed fatigue on the part of governmental and industry actors in the 1990s policy cycle. In fact, industry has initiated proposals to fundamentally address the American trade pressures that, if implemented, would crack the long-standing compromise and alienate forest workers from the policy subsector. We review these proposals below and also include a policy alternative that would bring an environmental goal into the timber pricing subsector.

Change in Instruments toward Competitive Bidding and Free Trade in Logs

Except for the Small Business Enterprise Program, the return of competitive bidding has, until recently, never been seriously entertained by members of the timber pricing policy regime. Competitive bidding as a policy instrument addresses one of the concerns of American lumber companies, though it would need to be accompanied by free trade in raw logs to avoid existing American complaints about subsidy issues.[91] What are the chances of the subsector policy regime actors seriously considering such a change? Just a decade ago industry was resolute in its support of the residual system, and organized labour was strongly against free trade in raw logs and a competitive bidding system that might remove appurtenancy requirements. However, the recent quota system behind the Canada-US Softwood Lumber Agreement and the effects of the Asian crisis have so frustrated BC coastal forest company executives that a number of them have now suggested moving toward a market-oriented competitive bidding system in the hopes that this move might reduce or eliminate American countervail pressure.

Before Weyerhaeuser purchased MacMillan Bloedel in January 2000, MB had taken the lead in this regard, circulating a white paper that proposed a scheme to open half of BC's fibre supply to a market-oriented competitive bidding process. The proposal is ingenious, recommending that existing tenure holders be paid to manage the new competitive bidding system on lands that fall within their current tenure arrangements.[92] Moreover, current tenure licensees would have the right of first refusal, in effect being allowed to outbid the highest bidder. The plan also calls for government to derive stumpage rates for noncompetitive tenures based on competitively bid stumpage rates. The white paper acknowledges that such a system would require existing appurtenancy requirements to be dropped from licence agreements because such requirements would hinder implementation of a free market-oriented, competitive bidding system. Similarly, the high-wages objective would necessarily be dropped from the timber pricing subsector because a competitive bidding policy instrument would not allow government to set a rate in accordance with the wages objective. (Wages and employment objectives would have to be addressed elsewhere in the broader forest policy sector.) The paper also acknowledges that to address all of the American industry's concerns, raw log export restrictions would have to be eased because these policies drive down the market price of lumber by reducing demand.

Despite these political hurdles, a market-oriented stumpage system is gaining support within the BC industry and from forestry critics. Recently, BC's Fletcher Challenge advocated a market system,[93] and the president of COFI also began arguing that BC should move toward a market system.[94] This position on the part of BC industry represents a radical change from its stance just a few years before. The collapse of the Asian market and the SLA constraints led to these proposals, which if implemented would challenge the existing historical compromise. Indeed, although industry has criticized most aspects of the Wouters forest policy review in spring 2000, it did support the review's call for the establishment of a competitive log market "within six months" that would allow for a market-based pricing system for logs.[95] The opposition Liberals, who appear poised to win the next election, similarly supported the stumpage reform ideas.[96] Wouters himself optimistically argued that, if established, "The new market model on the coast ... will make it more competitive, more dynamic and should make it more diversified; and with the free trade problem, allow better access into the US market."[97] If the choice was made at the decision-making stage of the policy cycle to proceed with such a system, important spillovers could be felt in the tenure system, which would also have to be altered.

Certainly, forest companies are gearing up for a new softwood lumber dispute battle in which a variety of options and strategies will be considered. In summer 1998, a coalition of forest company executives from British Columbia, Alberta, Ontario, and Quebec met with American lumber dealers, home builders, and consumer organizations to propose free trade in lumber between Canada and the United States.[98] In early 2000, British Columbia's coastal and Interior companies put aside their differences and formed the BC Lumber Trade Council (LTC), cochaired by Lignum's Jake Kerr, from the Interior, and Interfor's Bill Sauder, whose operations are concentrated on the coast.[99] In a similar move, Canadian companies in eastern Canada formed the Free Trade Lumber Council (FTLC).

Just how far companies want, or will be able, to move in a more market-oriented direction is unclear. The FTLC's strategy is to advocate ending the SLA and promoting some form of free trade in lumber between the two countries. However, the LTC's strategy was to renegotiate the SLA, cognizant that twenty years of dispute had yet to provide any long-term resolution. When the two Canadian forest company associations failed to agree on substantive matters, they at least agreed on principles, including asserting that any new initiative must recognize that "fundamental changes to Canada's public land model are not possible in light of the country's historical, cultural and philosophical beliefs regarding land ownership."[100] This statement seems to characterize the Canadian companies as ready to compromise but unwilling to advocate radical adjustments because of American pressure. Indeed, the FTLC has increased its coalition building with American home builders, retailers, and other consumer organizations, indicating they are now aiming to fight the CFLI on its home turf by making clear the effects of this ongoing trade action on the American consumer.[101] These are not the actions of an industry about to acquiesce to American countervail pressures.

Even if BC forest companies agreed to a market-oriented competitive bidding system, other obstacles remain. First, the BC government would have to be convinced of the need. Government actors have expressed interest in MB's proposal, and MB CEO Stephens indicated he was involved in talks on this issue with governmental officials.[102] However, there is little evidence that the current NDP government has the desire to implement such wide-ranging systems. Although the Wouters report supported the concept, the BC Ministry of Forests released a stumpage and tenure discussion paper in fall 1999 that failed to promote such a departure, instead raising questions but proposing very little.[103] British Columbia government policy as of June 2000 was to not support renewal of the SLA, which it argued caused too many problems for BC companies. At the same time, the government recognized that without some kind of negotiated settlement, a countervail duty is likely and would be harmful to BC lumber producers.[104]

Second, the NDP's labour ties, especially BC's Industrial, Wood and Allied Workers Union (IWA), has strongly opposed Stephens's plan, and it has instead developed its own sophisticated critique of stumpage and tenure policy. The IWA released a detailed analysis that claimed the SLA backfired, with softwood logs from private forest lands in BC and other jurisdictions "flooding" the American market.[105] The IWA's response is understandable because the wages objective would essentially be removed from the timber pricing subsector. IWA membership in this subsector would be at stake because the key benefits to labour from the existing administered system would be eliminated.[106]

These two hurdles could likely be overcome with the election of a Liberal government, but a more important third hurdle remains: competitive bidding would also require change in the tenure policy subsector, which we show in this book to be highly resistant to change. What these proposals do provide is a glimpse into what future significant change might look like and which external perturbations caused these ideas to be inserted into the policy agenda.

We must note that even if these proposals for a competitive bidding system were implemented, they would not change the timber subsector's goal of promoting a healthy forest industry. However, under such a regime the wages objective would have to be dropped because a competitive bidding system is incompatible with such an objective.

We turn now turn to proposals that would explicitly recognize ecological values in timber pricing, which would add a new goal to subsector dynamics.

Pricing for the Environment

The review above reveals that environmental groups have developed an increasing interest in timber pricing policy choices in the late 1980s and 1990s, but owing to the closed subsector, they have had limited influence. Environmental groups have also been hampered by their inability to propose any single alternative. Indeed, most of the criticism has been to use American timber company pressures as evidence that stumpage is too low. Consequently, most environmental groups have advocated that a market-oriented competitive bidding system be instituted to increase the price and to provide greater access to the forest resource.[107] As a result, environmental groups have mirrored American industry complaints about what type of policy instrument would be needed to correct alleged stumpage subsidies. In terms of this pressure, they have only focused on increasing the revenue objective within the current historical compromise rather than inserting new ecological goals. At the same time, support of the American pressure has placed environmental groups in a difficult position. That is, American pressure is now focused as much on removing BC's raw log export restrictions as on BC's stumpage rates.[108] But environmental groups have long advocated maintaining raw log restrictions because they believe such a policy increases value-added in the province, creates jobs, and renders the BC forest industry more diversified and less dependent on harvesting as an economic activity.

Indeed, the position to support competitive bidding and to maintain raw log exports has led environmental groups to be highly critical of MacMillan Bloedel's white paper, in part because they argue that the MB proposal leaves the door open for the removal of raw log export restrictions.[109] MB and other companies now appear closer to the position of the American timber lobby than do BC environmental groups.

At the same time, another idea has emerged from within the BC environmental community that would inject a new ecological goal into the timber pricing subsector. This idea revolves around the idea that stumpage prices should reflect a price differentiation for the type of harvesting undertaken, one that would reduce the price when more costly environmentally friendly harvesting took place. Indeed, Forest Policy Watch and environmental groups have publicized this issue, arguing that the "waterbed" effect of the CVPS means that, although a firm might get a lower stumpage rate on a selectively logged stand, the current system means that they would simply pay more elsewhere.[110] Under this type of change, the goal of environmental stewardship would exist alongside the goal of industry health, and it might also open up subsector membership to include environmental groups. Certainly, this type of change would reorient some environmental groups from simply arguing for constantly higher stumpage fees[111] to supporting lower stumpage fees where environmentally sensitive and more costly harvesting techniques were used. With the development of forest certification as an issue (see Chapter 3), companies might also support such a change.

In addition to any uncertain "green premium" companies might receive from logging according to certain "green" criteria, they would also be able to deduct related costs from stumpage, putting themselves at a competitive advantage compared to their counterparts in the United States and elsewhere that rely mostly on private forest land for their sources of fibre. But there remain three key drawbacks. First, it is difficult to distinguish "green" harvesting from other forms of harvesting, rendering cumbersome stumpage reduction calculations. Second, such a system could be vulnerable to new subsidy complaints from American forest companies. Third, any changes would inject new goals and objectives into the subsector, and they would necessarily expand membership to include environmental groups. The uncertain implications of this situation on objectives might cause existing development coalition members to avoid such changes. Fourth, environmental groups have failed Kingdon's test of providing a unified policy alternative, with most still advocating higher market-oriented prices.

Conclusion

Policy choices made in the timber pricing policy subsector reveal the importance of distinguishing among goals, objectives, instruments, and settings when explaining policy change. Although settings (that is, prices) fluctuated wildly in the 1990s, that decade's story is one of goals, objectives, and instrument stability. All changes made to settings in the 1990s were conducted in accordance with the long-standing historical compromise that developed among members of the development coalition. The closed nature of the subsector and the inability of the environmental coalition to formulate a cohesive alternative meant that new ideas were either nondecisions (for example, environmental accounting procedures) or rejected as "negative" decisions (for example, competitive bidding) in the decision-making stage of the policy cycle. Spillovers did occur from other sectors, especially from the forest practices subsector, where increased operational costs were related to stumpage rate settings. However, these changes were conducted within the confines of the existing compromise. No other subsector managed to alter the timber pricing subsector's goal, objectives, or instruments. The tenure subsector might have had such an influence, but we have explained in detail why tenure remained stable. And because the tenure subsector and AAC sectors did remain relatively stable, sustained yield and appurtenancy commitments that came with them virtually eliminated potential for broader change.

This chapter highlights the importance of understanding the causes of setting changes before ascribing them significance. Settings may be caused by fundamentally different reasons, and understanding the causes of setting change is fundamental to appreciating how significant a policy change may have been. This chapter also reinforces the need to distinguish short-term decisions made at one stage of the policy cycle from long-term subsector dynamics. In this regard, the chapter qualifies the hypothesis we raised in Chapter 1 that "significant changes in policy that go against the interests of business groups (or other dominant actors) are unlikely without a burst in public salience of new values." We have found that considerable fluctuations can occur in settings that may displace business profitability interests without any "burst in public salience of new values," but such a phenomenon occurs only in the short term and only to promote long-term subsector stability. In this regard, the story reinforces the importance of understanding how and when the business cycle intersects with the policy cycle for understanding how and why business objectives become dominant vis-à-vis governmental and labour interests.

Overall, the story of the 1990s is one of stability among subsector goals, membership, instruments, and limited but fluctuating objectives. But the late 1990s provides a window into what more significant policy change might look like and from where the pressure might arise. For the first time, business is proposing a different policy instrument that would alter the historical compromise, with the softwood lumber dispute and fluctuating timber prices given as the reasons for the need to make such a change. This proposal seems to indicate that even in a closed subsector relentless external perturbations may eventually cause enough fatigue on the part of key subsector members that changes in objectives and instruments could occur. However, the fatigue would have to be so great as to convince governmental actors to alter tenure and AAC subsector choices, and to take a position at odds with organized labour's interests.

If fatigue were so great that this were to happen, causal relationships between the timber pricing subsector and the tenure and the AAC subsector would be reversed, with choices made at the timber pricing subsector causing change in tenure and the AAC. Given our discussion about the reasons for policy stability in tenure and the AAC, such a response would face immense institutional hurdles, even if governmental actors supported such changes.

In short, unless a sustained economic crisis forces a change in other subsectors (particularly tenure and the AAC) or a crisis in the broader forest policy regime, we can expect durability in the timber policy subsector. External perturbations from the American timber industry will be limited to facilitating short-term increases in stumpage policy or to forcing creative solutions to reduce stumpage rates.

8
"Don't Forget Government Can Do Anything": Policies toward Jobs in the BC Forest Sector

Of all the areas of policy development in BC forest policy in the 1990s, the most dramatic initiatives were in the jobs sector. While governments around the world were deregulating, privatizing, and retrenching, the BC government launched massive policy initiatives to influence the labour market in the forest sector. The first initiative, Forest Renewal BC (FRBC), was an instrumental part of the Harcourt government's package of environmental reforms. To compensate for job losses resulting from land use plans, the government earmarked funds derived from a massive stumpage hike and plowed the money back into various schemes to retrain workers, help affected communities, restore damaged watersheds, and conduct enhanced silviculture, all under the guise of "forest renewal." In its first five years, the agency spent $1.6 billion and managed, in the short term at least, to mollify some criticism of its environmental policies. However, by 1999, with dwindling revenues and escalating criticism of its efficacy and efficiency, the program was being attacked even by labour leaders, and its future seemed precarious.

The second initiative, one of the cornerstones of Glen Clark's mandate, was the Jobs and Timber Accord. Although it involved fewer public funds, it was no less ambitious, promising to create 21,000 forest sector jobs in four years. The policy was a spectacular failure. By late 1998, with industry jobs losses mounting, the Accord's Web site was quietly taken down, a powerful if indirect measure of policy termination in the virtual age.

The two initiatives did amount to significant policy change in terms of the magnitude of the role the government played in the forest industry labour market. At its peak, FRBC spending rivalled that of the Ministry of Forests. Through the combined effort of FRBC and the Jobs and Timber Accord, there was a substantial increase in the proportion of forest sector jobs paid for by the government as opposed to by the private sector. Over $1 billion was spent on land-based programs of forest renewal and environmental restoration that would not have been spent without these policy changes. Whether or not the initiatives will have any lasting impacts on the forest sector is far more uncertain. The future of FRBC seems tenuous, and the Jobs and Timber Accord seems to have had a short life.

These initiatives are revealing of the balance of power between the state and business. They involved an extension of state power and capacity that is remarkable given the antigovernment ideological climate that dominated much of the democratic capitalist world in the 1990s. State control over the land base gave the government the raw legal authority to assert this power, and control over the government by a union-based party beginning in 1991 gave the government the will to extend its power. Favourable background conditions deflected what would otherwise have been formidable industry opposition. The structural change in product markets dramatically increased product prices and thus the total amount of rents available for distribution. Combined with the persistent threat of American retaliatory trade action, the background conditions promoted industry acquiescence. When the background conditions changed in 1997 – when a collapse in demand led to a downward industry spiral – the balance of power between industry and the state shifted, and the programs were reconfigured to better serve the needs to the embattled industry rather than the government.

Because it tackles these two policy initiatives, this chapter is structured somewhat differently. After a discussion of the nature of this subsector, the Forest Renewal and Jobs and Timber Accord cycles are addressed. In important ways, the Jobs and Timber Accord is best seen as a revision of the Forest Renewal initiative, growing out of the frustration of its slow implementation. It was a major policy initiative unto itself, however, and for that reason, it is treated separately. This organizational distinction should not distract from the close links between the two policies.

Forest Jobs Policy Subsector

The forest jobs policy subsector is distinctive in the central role played by labour, particularly the IWA, and the different institutional arrangements. Labour does play a key role in many other subsectors, particularly because of its close relations with the NDP government. But in most other cases, labour participates at a far more general level. In the case of jobs, organized labour has played a central role in designing programs and, in some cases, implementing them as well. In contrast, the role of environmental movement is far less direct. Some aspects of the jobs subsector have direct environmental implications (for example, the FRBC watershed restoration programs and the Jobs and Timber Accord revisions to the Forest Practices Code). However, because most of the issues here relate to the livelihood of forest workers, environmentalists have been less active and less influential.

Figures 8.1 and 8.2 provide information on trends in forest employment in BC over the past several decades. Figure 8.1 shows trends in jobs from 1983 to 1998. Over this time, industry employment varied between 70,000 and 90,000 jobs. In 1998, there were about 82,000 jobs in the sector, 33 percent in harvesting and logging, 18 percent is wood manufacturing, and 49 percent in the pulp and paper sector. The largest and by far most influential union in the sector is the Industrial, Wood and Allied Workers, or the IWA, which split from the International Woodworkers of America in 1987. In 1997, the IWA had 29,000 members in the forest industry in BC. The IWA represents workers in logging and sawmilling but not in the pulp and paper sector. Pulp and paper workers are represented by the Communication, Energy, Paperworkers (CEP) (9,300 members in 1994) and the Pulp, Paper, and Woodworkers (PPWC) (7,000 in 1994). Thus, only about half the workers in the forest industry are unionized. There is a significant regional difference in union rates. Harvesting on the coast is done by unionized employees of major companies; in the Interior, virtually all harvesting is done by nonunion contractors for the major companies.

Figure 8.2 gives trends in the labour intensity of the forest sector, measured by the numbers of workers per cubic metre of timber harvested. Despite the claims of many environmentalists, community activists, and scholars, the last several decades have not witnessed a decline in labour intensity. It is true that over the postwar period, labour intensity declined dramatically until the mid-1980s; by comparison, in 1965 the figure had been approximately 1.7.[1] This decline was driven by technological change in harvesting and milling, which was in part driven by high wage rates. Although it is not widely acknowledged, BC forest workers are among the highest paid in the world.[2] It is not clear why this decline in labour intensity has been reversed.

The institutional apparatus for forest jobs policy has also been different. Rather than the Ministry of Forests being the core government actor, separate organizations were created to implement these two initiatives. For Forest Renewal, a massive Crown corporation was created to administer the "super stumpage" funds collected. Although FRBC was given some independence, the government exercised control through appointed board members, legislative design, and, as we will see, the occasional threat of confiscation of funds.

For the Jobs and Timber Accord, the government appointed an advocate to oversee and implement the agreement. The position of the advocate is extraordinary. The salary of Garry Wouters, appointed in April 1998, is partly funded by the forest industry. The forest industry pays $53,000 of his $168,000 annual salary, and there is a board of directors for the Office of the Advocate that consists of the deputy minister of forests, the chair of

Figure 8.1

Trends in forest sector jobs, 1983-98

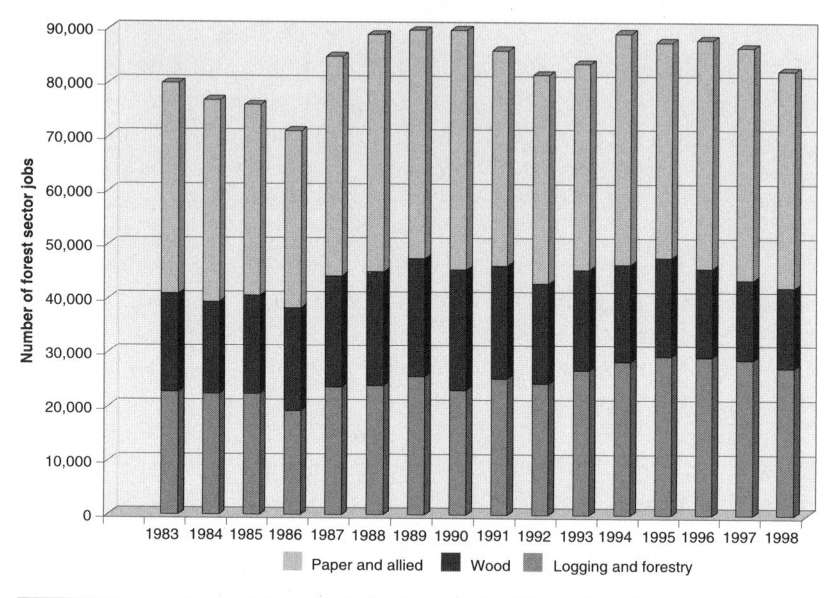

Source: Based on Statistics Canada data provided by the Ministry of Forests.

Forest Renewal BC, and two forest companies.[3] Because of its limited duration, though, the Office of the Advocate was never accorded many resources or much status until it was diverted to a broader forest policy reform in late 1999.

Forest Renewal

Agenda Setting

A complex array of forces brought Forest Renewal policies onto the government agenda. First, there were spillovers from other forest policy subsectors. Initiatives designed to increase the environmental sensitivity of BC forest policies – land use policies, the Forest Practices Code, and the Timber Supply Review – all exerted downward pressure on jobs. Second, the political stream was dominated by the governing NDP, which was at its core a labour party. On forestry issues, where there are frequently challenging distributional choices between jobs and environmental protection, the NDP was faced with dilemmas. If the NDP moved too far in the environmental direction without protecting workers, it would not be able to survive.

At the same time as these environmental policies were creating explosive tensions within the NDP coalition, several enabling conditions were

Figure 8.2

Labour intensity of BC forest sector, 1983-98

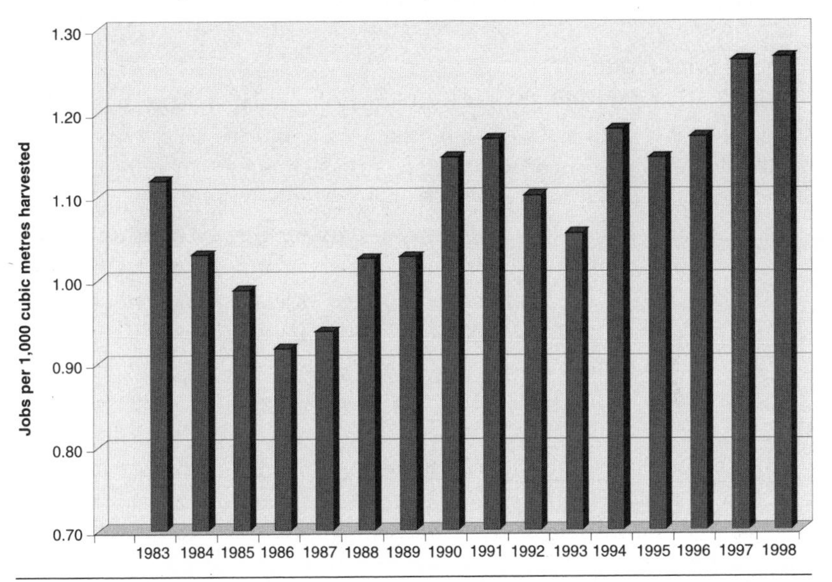

Source: Based on Statistics Canada data provided by the Ministry of Forests.

in the background. The softwood lumber dispute with the United States was at a critical stage, and BC and other provinces were looking for ways to deflect the argument that low stumpage fees were illegally subsidizing BC producers. In addition, the industry was at the height of its prosperity, pushed by strong demand and what many considered to be a structural change in lumber prices. The combined effect of these two background conditions was that the industry accepted that stumpage prices would need to increase to deflect American pressures, and, because of high lumber prices, the industry believed it could afford that increase.

These tensions were brought into focus by the intense land use conflicts on Vancouver Island in 1993 and 1994.[4] The decision to allow two-thirds of Clayoquot Sound to be logged provoked outrage among environmentalists in summer 1993 and put the government on the defensive. In response, an impressive grassroots mobilization of loggers and timber-dependent communities occurred, culminating in a massive demonstration in Victoria in March 1994 to protest the Commission on Resources and the Environment's proposed land use plan for Vancouver Island. This direct linkage between land use and the jobs subsectors solidified a particular problem definition within the forest sector: if the government was going to raise any costs for the forest industry, it had to keep the funds within the forest sector, broadly defined here to include workers, companies, and communities, to compensate for environmental policy initiatives. This problem definition was absolutely critical to getting the agreement of divergent interests at work in the sector.

Policy Formulation

The Forest Renewal initiative was uniquely formulated. The architect of the initiative was the Forest Sector Strategy Committee, a secretive multistakeholder body created by the Harcourt government in April 1993 to develop an industrial strategy for the forest sector. The nineteen-member committee had broad representation, heavily weighted toward corporate CEOs but also containing representatives of communities, labour, and environmentalists. The committee reported directly to a committee of deputy ministers, chaired by Doug McArthur, who was then deputy minister of the planning secretariat for the provincial cabinet.[5] Bringing a representative group together in this quiet forum created the potential to address related issues in a structured atmosphere. Despite their time commitments, the CEOs participated actively, in part because McArthur and Forest Minister Andrew Petter attended and contributed substantively, ensuring that the efforts of the committee would be taken seriously by the government.

This committee developed and elaborated the concept of increasing stumpage and putting the funds in an independent, statutorily protected entity, to spend on a wide range of "forest renewal" activities. For labour, it promised job protection for forest workers. For environmentalists, it promised new financial resources dedicated to environmental restoration and a way to deflect criticisms that their actions were destroying jobs. For industry, which was being asked to absorb a significant cost increase, it promised that the money raised would be reinvested in the forest sector to contribute to long-term productivity. For government, because the initiative addressed the needs of all three of these groups, it promised an end to the "war in the woods." It also provided a significant increase in resources available to government and an enlargement of the state role in the forest sector. With the committee setting the intellectual and political groundwork for the initiative, Premier Harcourt seized on the idea at a crucial juncture, just when his efforts to forge a new, more sustainable forest policy seemed to be jeopardized by escalating labour opposition.

Decision Making

Faced with a potentially fractious conflict at the 1994 party convention in late March, Harcourt announced his bold pledge that "not one forest worker will be left without the option to work in the forest as a result of a land-use decision."[6] The government backed up this commitment by introducing the Forest Renewal Strategy in April 1994, three weeks after the loggers' protest in Victoria and two months before the government issued its first major decision on the CORE Vancouver Island plan. The plan would double stumpage payments and place the funds under the control of an independent government entity, Forest Renewal BC. This agency was required to reinvest the money in the forest sector through forest restoration, intensive silviculture, worker retraining, value-added manufacturing, and community redevelopment. Annual revenues were projected to be approximately $400 million and revenues over a five-year cycle to be $2 billion.[7]

The decision to create a statutorily protected Crown corporation to administer the funds was critical to obtaining agreement, especially from industry. Industry was unwilling to support the increased costs without an ironclad guarantee from the government that the money would remain in the forest sector and not be absorbed into general government revenues. History supported the forest sector scepticism, as a number of previous initiatives to establish silvicultural funds were quickly dismantled because the government of the day could not resist clawing the money back into general revenues.[8] In establishing the new program, the government did everything it could to assure industry, in both the structure of the legislation and the statements by relevant ministers. Cabinet Minister Dan Miller stated: "There won't be a politician next week, next month, next year, or 20 years from now who will dare to put their hands into that pocket of money."[9]

The agreement was announced with great fanfare and excitement, and with explicit endorsement from labour, industry, and environmentalists.[10] The plan seemed, at long last, to promise an end to the war in the woods. The economic consequences of environmental restrictions could be alleviated by using the fund to employ dislocated workers in forest renewal, with the promise of more intensive silviculture producing more jobs in the woods in the long run.

The goals of Forest Renewal are best described as promoting the resolution of forest-environment conflicts by providing a mechanism to compensate workers displaced by environmental reforms. On close inspection, Harcourt's promise of no jobs lost was carefully crafted to limit the extent of the government's formal commitment. Note that Harcourt promised no job losses from "land use decisions," leaving aside the Forest Practices Code and Timber Supply Review that were also putting downward pressures on jobs. Nevertheless, the government was hopeful that enough funds would be available to protect those workers as well.

The objectives of Forest Renewal were outlined in detail by the government. There were two major priorities: "renewing our forests" and "creating more value, strengthening communities." Under the banner of forest renewal, the government listed the following:

- improving reforestation and silviculture to ensure healthy, productive second-growth forests
- cleaning up environmental damage, particularly watersheds damaged by past logging.

Under the rubric of "getting more jobs and value from our forests," the government stated these objectives:

- assisting more value-added companies to start up, expand, and develop new markets
- providing new training for forest workers, to prepare them for new job opportunities in forest renewal and value-added manufacturing
- enhancing First Nations' participation
- strengthening communities through economic development and diversification.[11]

The central instrument of Forest Renewal was spending the funds generated by the new stumpage revenues through the new organizational mechanism of Forest Renewal BC. The revenue formula was designed to provide approximately $400 million per year, creating a massive pot of new money to help buy peace in the woods. (For reference, the expenditures by the Ministry of Forests in the 1995-6 fiscal year were $690 million.[12]) The initial government decision did not contain any guidance about how the spending would be allocated to the various objectives listed above. It did, however, require that spending be "distributed fairly throughout all regions of the province."[13] FRBC was to be governed by a cabinet-appointed board made up of representatives of government, the forest industry, workers, communities, First Nations, and environmentalists (see Table 8.1 for the composition of the original board). The board's only chair has been Roger Stanyer, who before taking a position as a senior advisor to Harcourt was a logger on Vancouver Island and then an IWA staff member. In the public documents released when the decision was made, a great deal of emphasis was placed on stakeholders running the organization. For example, in a statement reminiscent of Charles Lindblom's famous dictum that the measure of a good policy was agreement among interest groups,[14] the investments were said to be effective because "all partners will take part in making decisions on where the dollars are invested."[15] The legislation required the establishment of five committees that would advise the board on spending decisions in each of the major areas of the FRBC mandate. FRBC was required to submit an annual business plan to the minister of forests and the legislature.

The decision to embark on Forest Renewal BC was monumental. It was brilliant in concept: the government got industry to agree to a significant

Table 8.1

Original Forest Renewal BC Board of Directors

Gerry Armstrong, deputy minister, Ministry of Forests

Peter Beulah – Penticton, president, Greenwood Forest Products; president, Interior Value Added Wood Association

Ric Careless – Gibsons, western regional director, BC Spaces for Nature

Desmond Gelz – Prince George, vice president, Northwood Pulp & Timber Ltd.; director, Canadian Wood Council

David Haggard – Port Alberni, president, IWA 1-85; member, BC Federation of Labour Executive Council

Ann Hillyer – Victoria, staff counsel, West Coast Environmental Law Association; member, National Pulp and Paper Round Table

John Kerr – Vancouver; chair and CEO, Lignum Group of Companies

Joanne Kineshanko – Lumby, mayor, Lumby; Kineshanko Logging Ltd.

Doug McArthur, deputy minister to the premier, cabinet secretary

David McInnes – Vancouver, chair of the board, Weyerhaeuser Canada Ltd.

Garry Merkel – Kimberley, vice-chair, Columbia-Basin Trust

Brian Payne – Vancouver, western vice-president, Communications, Energy and Paperworkers Union of Canada

Andrew Petter, MLA, Saanich South, BC minister of forests

Jackie Pement, MLA, Bulkley Valley-Stikine, BC minister of transportation and highways

Moe Sihota, MLA, Esquimalt-Metchosin, BC minister of environment, lands and parks

Roger Stanyer – Duncan, chair of the board of directors, Forest Renewal BC

Ralph Torney – Victoria, vice-president, Truck Loggers Association, president, R.F.T. Trading; vice-president, Husby Forest Products Ltd.; president, Canadian Air-Crane Ltd.

George Watts – Port Alberni, First Nations Summit representative; chief of the Tseshaht

Source: Forest Renewal Annual Report, 1994-95.

cost increase, creating the financial means to cover at least some costs created by the government's environmental initiatives. The government was able to keep its precarious coalition of environmentalists and labour together for the time being, without provoking industry opposition. The policy was a significant victory for labour, which obtained access to the financial means to compensate those displaced by environmental policies.

Environmentalists were willing to support the initiative because it promised the means to pay for environmental restoration and because it mollified political opposition to environmental initiatives in the forest sector. But the biggest winner had to be the government, which was able to dramatically expand its financial resources in an era of pervasive fiscal austerity. Essentially, the initiative involved the government seizing control over rents that otherwise would have gone to industry, placing them in a fund for government to allocate according to its own objectives.

Industry lost to the extent that it was forced to agree to significant cost increases. A report on costs increases from 1992 to 1997 concluded that the stumpage increases of 1994 increased delivered wood costs by 33 percent.[16] Industry's opposition was reduced by the government's commitment to protect the fund from general revenues and by the fact that the policy promised to invest in increasing the future productivity of forests, in effect socializing in part the costs of producing the future forest from which they would benefit. Nonetheless, the fact that control over those spending decisions would be transferred to government bureaucrats reflected a clear distribution of power away from industry. Such a dramatic move would not have been possible without facilitating background conditions: namely, the apparent structural shift in product prices, the accompanying boom in the BC forest industry, and the need to address the demands of the tenacious American softwood lumber lobby.

Implementation

Brilliant in concept, Forest Renewal has had extraordinary difficulties in implementation. It has proven far more effective at collecting money than spending it or spending it wisely. Figure 8.3 shows trends in FRBC revenues, spending, and equity. Revenues for the first year were $462 million, and they stayed at or above the originally projected annual revenues until the 1998-9 fiscal year, when the market downturn reduced harvest levels and thus stumpage payments, and the government adjusted the stumpage formula to reduce the amount paid per unit harvested (see Chapter 7). Revenue for the 1999-2000 fiscal year was $268 million (less than half the peak), and it is projected to be about $250 million for the next several years. Spending started much more slowly, with only $40 million in the first year and $158 million in the second. Under intense pressures from all parties to begin to deliver, spending increased dramatically over the next several years, peaking at a staggering $646 million in the 1997-8 fiscal year. The business plan for 2000-1 projects spending of $318 million, less than half the peak.[17]

Figure 8.4 shows how the spending over the first five years was allocated by function. The largest area was land and resources, with environment coming a close second. These two functions accounted for 82 percent of all spending. The programs on workers, communities, and value added were small in comparison. Within the land and resources program, the biggest category of spending was enhanced forestry: $310 million, or 43 percent of the land and resources envelope. Enhanced forestry is silviculture beyond the basic reforestation required by law, including activities such as spacing trees, removing competing brush, pruning branches, and fertilizing stands. Reforesting costs totalled $108 million for areas that were not properly reforested in the past and another $141 million for inventories of forest resources. The bulk of the work in the environmental program focused on restoring watershed damaged by past logging ($302 million, or 49 percent), with the next largest category of spending being inventory work on nontimber values, such as wildlife habitat requirements.[18]

The first major crisis for FRBC occurred in summer 1996, two years after its birth. FRBC was bulging with money it could not figure out how to spend, and the newly elected Clark government was mired in a political crisis over deficit spending. As a result, the government decided to claw a sizable fraction of the money back into government revenues. All parties to the initial agreement announced their sense of betrayal. Peter Bentley, the chair of Canadian Forest Products Ltd., who had stood next to Premier Harcourt when he announced the creation of the FRBC in 1994, stated: "It is a complete double-cross ... I would say the government cannot be trust-

Figure 8.3

Forest Renewal BC financial trends, 1995-2000

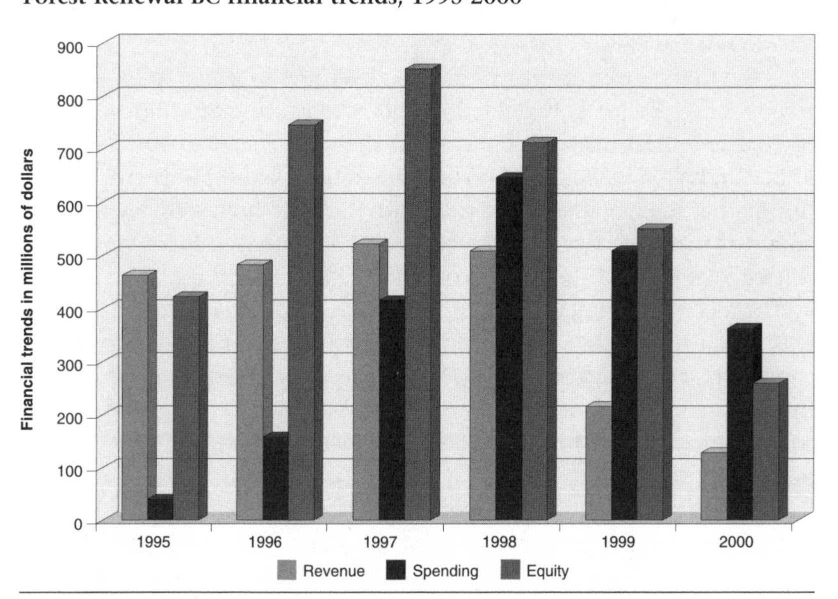

Source: Forest Renewal BC annual reports and business plans. Figures for 2000 are projections.

Figure 8.4

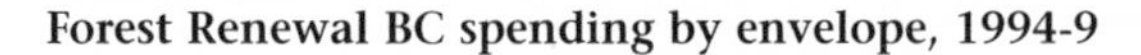

Forest Renewal BC spending by envelope, 1994-9

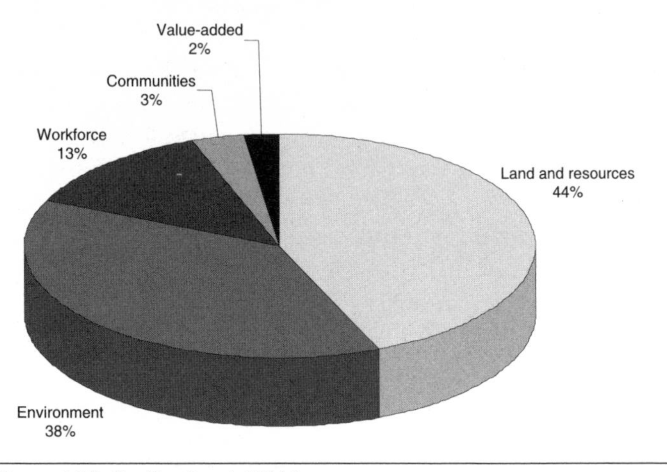

Source: Forest Renewal BC, *Five-Year Report*, 1994-9.

ed."[19] In the event that there was any misunderstanding about the nature of power in the BC parliamentary system of government, Forest Minister David Zirnhelt uttered his now infamous remark: "The government can change the law. Don't forget the government can do anything."[20] Although not far off the mark as a statement about how the BC government operated, the comment was stunning for its callous disregard for the delicate bonds of trust that had developed among divergent interests as a result of the creation of the FRBC.[21]

As the Clark government was mulling over its budgetary options, FRBC took another major public relations hit when a former senior Ministry of Forests official made a public claim that 35 percent to 40 percent of FRBC funds were being spent on administration. Forest Minister Zirnhelt, noting that the maligned FRBC administration did hold well-paying jobs, admitted there was a serious problem and committed to addressing it.[22]

Recognizing the consequences of reneging on the deal for forest politics and other issues facing the province, Premier Glen Clark retreated in early 1997 from his threat to transfer the funds to general revenues.[23] In the short term, the issue was resolved far more quietly when the government, in its March 1997 budget announcement, had FRBC pick up the tab for $100 million of the Ministry of Forests silviculture budget.[24] FRBC clearly got the message and ramped up spending dramatically.

As spending increased, however, criticisms began to emerge about how the money was being spent. Colourful anecdotes began making their way into the media: the wrong logging roads being deactivated and having to be replaced with new roads; displaced forest workers being paid to learn to play the saxophone or golf. These sensational stories distracted attention from the far more weighty concern that the huge investments in silviculture funded by the agency were being squandered on activities that had little payoff in increased volume or value.

Just as criticisms of the wisdom of FRBC spending were beginning to intensify, the amount of money available to spend began to plummet. Spending in the 1998-9 fiscal year dropped to $463 million (compared to peak spending of $646 million in 1997-8), and it is projected to decline further to about $330 million in the 1999-2000 fiscal year.[25] This combination of increasing criticism and decreasing resources forced the agency to reconsider its mandate. In February 1999, FRBC announced a significant reduction in staff and planned spending along with a shift in priorities. Agency staff was cut from 195 to 117 (a 40 percent reduction). Support to workers and communities (already only 16 percent of funding over the first five years) was also scaled back, limited only to "community and workers in crisis." The agency would focus its efforts on enhanced forestry, environment, and value-added industries, but the amount spent on the environment was expected to decrease.[26]

The new priorities were reflected in the 1999-2000 business plan. The funding expected to go to the environment envelope was being reduced to 26 percent of the (much smaller) 1999-2000 total, from an average of 38 percent over the first five years and 33 percent of the 1998-9 total.[27] Environmentalists, who supported the creation of Forest Renewal BC, expressed alarm at the cutbacks to both funding for watershed restoration and inventories of wildlife and other nontimber values.[28] In addition, the agency announced that it was dipping into the reserves set aside for hard times, the so-called rainy day fund.[29]

Tensions mounted as the concerns over the wisdom of FRBC spending drew the attention of Auditor General George Morfitt, who conducted a major audit of FRBC in spring and summer 1998. Even before the results were known, the agency began a defence against the expected criticism by shifting the standard of evaluation. Forest Minister Zirnhelt, always ready with the colourful quotation, stated: "You can't measure the efficiency of some things. The bean counters will always argue that if you put the money in the bank, you'll have more money. We're looking at a forest a hundred years from now. That is what I say to the critics."[30] Chief Executive Officer Lee Doney had the same message: "A straight financial return on some silviculture investments, you'd probably do better putting the money in the bank. But that's not why you do silviculture. You do it because you are hoping a hundred years from now there is going to be a market for that wood."[31] What they did not say here, of course, is that the dominant objective of the silvicultural strategy was to keep people employed in the forest industry.

Denouncing government auditors as bean counters and shifting the criteria for evaluating investments to some vague, unlegislated objective did little to quell the rising tide of criticism, even among those who would be expected to be its strongest supporters. In May, NDP MLA Gerard Janssen told the *Vancouver Sun,* "I'm trying to put together how you can spend more money, create fewer jobs, and do a heck of a lot less work."[32] Doug Smyth, an IWA researcher, argued that FRBC should be replaced with a more incentive-based systems for firms.[33] Don Cochrane, the NDP government's first Forest Jobs commissioner, wrote a blistering report, released in summer 1999, blasting the FRBC worker transition program as being ineffective and far too cozy with the IWA. Local union halls were apparently used to administer the program.[34] In July 1999, the *Vancouver Sun* ran a four-part series called "Lost in the Woods," illustrating widespread opposition to the operation of FRBC. Garry Wouters, the province's Jobs and Timber Accord advocate, is quoted as saying the government was looking for a new model.[35]

FRBC has produced summaries of outputs, and journalists have produced other useful aggregate information. According the FRBC, the $1.6 billion worth of spending from 1994 to 1999 produced the following outputs:

- enhanced silviculture on 311,000 hectares of land, increasing BC's potential future timber supply by 12 million cubic metres
- restored 150 damaged watersheds
- trained 25,520 workers
- created an annual average of 4,500 jobs, with an additional 1,000 jobs total created through community economic diversification
- reduced core administrative costs to 5 percent of total spending by 1999.

What is missing from this self-reporting is an assessment of the efficiency or effectiveness of these efforts. The auditor general's report, finally released in fall 1999, is critical of the efficiency and effectiveness of the corporation's core programs. The report's criticisms were soft-spoken but nonetheless pointed. It stated: "The corporation spent about $1.2 billion without a strategic plan and clear objectives to guide program investments."[36] With regards to the silviculture program, the report concludes that "significant improvements are required in several areas before Forest Renewal BC can demonstrate that it is spending wisely." One clear consequence of FRBC initiatives is that the costs of silviculture increased significantly, though the magnitude of the increase is disputed. Historically, much of the silvicultural work in the province had been done by a nonunionized workforce, working on piece rate. "Treeplanters" have become a cultural fixture in BC life, a fascinating combination of rural residents, counterculture enthusiasts, and university students looking for a quick infusion of cash. Through FRBC, and the Jobs and Timber Accord initiative that created New Forest Opportunities, some of this work, especially on the coast, was given over to displaced forest workers who were given IWA salaries. It was inevitable that costs would increase. As the auditor general's report concludes, "adding social objectives to Forest Renewal BC's mandate (such as the creation of partnerships and establishment of New Forests Opportunities) has meant that services purchased by the corporations are not at least cost."[37]

To summarize the implementation record of FRBC, the agency has been highly effective at collecting money, and it became effective at spending money. That spending has helped provide employment for forest workers and support for affected communities. But it is doubtful that the spending was maximally effective in achieving the mix of objectives the program was established to pursue. As revenues decline, and criticisms about its performance intensify, the future of the FRBC is in grave doubt.

What do the six criteria for successful implementation outlined in Chapter 1 tell us about how to explain this pattern?

(1) *The enabling legislation or other legal directives mandate policy objectives that are clear and consistent or at least provide substantive criteria for resolving goal conflicts.* FRBC was given ambitious objectives, but there was little guidance on how to allocate resources among those objectives. In the beginning, when the corporation couldn't spend money fast enough, this situation was not a problem, but it became more intense as resources dwindled.

(2) *The enabling legislation incorporates a sound theory identifying the principal factors and causal linkages affecting policy objectives and gives implementing officials sufficient jurisdiction over target groups and other points of leverage to attain, at least potentially, the desired goals.* FRBC was established to pursue a range of objectives, each with distinctive causal theories involved. Overall, the agency's approach was based on the belief that money could solve a lot of the forest sector's problems. What was lacking was clear causal linkages between spending and desired outcomes. For silviculture, nearly $1 billion has been spent, much of which, by the agency's admission, does not meet the criteria of desirable investments based on a rate of return analysis. For worker training and community assistance, there is a considerable amount of controversy about what sorts of interventions are effective in diversifying communities and helping structurally displaced workers.[38]

(3) *The enabling legislation structures the implementation process to maximize the probability that implementing officials and target groups will perform as desired. This condition involves assignment of authority to sympathetic agencies with adequate hierarchical integration, supportive decision rules,*

sufficient financial resources, and adequate access to supporters. Financial resources and legal mandate were certainly not the problem: FRBC had both in abundance. Knowing what to do and how to do it were much bigger problems.

(4) *The leaders of the implementing agency possess substantial managerial and political skill, and they are committed to statutory goals.* FRBC had definite management problems, especially in the beginning. It had the enviable task of spending massive sums of money on vague legislative goals, but the agency lacked the managerial skill and resources to establish an effective administrative regime. By the time the agency seemed to have gotten its administrative house more in order, its funds were drying up. Overly cozy relationships with IWA offices aggravated the tendency to skirt effective procedures and fiscal discipline.

(5) *The program is actively supported by organized constituency groups and by a few key legislators (or a chief executive) throughout the implementation process, with the courts being neutral or supportive.* FRBC had widespread support early on, but as problems mounted, so did the criticism, and the future of the agency now seems precarious.

(6) *The relative priority of statutory objectives is not undermined over time by the emergence of conflicting public policies or by changes in relevant socioeconomic conditions that weaken the statute's causal theory or political support.* The ability of the agency to pursue its mandates has certainly been influenced dramatically by the downturn in the forest economy that reduced the agency's revenues.

FRBC could solve any problem that spending could address. The problem was that the agency lacked the knowledge base and managerial skill to direct the funds effectively, and many of the most fundamental forces at work in the BC forest economy were structural, economic, and global, which were not easily rectified by ramping up government spending.

We now turn to an analysis of the Jobs and Timber Accord. After that, we will return to more general analysis of policies toward forest jobs in BC.

Jobs and Timber Accord

Agenda

The agenda-setting process for the Jobs and Timber Accord was less complex than many other cases in this book. The two major forces at work were a surplus of unspent FRBC money and an aggressive new premier with an exceptionally activist approach. This initiative emerged in early 1996 when the main problem with the FRBC was that it could not spend money fast enough. To that extent, the Jobs and Timber Accord arose as an effort to refocus Forest Renewal. More important, the Jobs and Timber Accord resulted from the personal initiative of Premier Glen Clark, immediately after he took over from Mike Harcourt. Clark, more union-oriented and enthusiastic about government intervention in markets, wanted a major policy initiative to act as the centrepiece of his party's campaign in the upcoming election. In March 1996, in his first major announcement about policy initiatives going into the upcoming election, Clark told the industry that it would need to create 21,000 more jobs to maintain cutting rights. When asked about industry's reluctance to join the effort, Clark explicitly threatened the industry: "Frankly, I'm not too worried about it because we are the landlord, we are the owners of the forest ... I indicated to the industry that the government is committed to this, we are serious about it, we want to work together to do that. Failure to work together means the government would act unilaterally."[39]

The problem definition promoted by the NDP government was that the industry was not doing enough to provide forest jobs with the public resource they were exploiting. As the summary document of the accord stated: "Public forest lands belong to the people of British Columbia, and they have a right to expect more jobs and other social benefits from every tree cut on public lands."[40] Clark explicitly referred to the need for a new social contract, where the industry would do more in exchange for its rights to harvest Crown timber. The 21,000 figure was derived by comparing BC's number of jobs per volume of wood harvested with that of Washington and Oregon.

Although the industry could not contest its dependence on Crown timber supplies, it was taken aback by how explicitly Clark referred to the landlord-tenant relationship, as if he were giddy in brandishing his new-found sword. It focused its challenge on the argument that it wasn't providing the same number of jobs per cubic metre as comparable jurisdictions. The Council of Forest Industries commissioned a study that argued that once corrections were made to take into account the differences among the jurisdictions, the differences in labour intensity disappeared.[41]

Policy Formulation

Despite its importance to Clark, the initiative became stymied in difficult negotiations with the industry. The Forest Sector Strategy Committee, despite its success in producing the innovative FRBC strategy, was never involved in the formulation of the Jobs and Timber Accord. Everyone had understood that the committee would be, but Clark and his advisors closely guarded the negotiations. Then the Asian economic crisis exploded and the export markets dwindled, and the balance of power between the government and industry shifted. The initiative opened with Clark trumpeting the power of the Crown through its landlord status, but as time went on, the government felt the escalating pressures of addressing the cost structure of an industry in crisis. In a meeting with senior government officials, industry put its cards on the table with an alarming report entitled *Industry on the Brink,* which emphasized the emerging crisis within the industry and the role of recent government policies in aggravating the problem.[42] As a result, what might have become a formidable policy instrument involving legal commitments linking job targets to cutting rights became little more than an overinflated announcement by a charismatic leader. There was little surprise when it collapsed within a little more than a year.

When it became apparent that addressing the industry's cost problems involved reforming the Forest Practices Code, environmentalists insisted on being involved in the deliberations. The government struggled to address this problem, but it was ultimately unable to do so. The government pursued a divide and conquer strategy. It denounced Greenpeace for its aggressive tactics, both on paper in its report *Broken Promises* and on logging roads in the Central Coast, where it began blockades in spring 1997 (see Chapter 2). The government said it would not deal with Greenpeace, and the other environmental groups said that they would not sit down with the government unless Greenpeace was invited. As a result, environmentalists did not play a formal role in the development of the accord.[43] Meanwhile, government and industry sought to minimize the impact of environmental opposition by doing their best to "decouple" the Jobs and Timber Accord from the upcoming Code revisions.

Decision Making

The agreement was finally announced on 19 June 1997, over a year after Clark announced his intentions, with the great fanfare and hyperbole that came to characterize Clark's leadership style. The day before the announcement, he told the legislature: "It's the most ambitious package of job creation in the history of Canada."[44] When the formal announcement was made in Prince George, the premier made his commitments extremely explicitly, in terms of both broad goals and specific objectives. Regarding goals, the premier stated: "The Jobs and Timber Accord is the next step in my government's long-term plan to renew our forests and sustain the communities that depend on them. I am determined, in partnership with industry, forest workers and communities, to put the same energy and commitment into creating jobs as we put into protecting the environment."[45] Regarding objectives, the initiative promised to create 22,400 new direct forest jobs, as follows:

- *6,500 jobs from the small business and secondary industry:* These jobs were to come from two sources: government producing the full cut allotted

to the Small Business Enterprise Program, including the accumulated backlog, and an increase in the amount of fibre available to small, value-added manufacturing enterprises due to requiring major licences to offer 16 percent to 18 percent of lumber to secondary manufacturing businesses at fair market prices. Value-added firms have long complained that the vertically integrated nature of the BC forest industry made it very difficult for them to get access to adequate wood supplies.
- *5,000 new jobs from renewing our forests:* These jobs come from a commitment from Forest Renewal BC to dedicate $300 million per year of its accumulated funds to reforestation and stand tending, enhanced forestry, watershed restoration, and other projects.
- *5,900 new jobs from forest companies:* The industry committed to creating 2,000 direct jobs through new investment and developing higher-value products. The remaining 3,900 jobs were to come from the government commitment to close the gap between allowable cut and actual harvest levels (in part by simplifying the Forest Practices Code) and by using FRBC to invest in future timber supply increases.
- *3,000 new jobs from new work arrangements:* These jobs were to be created through a government commitment to work with industry and unions in the forest sector in the collective bargaining process to reduce overtime. The government offered up to $20 million to offset payroll costs for the additional employees.
- *2,000 new jobs from the creation of Fisheries Renewal BC:* FRBC money would be used create a parallel Crown corporation to restore salmon streams damaged by past forest practices.

In addition to these 22,400 direct jobs, the government claimed that 17,400 indirect jobs would be created through a multiplier effect, resulting in a total of 39,800 new jobs by 2001.[46]

In terms of policy instruments, the threatened link between job creation and cutting rights was essentially severed by the time the agreement was made. The announcement stated that the "job targets will be achieved through a combination of incentives and compliance measures. Companies which meet their obligations under the accord will be eligible for Forest Renewal BC funding and new Innovative Forest Practices agreements. They will also be eligible for exemption from the five percent 'take-back' of allowable cut that applies to the sale or transfer of a license. Companies failing to deliver jobs will be ineligible for these benefits, while government may take additional steps to ensure compliance with job targets."[47] The sticks had been put away and the carrots set out.

Softening the threatening tone was essential to getting the industry to support the accord, which despite its label was not signed by any industry representatives.[48] It was, however, informally agreed to by seventeen firms accounting for 80 percent of the province's cut. Forest company CEOs acknowledged that they understood that if they did not respond to the incentives in the accord, Clark might take more dramatic action down the road. Lignum CEO Jake Kerr, who led the industry negotiations on the accord, said, "Glen Clark is not a shrinking violet and if things don't go well, then I assume he has more cards to play, and he said so this morning."[49] To signal clearly how the scheme would work, Clark awarded Kerr's company the first Innovative Forest Practices Agreement on the same day the accord was announced.[50]

What does the accord say about the relative balance of power among the various actors in the forest policy regime? Clark clearly saw it as a vehicle to exercise the muscle of the government, supported by its institutional status as landowner and by its financial status as banker sitting on large quantities of underused FRBC funds. But when the content of the accord is scrutinized, the industry does not seem to have lost much. The major firms did agree to allocate a certain fraction of their cut to value-added production, reflecting a reallocation of benefits within the forest industry. But they were able to count their value-added activities toward that reallocation. Their only other commitment was the vague promise of new investment to create 2,000 jobs (less than 10 percent of the total number of jobs promised by the deal), but a look at the fine print shows that was not a costly commitment for industry. It was "subject to the tenure holders' satisfaction as to commercial feasibility in the context of an economically viable forest industry."[51] Thus, industry committed to little more than agreeing to create jobs if it were profitable to do so. Far more jobs were to be created by the government providing direct benefits to industry with their commitments to FRBC funding and to close the gap between allowable cuts and harvest levels. (At the same time, the government was changing policy in other subsectors to address industry concerns about costs. See Chapters 3 and 7.)

Unionized workers came out the big winners, with the general commitment to increase jobs and the specific commitment to create a new agency to ensure laid-off union workers got first crack at coastal silvicultural jobs. Environmentalists were losers because Code requirements were relaxed somewhat. But as Chapter 3 argues, those losses were relatively minor.

Implementation

The accord was full of bold promises, remarkable for the specificity of its targets. What has been accomplished? In April 1998, a Jobs and Timber Accord advocate, Garry Wouters, was appointed to coordinate government efforts. The new forest worker agency promised by the accord was created. After collective bargaining sessions with the IWA, New Forest Opportunities Ltd. was established as a subsidiary of FRBC to put displaced forest workers back to work in enhanced silviculture, watershed restoration, and recreation.[52] Several additional Innovative Forest Practices Agreements have been reached.

Despite its apparently defunct status, Wouters's office produced an annual report that was scheduled to be released in late September 1999. Its findings are presented in Table 8.2.

The report shows that between March 1996 (the baseline provided by the agreement) and March 1999, the industry lost 7,500 jobs, an obvious failure given the objective of creating 22,400 new jobs. However, the report argues that the initiatives of the accord did produce a substantial number of jobs, and it claims credit for producing 12,500 jobs. (The report does not add up the various totals.) According to the report, two of

Table 8.2

Results from the first two years of the Jobs and Timber Accord

Jobs and Timber Accord commitment	1998-9 annual report results
6,500 new jobs created by delivering the full small business cut, by reapportioning more of the small business cut to value-added manufacturers, and by requiring major licensees to offer 16-18 percent of lumber to secondary manufacturing businesses at fair market prices and to direct Small Business Enterprise Program sales to value-added industries	6,600 jobs created. 4,800 jobs created from small business sales and reapportionment. 1,800 jobs created from transfers from the major forest companies. When including value-added work by the majors, the amount of lumber available for value-added did reach the percentage targets, but it did not have the expected impact on the level of fibre transferred
5,000 new jobs from Forest Renewal BC silviculture investments	6,100 jobs created in 1997-8; 5,100 jobs created 1998-9; 1999-2000 forecast is for 2,800 jobs
2,000 from industry commitment to new investment and developing higher-value products	800 jobs created by new industry investment
3,900 jobs from government commitment to close the gap between allowable cut and actual harvest levels	Despite government success in increasing the standing timber inventory, the market downturn led to a reduction in harvest levels and accompanying loss of jobs
3,000 new jobs from new work arrangements	No jobs created here, due to lack of interest by companies and unions
Total: 20,500 jobs	Total: 12,500 jobs

Source: Office of the Jobs and Timber Accord Advocate, *1998/99 Annual Report,* September 1999.

the five initiatives were successful. Delivering the small business cut and reallocating fibre to value-added manufacturers created 6,600 of these jobs, exceeding the target in this area by 100 jobs.[53] FRBC, which was targeted to create 5,000 jobs, created 6,100 in the first year and 5,100 in the second. (FRBC jobs are projected to decline to 2,800 jobs in 1999-2000.) The industry investment initiative is listed as partially successful, credited with creating 800 new jobs compared to the target of 2,000. The report takes an exceptionally optimistic interpretation of this figure, of course, because it declines to subtract the considerable number of jobs lost as a result of divestment by firms (that is, mill closures and other actions). The two clear failures were the effort to create new work arrangements and the commitment by the government to close the gap between harvest levels and the allowable cut. There were no corporate or union takers for the new work arrangements, despite the subsidies offered by the government. And although the government was successful in significantly increasing standing timber inventory, the decline in market demand meant that there was a reduction, not an increase, in harvest levels.[54]

The initiative said it would produce 22,400 direct jobs by 2001. Taking out the commitment to fisheries, the goal for forest jobs was 20,400 jobs. Adding up the numbers in the report, the government is claiming that the initiative created 12,500 jobs. Because forest jobs have declined by 7,500 since March 1996, the implication is that but for the Jobs and Timber Accord, industry jobs losses would have been 20,000 jobs. The term of the accord will run until 2002. Is it conceivable that the government could claim that it met its original targets? Yes, but only if the government claims credit for events that it is refusing to take blame for. If the market turns around and harvest levels increase, the government will claim credit for jobs created as a result of closing the gap between the AAC and harvest levels, only some of which will be attributable to government policy changes. Additional jobs might be created if there are any significant new corporate investments. Although the government could put together numbers to claim success, this action is unlikely. There is no apparent desire by labour or business to pursue the alternative work arrangement provision, and the report notes that FRBC spending, and therefore jobs created, is projected to decline significantly.

Why was the accord so unsuccessful in meeting its jobs targets? Given the conditions for effective implementation outlined in Chapter 1, the fundamental reason clearly seems to be that the policy instruments adopted did not permit sufficient control over the most important factors determining jobs levels in the industry. The government was successful at producing jobs from initiatives that it had direct control over. It could ramp up harvesting by the small business program and use the vast resources of FRBC to create jobs (in part by providing them to displaced forest workers). But there was nothing in the accord to force the industry to create jobs and nothing to prevent the job losses resulting from the market turndown.

As an initiative to assess progress toward measurable goals, the accord seems to have died, despite the efforts of the Office of the Advocate to breathe some life into it. But some of its institutional machinery is still in place and having an impact on forest policy. The accord continues to be used as an umbrella for new forest policy initiatives. For example, when the province launched its community forest pilot project in late 1997 (see Chapter 4), it did so under the auspices of the accord.[55] In fall 1999, when the government finally announced its major forest policy review, the task of coordinating the effort was given to Garry Wouters.[56] Despite the apparent disappearance of the accord from active policy status, no effort was made to change Wouters's title.

Conclusion

These two initiatives have resulted in significant change in the role of the government in the forest sector labour markets. The goals and objectives were lofty and ambitious. For FRBC, the key goal was to create the means to compensate those dislocated by land use and other environmental initiatives, thus dramatically increasing the potential for finding mutually acceptable solutions. As a result, developments in this subsector facilitated the policy changes in other subsectors, most notably land use and forest practices. For the Jobs and Timber Accord, the goal was to go beyond compensation to facilitate a structural change so that more jobs were produced from each unit of timber harvested. The key instrument in both cases was government spending. The Jobs and Timber Accord hinted at regulatory changes for uncooperative parties, but the initiative itself was based on the promise of additional spending. The policies were extremely liberal with the use of the spending instrument: FRBC expenditures totalled $1.6 billion over its first five years.

These initiatives are highly revealing of the balance of power within the forest sector. They involved a dramatic enlargement of state power and capacity, and the allocation of significant benefits to the government's allies in the labour movement. State control over the land base gave it the raw legal authority to assert this power, and control over the government by a union-based party beginning in 1991 gave it the will to extend its power. Favourable background conditions deflected what would otherwise have been formidable industry opposition. The structural change in product markets dramatically increased product prices and thus the total amount of rents available for distribution. Combined with the persistent threat of American retaliatory trade action, the background conditions promoted industry acquiescence.

This configuration of forces was not durable, however. There was a dramatic shift in background conditions in 1997 as a collapse in demand created a crisis in the industry. The balance of power between industry and the state then shifted, and the programs were reconfigured to better serve the needs of the embattled industry rather than the government. The Jobs and Timber Accord, originally envisioned as creating a direct link between industry job creation and access to Crown timber, was transformed into refocused government spending and financial inducements for industrial cooperation. Meanwhile, FRBC spending was reconfigured to better support industry's needs (and in other subsectors costs were being reduced through stumpage cuts and regulatory reform). One anecdote is revealing of how, overall, FRBC affected the industry. Bill Dumont, chief forester at Western Forest Products, claims that his firm paid $66 million in increased stumpage and got $44 million back in FRBC spending.[57] Although these figures may or may not be representative, they seem to reasonably depict the magnitude of impact on the industry.

In his infamous quote, Forest Minister David Zirnhelt said, "Don't forget the government can do anything." This analysis of policies directed toward forest jobs in BC shows that the government could indeed do a great deal. In the BC parliamentary system, the institutional power of the government of the day is virtually unlimited, and in the forest sector, control of the land base gives the state additional tools to exercise power. When cash is available, as it was in the heady market days of the early 1990s, the spending instrument can produce significant benefits for recipients – in this case, mostly forest workers displaced by changes in policy and market structure.

But this analysis also reveals the limits to what government can do. Leave aside for now the limits on government actions erected by the party in power's desire for reelection and thus its need to maintain sufficient support from the electorate. More telling in this case is that, in a market economy, the prime driver of jobs is still private investment. And when government policy or business cycles undermine business confidence, investment dries up, or moves on, and jobs disappear. As a result, the government's dependence on the private sector for investment will always be a profound constraint on its power, a constraint whose magnitude waxes and wanes with the business cycle. Moreover, while Forest Renewal BC certainly had the financial capacity to do great things, it did not have the analytical or managerial capacity to take the best advantage of its remarkable fiscal opportunity. As the report of the auditor general confirms, the agency had both the will and the financial means to execute bold plans but lacked knowledge and managerial or fiscal discipline.

As the 1990s drew to a close, however, those grand schemes appeared to be withering on the vine. The fundamental question is now whether the extraordinary opportunity created by Forest Renewal has been squandered. Have the efforts truly contributed to the renewal of a forest industry capable of competing effectively in global markets in the twenty-first century? More profoundly, might the people of British Columbia have been better served by using those resources to foster a transition to an economy based less on industrial forestry, a hazardous, ecologically disruptive commodity sector perpetually buffeted by the unpredictable swings of fickle global markets? Even asking the question is equivalent to heresy in the province, and that fact is testimony to the enduring legacy of a provincial economy dominated by the forest industry.

9
Conclusion: Change and Stability in BC Forest Policy

As the previous chapters demonstrate, the 1990s witnessed an exceptional level of policy activism in the forest sector. Our aim has been to describe and explain what came of the reform initiatives launched by the NDP governments since they took power in 1991 and to draw from our findings lessons that would advance the development of public policy theory. We have focused on two fundamental questions: How much change? And why? In pursuing these questions, we have been guided by several core convictions.

First, we knew that to understand the dynamics of public policy in an issue as complex as this one, we would need a disaggregated approach, one that would focus attention on developments within particular issue areas or subsectors. It made sense to organize the book around separate analyses of tenure, Aboriginal issues, pricing, timber supply regulation, land use, and forest practices. This approach has the added benefit of allowing us to examine the relations among subsectors and between the subsectoral and sectoral levels. Second, we believed that providing a clear description of what did and did not happen requires careful attention to the measurement of policy content. In so doing, we distinguished four aspects of policy content: goals, objectives, instruments, and settings.

Third, we knew that a full understanding of policy developments would not be possible without examining all aspects of the policy cycle. We set out to evaluate developments in each subsector at five different stages: agenda setting, policy formulation, decision making, implementation, and evaluation. Although we anticipated some difficulties in deciding where one stage ended and another began, we believed that deployment of this model would sharpen the account of policy evolution. In particular, we knew that it would be impossible to gauge the full extent of policy change without carefully examining the implementation stage.

Finally, recognizing the need for a broad explanatory framework, we adopted the policy regime framework, which views policy as determined by interactions among public and private sector actors, all pursuing particular interests within shifting institutional and ideational contexts. Policy outcomes are seen as determined by interrelated changes in regime components (actors, institutions, and ideas) and background conditions (including, for example, economic conditions, public opinion, and other exogenous factors). Changes in these regime components and background conditions promote change in policy. But forces promoting change also need to overcome the inertia of the existing policy framework, resulting from long-institutionalized courses of action or resistance from powerful regime actors who benefit from the status quo. The general shape of policy outcomes is determined by the relative balance between these forces for change and stability at any given time.

In this concluding chapter, we begin with a brief summary of the forces for change emerging across the forest policy sector in the 1990s as well as the obstacles those forces confronted. We also summarize the results of the analysis of policy dynamics in each subsector. We then move to an explanation of the similarities and differences across subsectors, including an examination of the relationships between them. And in the final section, we survey some general trends buffeting the sector as we move into the new century, suggesting that the policy reforms of the 1990s failed to create any long-term stability in the sector.

Regime Dynamics: Sectorwide Forces for Stability and Change

In the late 1980s and early 1990s, an apparently powerful combination of changes in regime components and background conditions led us to expect substantial policy change across the sector. First, environmental groups became stronger and more sophisticated.[1] They built new alliances with scientists, First Nations, and activist allies across the border in the United States and abroad. As a result, environmental groups fundamentally altered the composition of the provincial forest policy community. Second, new ideas about ecological sustainability and biodiversity emerged, challenging the liquidation-conversion paradigm that has historically dominated the forest sector in the province.[2] Third, new institutional innovations in shared decision making and American-style legalism, along with the growing popularity of more decentralized, community-oriented forms of forest management, seemed to threaten the dominance of the model of executive-led bilateral bargaining that characterized earlier forest policy making.

In addition to these changes within the forest policy regime, profound shifts in background conditions altered the balance of regime forces. In the realm of public opinion, the "second wave of environmentalism" that swept across the developed world in the late 1980s had major impacts on politics in BC, especially on the overlapping issues of wilderness preservation and forest policy. The burst in salience of these issues in public opinion gave environmentalists a valuable resource with which to influence politicians.[3]

The election of the NDP in 1991 transformed the government from one openly hostile to environmental initiatives in the forests to one dedicated to bringing about "peace in the woods." The NDP pursued a bold package of policy reforms designed to appeal to urban environmentalists as well as the party's more traditional supporters in the labour movement. Anxious to hold and expand support among environmentally conscious, urban middle-class voters, Harcourt and his team listened carefully to the party's green caucus as well as to its traditional labour supporters. The result was a platform containing a long list of forest reform proposals. These reforms included pledges to double park land, end land use conflict, settle land claims, increase forest employment, introduce legislation to regulate forest practices, encourage value-added manufacturing, and establish a royal commission review of the tenure system and other aspects of policy.

Finally, forest products markets also began to shift in directions conducive to policy change. Downward pressures on timber supply in the US Pacific Northwest contributed to escalating prices for BC products, creating a major period of boom for the industry by 1993.[4] Given past evidence supporting the hypothesis that the industry's political clout declines during periods of prosperity, there was thus reason to believe that the NDP initiatives would confront less effective opposition from industry.

This constellation of shifts in regime components and background conditions suggested a policy sector ripe for significant change, as these forces combined to create a powerful new environmental challenge to the power of the old development coalition. Consisting of the forest industry, forest unions, and the Ministry of Forests, this coalition had long centred on a common interest in defending the liquidation-conversion policies at the heart of sustained yield and integrated resource management initiatives introduced after the Second World War.

Although this environmental challenge was formidable, it still confronted daunting obstacles. First, the timber-oriented forest policies established by the province after the Second World War set the province down a policy path that would prove difficult to alter.[5] Second, actors benefiting from the status quo retained considerable power resources. Although some industry leaders had begun to see the need for farsighted compromise with these new environmental forces, many industry officials and the major forest industry associations opposed or had serious reservations about much of the new government's agenda. The companies' control over numerous levers affecting overall economic health, and thus their ability to influence the new government's reelection prospects, meant that their arguments would continue to receive a careful hearing in

Victoria even with an NDP government in office.[6] In addition, company leaders' positions on many policy questions were echoed by the IWA, long one of the NDP's strongest backers. A situation marked by company influence on Social Credit governments was being transformed into one marked by IWA influence over the NDP.

Third, the strengths of the environmental coalition were matched by some weaknesses. Most important, from the outset this coalition encompassed a diverse assortment of groups. Some groups focused on protecting precious areas, others pushed aggressively for a dramatic diminution of extractive activity across the forest land base. Moreover, there was no single idea (or idea set) that would serve to unify the fragmented assortment of advocates for change and help them win support for new initiatives from the broader public. Although there was no shortage of interesting ideas floating around, most addressed only small portions of the problem set perceived to be afflicting the sector. As noted, the main challengers coalesced around the concept of sustainability, but many of the ideas advanced in attempts to elaborate its meaning were ambiguous, poorly formulated, or untested. To phrase it in terms of one of John Kingdon's metaphors,[7] because there was no single unifying idea, it seemed unlikely that the policy story would take on the features of a bandwagon rolling through the sector and across the decade.

Finally, although by 1991 the pro-change environmental coalition had managed to undermine the legitimacy of the development coalition, the technical nature of many facets of the policy domain meant that policymaking authority – particularly authority over policy implementation – would remain in the hands of a community of experts who, by and large, supported the status quo. Environmentalists and other critical outsiders had sharply increased their capacity for effective oversight of many dimensions of forest policy. Nonetheless, thorough and lasting policy change would require major measures to open up complex policy development and implementation processes dominated by professional administrators and foresters in government and industry. These measures would require reconsideration of many traditional assumptions about the accreditation of expertise as well as government acceptance of arguments about the need to subsidize informed public participation in various parts of the sector. The NDP seemed positively disposed to a range of policy reforms, but its willingness to promote deep process changes was in doubt.

At the beginning of the 1990s, then, some significant change-inhibitors were at play alongside some obvious change promoters. The complexity of this recipe reinforced our initial expectation that change dynamics were likely to vary across subsectors. We realized that any attempt to sort out just what combination of forces shaped policy outcomes would require careful subsector-by-subsector analysis of policy dynamics along with full consideration of how developments in one subsector affected those in the next.

Characterizing and Explaining Change at the Subsector Level

As expected, we found considerable variation across subsectors in the extent and nature of policy change. The task of comparing across subsectors is complicated because no general measure of change can be applied across areas. Inevitably, the characterization of change requires that the case be examined in historical context or compared with other potential outcomes revealed by proposals of various actors in the dispute or by the policy experience of other jurisdictions.

Table 9.1 broadly characterizes the four different components of policy content in each subsector (goals, objectives, instruments, and settings) and presents a general assessment of the degree of change. In two cases, timber supply and tenure, there has been virtually no change at all, despite substantial pressures. We found clear evidence of change in four cases: land use, forest practices, jobs, and pricing. In each of these cases, however, change was severely constrained. We have witnessed nothing like a paradigm change, where fundamentally new goals or objectives have emerged. In one other case, First Nations, the likelihood of fundamental change seems substantial but has not yet been fully implemented.

Table 9.2 shows the position of different actors in each subsector policy community. Some subsectors are far more open to a diversity of interests than others. Not surprisingly, the Ministry of Forests and the forest industry are at the core of almost all of the subsector policy communities. The more open the policy community to other interests, the more likely we are to have witnessed change.

The land use subsector produced the clearest evidence of change. Here we found a significant response to the environmental coalition's protected areas agenda, with major changes to goals, objectives, instruments, and settings. The most important indicator of this level of change has been the government's commitment to double park land to 12 percent of the province, an objective that it has essentially attained. The huge area added to the parks system during the 1990s should undoubtedly be highlighted in assessments of the Harcourt and Clark governments' substantive accomplishments. Indeed, because of progress in this subsector, BC will continue to receive favourable notice from those searching the globe for signs of late-twentieth-century green policy advances. More process-oriented evaluations will no doubt also reflect positively on the shifts in land use planning approaches engineered by the architects of CORE and the LRMP process.

However, the limits on the degree of change in this subsector are also evident in slow progress on both ecosystem representation and the establishment of Special Management Zones as "truly special." Nonetheless, by moving toward multiple zone categories, the government did endorse the notion that the land use components of biodiversity conservation policy must involve more than just designation of protected areas.

Land use is also the area where the policy community has proven to be the most open and diverse. Environmental groups not only had an immense role in placing wilderness protection on the government agenda but also became a vital part of the policy formulation process through their roles at the CORE and LRMP tables. The degree of policy change in this subsector is testimony to the resourcefulness of the BC environmental movement. It effectively capitalized on its public support and on its influence within the NDP. As well, we have noted the evidence supporting the wisdom of the movement's efforts to build alliances with groups and individuals based outside the province. The outer limits of the environmental coalition's influence, however, are clearly revealed in the more limited success on ecosystem representation and Special Management Zones.

In the forest practices subsector, important shifts toward more emphasis on environmental protection also took place. Goals and objectives shifted toward greater concern with environmental values, there was a dramatic transformation of the formalization of policy instruments as the Forest Practices Code was enacted with rigorous planning requirements and tough enforcement penalties, and settings in some areas were tightened. However, the magnitude of change was limited by the government's decision to cap the impact of the Code on allowable harvest levels at 6 percent. This cap, which spawned various subsidiary numbers games, is a very appropriate indicator of the limited degree of change in this subsector. Despite indications that the Code marks a dilution of the old timber-extraction policy goal, and despite indications that costs have shifted onto the industry, we do not see these changes as indicating fundamental change.

The policy community in the forest practices subsector was also expanded considerably but not nearly so much as in the case of land use. Environmentalists were effective in shifting the government's agenda. Later, when the government sought to relax the Code, the environmental coalition managed to limit the amount of retrenchment. It was not, however, able to penetrate the process at the formulation stage. Nevertheless, a more prominent role for the Ministry of Environment, Lands and Parks in formulation and decision making did shift the balance of interests represented in the core policy community somewhat. And the new Forest Practices Board opened the implementation and enforcement of the Code to greater public scrutiny.

The changes that did occur were the result of a powerful combination of domestic opinion pressures, a new and more environmentally oriented

Table 9.1

Summary of changes in policy content

Subsector	Goals	Objectives	Instruments	Settings	Overall characterization of changes
Land use	Peace in the woods	12 percent protected areas, worker compensation	Shared decision making, regional land use tables (CORE, LRMPs), LUCO, and multi-category zoning	Specific protected area/SMZ decisions in the CORE and LRMP areas	Significant
Forest practices	Appropriate balance of forest values	Greater environmental sensitivity, limited by 6 percent cap	Planning, regulations – significant increase in formalization	Change moderated by 6 percent cap	Moderate
Tenure	Create property rights facilitating economic development – no change	Promote economical operation of forest industry facilities – no change	Variety of lease arrangements, no significant change in 1990s	No major changes; minor reforms (for example, community forests)	No change, some experimentation
First Nations	Respect changing legal definitions and requirements of Aboriginal rights and title – fairly new	Provide mechanisms for consultation and comanagement of lands and forests subject to claims	New treaties, interim measures agreements	Variety of land title and comanagement arrangements/established – new	Significant but covering only modest areas of forest at present; extension to larger areas expected in

Timber supply	Promoting economic development	Update analyses in a more rigorous and transparent process	Yield regulation, cut control – no change	Some change within regions but no change provincewide	Marginal
Pricing	Forest sector health – no change	Profitability, jobs, government revenues – no change	Administered pricing – no change	Tripling of stumpage, reduced to a doubling	Moderate
Jobs	Peace in the woods through compensation – significant change	Forest renewal, value added, worker and community support – significant change	Spending through FRBC and JTA	Significant increase in spending	Significant

Table 9.2

Policy community differences across subsectors

Subsector	MOF	MOELP	Business	Environ-mentalists	Labour	First Nations	Communities
Land use	Core	Core	Core	Core	Core	Core	Core
Forest practices	Core	Core	Core	Core	None	None	None
Tenure	Core	None	Core	Periphery	Core	Periphery	Core
First Nations	Core	None	Core	None	None	Core	Periphery
Timber supply	Core	Periphery	Core	Periphery	Periphery	Periphery	Core
Pricing	Core	None	Core	None	Core	None	Periphery
Jobs	Periphery	None	Core	None	Core	None	Core

government, favourable market conditions, and the internationalization of environmental pressure tactics. But the 6 percent cap is a powerful reminder of how forces at the heart of a critical subsector (described in more detail below) can spill over to constrain the magnitude of change in other subsectors within the same broad policy area.

In the jobs subsector, we documented significant change in the government's role in the forest sector labour market. The government asserted a much more active role in pursuing the objectives of forest renewal, the promotion of value-added industries, and worker and community support. Using the spending instruments established with the creation of FRBC and the Jobs and Timber Accord, the government substantially increased the state role in subsidizing forest industry jobs. Whether these initiatives have any lasting impact on the sector remains uncertain. FRBC's future seems tenuous, while the Jobs and Timber Accord has withered on the vine.

As a new subsector, a distinctive policy community emerged in the jobs area. New organizations were created, including Forest Renewal BC and the Jobs and Timber Accord advocate. Labour, industry, and government actors all played key roles in this community. Despite having a position on the FRBC board, environmental groups did not play as central a role here as they did in land use or forest practices.

Developments in the jobs subsector reveal some important regime dynamics. These initiatives involved a major extension of state power and capacity. State control over the land base gave it the raw legal authority to assert this power, while NDP control over the reins of government gave it the will to extend its power. But these changes would not have been possible had favourable background conditions not deflected what would otherwise have been formidable industry opposition. The structural change in product markets in the early 1990s dramatically increased product prices and thus the total pool of rents available for distribution. Combined with the persistent threat of American retaliatory trade action, the background conditions promoted industry acquiescence. In 1997, the background conditions changed when the industry went into crisis. As a result, the balance of power between industry and the state shifted, and more industry-friendly policy changes were introduced.

The timber pricing subsector witnessed moderate change. There has been no change in the general goal of the subsector of promoting forest industry health. There has also been no change in the hierarchy of three objectives at work: industry profitability first, jobs and wages second, and government revenues third. The policy instrument of administered pricing has also been remarkably durable. Significant change did take place only in instrument settings, in this case stumpage rates, as the 1994 changes tripled stumpage charges. Changes in 1998 reduced rates significantly, but they were still twice as high as when the NDP government came to power in 1991.

The narrow pricing policy community, consisting of the Ministry of Forests, forest companies, and labour unions, has proven to be resilient. The changes that did take place were largely initiated to preserve this closed subsector's historical compromise, one in which objectives were limited to those promoted by members of the development coalition. Shifts in background conditions – the forest business cycle – can affect which of these three groups' objectives is promoted by policy at a particular time. When the forest sector is booming and profits are healthy, policy can afford to serve objectives lower down on the hierarchy, as happened in 1994, when stumpage hikes redirected rents to labour and government revenue. When the forest industry slumped later in the 1990s, a correction was necessary to ensure at least some level of profitability. Spillovers from other subsectors were also important in limiting change. The lack of change in tenure policy undermined proposals to move toward market pricing, and the central place of the timber supply subsector has kept changes to new goals, such as pricing for environmentally sensitive logging, off the agenda.

In the First Nations subsector, the potential for significant change is very high but thus far has been limited to small portions of the province. Very significant changes have occurred in provincial and federal policies toward First Nations in the 1990s, particularly with regard to the pursuit of Aboriginal title through a formal treaty process. There are clear signs that these changes are beginning to spill over to affect forest tenure and other aspects of provincial forest policy. In addition to investing a large allocation of government resources (and a large allocation of its political capital) in the treaty process, the NDP government also pioneered interim measures agreements as a device for expanding First Nations control. Although the full implications of the recent court decisions supporting the continued existence of unextinguished Aboriginal title in the province remain uncertain, outcomes here will likely continue to include the transfer of some lands to Native control and the adoption of co-management provisions for larger areas. Both responses will obviously have major impacts on tenure and other dimensions of forest policy, while altering the lineup of policy actors active in the sector.

The First Nations policy community is distinctive, consisting at its core of the federal and provincial governments along with the relevant First Nations representatives in each treaty area. Because of their control over harvesting rights in areas subject to land claims, forest companies also play an important but subsidiary role. Environmentalists and labour play only a minor role through alliances with First Nations governments. The changes that have taken place thus far are largely a result of a successful strategy of venue shifting by First Nations. Frustrated by the political process, First Nations have made episodic appeals to the judicial arena, where the courts have gradually altered and expanded traditional interpretations of the definition of Aboriginal title. These judicial decisions have dramatically strengthened the political resources of First Nations, who without favourable jurisprudence would be limited to the strategies of moral appeal and economic disruption.

The timber supply subsector has experienced only marginal change. The subsector's core goal of promoting economic development has remained intact. The mix of problem definitions and statutory objectives guiding policy became even more complex and contradictory, but no new goals emerged, even though the Timber Supply Review was officially listed as a policy initiative that would contribute to meeting the goal of sustainable development. There has been no change in the basic instrument of timber supply regulation, the yield regulation model. In terms of the policy settings of the levels of allowable annual cut, some tenures have seen significant reductions, but overall these reductions have been balanced by more aggressive cutting in low value and low quality stands to leave the provincial AAC virtually unchanged.

The timber supply policy community has been relatively closed and surprisingly resistant to change. Traditionally, the community consisted of the Ministry of Forests, particularly the chief forester and his staff, and relevant licensees. Efforts by environmentalists to penetrate the community have by and large been frustrated, despite the quantum leap in the transparency of AAC determinations. This more open process was cold comfort for those who wanted a different outcome and not just a different process. The outcome of the first round of the Timber Supply Review (TSR1) frustrated the environmental coalition's deeply held belief that a more rigorous application of the existing yield regulation model, combined with the input of more realistic data about regeneration and accessibility, would necessarily result in dramatic reductions in timber supply.

Environmentalists have had difficulty engaging with the technically complex discourse of timber supply planning. Moreover, they have had an uphill battle in the struggle against the weight of history. Although the postwar liquidation-conversion paradigm seemed to envision an inevitable falldown to a lower, more durable harvest level, the social, economic, and political resistance to reducing allowable cuts has been intense. Given the dependence of many areas of the province on the steady flow of affordable fibre to mills, it has proven extremely difficult to alter the trajectory of the inherited policy path. These powerful constraints on change in the timber supply subsector have radiated outwards to limit the changes possible in other subsectors.

The tenure subsector also experienced no significant change. Here, we witnessed "arrested" policy cycles resulting in long-term policy stability. Goals and objectives remained focused on property rights to facilitate industrial development. Over the years, the complex instruments of different forms of lease arrangements have been renamed, and some of their conditions changed, but what is most remarkable is the lack of change in the 1990s. The one exception to this pattern is community forests, where there were some innovative pilot projects late in the decade. Although potentially significant as a model for future changes, these community forests currently account for less than 0.1 percent of the province's forest land.

The tenure policy community has traditionally consisted exclusively of the government and the forest companies, though unions have occasionally had their say about arrangements that link licences to commitments to particular mills. Beginning in the 1970s, this community began to break down and lose some of its legitimacy, but advocates for change have made no substantial progress. Proposals for tenure change have repeatedly made it onto government agendas, and options formulated, but at the decision-making stage of the cycle, policy makers have chosen to maintain the status quo.

This lack of change results in large part from the institutionalization of past tenure policy decisions, which has entrenched in law the right of licence holders to compensation in the event of significant alterations to their tenure arrangements. Significant changes in any new direction seem blocked by resistance from powerful forces. Proposals to dramatically expand community control are thwarted by the businesses that benefit from current arrangements. Proposals to privatize forest land, or at least dramatically expand the security of tenures, have been met with outrage by environmental and community advocates. As in the case of the timber supply subsector, the stability of the tenure subsector has also weighed down the forces for change in other subsectors.

This disaggregated review of the BC forest policy sector reveals a mixed pattern of change and continuity. The core sectors of tenure and timber supply have witnessed the least change. There has been substantial environmental progress in land use and to a lesser extent in forest practices. Environmental values have acquired a more central place in BC forest policy, but they have not displaced timber production as the dominant value. The jobs subsector witnessed a dramatic increase in the government role in the labour market, though the long-term efficacy of those reforms is doubtful. The pricing subsector witnessed some change, at least in the form of settings. The First Nations sector may be on the verge of significant change, but thus far change has been limited. Each subsector has a unique pattern of changes in goals, objectives, instruments, and settings. As a result of the structure of the sector, and the linkages and spillovers between sectors, the subsectors experiencing the least change acted as a powerful drag on the forest sector as a whole.

Explaining the Subsector Similarities and Differences

How can this complex pattern of change and stability be explained? Explanations of the policy changes that did occur confirmed the importance of shifts in regime components and background conditions. In terms of regime components, new actors were clearly fundamental to the changes that took place, in particular the rise of the NDP to power and the increasing resources and sophistication of the environmental movement. The election of the more environmentally oriented NDP was fundamental to all the changes that took place here. In forest practices and land use, outcomes must in large part be attributed to the NDP's strong commitment to change. In pricing and jobs, the NDP's use of the state to support the labour component of its coalition ensured change. The 1991 election was preceded by a major resurgence of the environmental movement, resulting in more sophisticated groups with more resources. Environmentalists were able to elbow their way into the forest practices and land use policy communities, resulting in a significant expansion of the interests reflected in policy outcomes. First Nations groups also increased their resources and their ability to represent themselves effectively in various arenas.

These actors pursued strategies consistent with the expectation of the framework outlined in Chapter 1. Within a given institutional context and ideational context, actors adopt the strategies most likely to advance their interests. For example, believing that a greener forest policy would help expand their popularity in influential urban and suburban ridings, the NDP wanted to enact and implement many of the environmental initiatives in its 1991 election platform. But we also noted how actors, when they believe existing institutions and ideas to be biased against them, actively try to reshape their contexts to promote their interests.[8] Governments, of course, have the most direct control over institutional structures, and the NDP created several new institutional arrangements to facilitate changes it viewed as politically profitable. Forest Renewal BC, CORE, and the Forest Practices Board are all good examples.

But nongovernmental actors also have an interest in shifting institutional venues. When the government showed an interest in devolving some land use decision-making authority into the hands of regional stakeholders, environmentalists worked hard to participate effectively in the new forums. Frustrated by the distribution of interests within the narrow confines of provincial politics that dominated BC forest policy, environmentalists undertook an extraordinary strategy of shifting the venue from the provincial government to the international marketplace. This strategy has proven to be enormously successful, largely because it hit the industry where it counts: at the bottom line. With their profits and market share threatened, the industry was forced to accede to a greening of forest policy. In the first half of the 1990s, this situation facilitated the adoption of the Forest Practices Code and progress in the Protected Areas Strategy. Toward the end of the decade, escalating international market pressures in fact pushed companies to go beyond provincial regulations and adopt more environmentally sensitive forest practices. Similarly in their efforts to achieve Aboriginal title, the influence of First Nations had proven relatively limited within the confines of the federal-provincial treaty process. So they shifted the policy venue to the courts, where new interpretations of the meaning of Aboriginal title strengthened First Nations when they went back to the bargaining table. Not all strategic innovations have proven profitable however. When environmentalists tried to import American-style legalism, for example, the effort floundered because of the discretionary nature of Canadian law.

In addition to changing venues, actors also pursue idea-based strategies such as attempting to reframe problem definitions in a manner more amenable to their interests.[9] Idea-based strategies have been critical to the success of the environmental movement in BC. Most important their relentless assault on the canons of the liquidation-conversion paradigm undermined the legitimacy of the old-style policies.[10] Drawing inspiration from developments south of the border and in the international realm, environmentalists introduced British Columbians to the concepts of sustainability, biodiversity, and ecosystem management. The NDP government framed its policies in the rhetoric of sustainability, renewal, and multistakeholder consensus, glossing over the persistence of conflict as it pursued approval of its policies.

The policy changes that we have described were clearly influenced by the emergence of new regime actors and strategies, but they would not have been possible without significant changes in background conditions, particularly changes in elections, market conditions, and public opinion. As noted, the changing electoral landscape was fundamental in that it brought a new, greener party to power. The shift in market conditions was also critically important. Consistent with the proposition stated in Chapter 1, the boom period of the early 1990s reduced the power of business to defend itself against cost increases in all four of the subsectors where change has been apparent. The fact that policy retrenchment followed the downturn in business conditions in 1997 provides further support for this proposition.

The shift in public mood within BC to greater concern with environmental values in the forest also strengthened the demands of environmentalists by forcing election-minded politicians to take heed. This fact provides some support for another proposition stated in Chapter 1 – that significant changes in policy that go against the interests of business groups (or other dominant actors) are unlikely without a burst in the public salience of new values. The increased salience of environmental values clearly contributed to changes in the land use and forest practices subsectors. Moreover, the internationalization strategy pursued by environmentalists depends on greater attention to environmental values among international consumers of BC forest products. However, the explanatory power of the proposition is somewhat limited by evidence from the pricing and jobs sector. In these cases, change was not the result of shifts in public opinion as much as it was of the specific objectives of the government in power, in the context of extremely favourable market conditions. In the Westminster system, significant policy change can occur without broad-based public pressures.

Spillovers among policies at the sectoral and subsector levels also promoted change in some cases. The government's environmental agenda in forestry threatened jobs, creating powerful pressures for change in the jobs and pricing subsectors. Changes in the pricing subsector permitted expanded government revenues facilitating the initiatives in the jobs subsector. Changes in federal and provincial policies toward Aboriginal land claims have also begun to change the BC forest sector, and those claims are likely to bring about more significant change over the next decade.

Although forces promoting policy change had a clear impact on a number of subsectors, they also encountered resistance from a variety of forces. The analysis returned time and again to three closely interrelated explanatory concepts: path dependence, critical subsectors, and the structural power of the forest industry. First, path dependence plays a significant ongoing role across the sector. Past decisions become institutionalized, leaving a broad array of government and nongovernment actors committed to maintaining the inherited policy path.[11] These policy legacies come to represent major constraints on policy change. Perhaps the best example is timber supply, where decisions made when the sustained yield paradigm was established after the Second World War set the province on a path that has been and will continue to be extremely costly and disruptive to reverse.

In some instances, the effects of this resistance were not evident until the decision-making stage. In tenure policy, for example, alternative ideas were widely debated in the 1980s and 1990s, but attempts to translate these ideas into policy were blocked by a set of forces closely associated with the legacies of crucial tenure policy decisions made earlier in the century. In other cases, the effects of policy legacies were evident in successful efforts to exclude alternative ideas at the beginning of the policy cycle. In the timber supply subsector, for instance, a tight policy community dedicated to maintenance of traditional policy assumptions and problem definitions managed to keep alternatives off the agenda.

The second major impediment to change arises from the constraints rooted in the relationships among subsectors. The analysis provides considerable evidence of how influences associated with certain subsectors spill over to impede or limit change in others. These findings suggested the concept of the "critical subsector," or as it was put earlier, the notion that decisions and policies taken in such a subsector will invariably prevail over contradictory proposals under consideration in other subsectors, thus imposing a unified direction on the sector as a whole. This notion seems to apply particularly to the role of the timber supply subsector. Those responsible for timber supply policy not only resisted demands for change from the forest practices and land use subsectors but also managed to exert a constraining influence on the extent and type of change taking place in those and other subsectors. The influence of the critical timber supply subsector was evident in the decisions to impose caps on the allowable cut impacts of the Code and the Special Management Zones. As Chapter 7 makes clear, fundamental changes in that sector are also powerfully constrained by the nature of the tenure system.

The third fundamental constraint to change is the structural power of business. At the heart of the market economy are business decisions on investment and divestment: where to build mills, what type of mills to build, at what capacity, with what kind of labour and technology, and so on. To build and maintain these facilities, forest companies need to make an adequate level of profit to attract capital. This fundamental requirement places profound constraints on the degree of change even the most activist government can achieve. In Canada's parliamentary system, majority governments do have a great deal of freedom to enact policies of their choosing during their mandates. But all governments have a fundamental interest in reelection, which means that deliberations on the costs, risks, and benefits of policy options are always strongly influenced by worries about maintaining business confidence. With bold schemes such as Forest Renewal BC, the NDP government did test the limits of what the industry would tolerate. In other respects, though, the government was definitely constrained by concerns about damaging investment and employment levels as well as its reelection prospects.[12]

These three constraining forces – path dependence, critical subsectors, and the structural power of business – are closely related. The structural power of forest companies in the province is a large part of the foundation of the liquidation-conversion juggernaut. In turn, the powerful inertia of the policy path established at a much earlier stage of provincial forest policy history promotes continued industry power. In general terms, the entire development coalition – forest companies, their workers, those workers' communities, and the state – all have a strong vested interest in the continuation of the tenure, pricing, and harvest control policies put in place to guide and facilitate the liquidation-conversion project. In specific terms, all these interests depend on a continued flow of reasonably priced old-growth fibre to the province's manufacturing facilities. As the capacity embodied in these mills and other parts of the industry expanded, more and more investors, workers, and community supporters came to share this dependence. Their intense feelings about the importance of not killing (or disturbing) the goose that lays the golden eggs tends to be shared by government actors who see their jobs, political health, and cherished programs depending on continued flows of stumpage and taxes into provincial coffers. The breadth of the constituency favouring continuation of the liquidation-conversion project, and the intensity of their commitments, are at the root of the forest industry's power. Explanations of this power that give primacy to the effects of corporate wealth ignore the extent to which the economic lives of people across the province depend on a continued flow of fibre to and through harvesting and manufacturing facilities.

In addition, business power is closely related to the operation of the critical subsectors. It is surely no accident that the two subsectors identified as having a powerful constraining influence on other subsectors – timber supply and tenure – are those most directly related to controlling the core business activities of the sector. Tenure determines the property relations, while timber supply determines the flow of fibre, which in the end is what produces the profits, wages, and tax revenues that make the companies, workers, communities, and government so dependent on this particular policy path.

All these primary change impediments can be linked to secondary ones. The success of actors seeking to continue on an established policy path, for example, will depend on their ability to restrict membership in the policy community, stigmatize (or otherwise exclude) alternative ideas, and maintain the legitimacy of key institutional arrangement and assumptions. The analysis of factors obstructing or limiting change highlighted a number of features of the political strategies involved, including the reminder that the technical complexity of some policy areas works to the advantage of those opposed to change.[13] Again, this situation was particularly evident in the timber supply case, where environmentalists failed in their efforts to frame the issue to garner the same sort of public attention they were able to attract to land use and the more evocative aspects of forest practices. Similarly, the bewildering complexity of the province's tenure policies has also frustrated issue expansion in that subsector.

Environmentalists and advocates for forest policy change have thus confronted a daunting set of obstacles. Could environmentalists have extended their policy gains beyond the land use and forest practices subsectors (and won greater gains in those subsectors) if they had played their strategic cards differently? We doubt it. The environmental coalition invested its political resources wisely, demonstrating considerable ingenuity and resiliency in its response to the obstacles enumerated here. The institutional and ideational obstacles faced by BC environmentalists come into sharp relief when we compare their situation with that faced by their counterparts across the border in the Pacific Northwest. The American institutional system gives environmentalists much more promising opportunities for venue shifting to the courts or to national publics. In the United States, once the courts swing into action, it is possible for a relatively small group of well-funded and litigious environmentalists to put a serious spanner in the industry's works and thus to increase their bargaining power.[14] Environmentalists have adapted well to British Columbia's bleaker institutional terrain, using a combination of high-energy activists and effective appeals to a wider international audience.

The environmental coalition would clearly have fared better, however, if after successfully undermining the old liquidation-conversion paradigm, it had been able to win broad support for a new and compelling forestry vision. The concept of sustainable development became extremely popular in the late 1980s, but it lacked sufficient specificity to mobilize broad support. Abbreviating it to "sustainability" held promise for some advocates, but any concept that can be used to support protection of both ecosystems and forest industry jobs creates obvious problems for those promoting fundamental change. Perhaps the most compelling alternative paradigm to emerge has been "ecosystem management." As its name suggests, ecosystem management has as its primary objective the maintenance of ecosystem integrity, usually defined as sustaining ecosystem structure, function, and composition. Commodity outputs are allowed to the extent that they are not incompatible with maintaining ecosystem integrity in this sense. This concept has found its way into federal forest policy in the US Pacific Northwest,[15] but to date in BC its influence has been minimal outside of the area considered by the Clayoquot Sound Scientific Panel.

Why did the idea of ecosystem management gain a foothold in the Pacific Northwest but not in BC forest policy generally? In the United States, the idea took hold because it satisfied the demands of a rigid legal framework – a combination of the Endangered Species Act, the National Environmental Policy Act, and the species viability regulations under the National Forest Management Act – that gave environmentalists the tools to force it on reluctant policy makers. The nature of American landownership, where business interests could still be realized on vast private lands, also reduced resistance to the adoption of ecosystem management principles on federal lands.[16] In Clayoquot Sound, the provincial government was willing to accede to the idea because of the intense political pressure the government faced in that area and because the government believed it could easily be contained to that one location in the province.[17] In the broader BC forest sector, ecosystem management failed to take hold because it was too threatening to powerful business interests and their advocates within the government.

In summary, the variation across subsectors in the degree of change is explained by the relative balance of change promoters and inhibitors in each subsector as well as the links between the critical and other subsectors. In timber supply, the pressures to reduce the cut immediately confronted the power of the industry and all of its supporters in unions and communities throughout the province. Tenure change has been impeded by a combination of the government's reluctance to give up control over forest land, resistance from economic actors that benefit from current arrangements, and a populist attachment to public ownership. In both cases, the technical complexity of the issues made it more difficult for environmentalists to mobilize domestic public support or international market pressures for change.

In jobs and pricing, the government had substantial room to manoeuvre. It seized these opportunities by expanding its capacity to pursue its policy goals. But even here the magnitude of change was limited by government's dependence on the health of the forest industry. When the business cycle turned down in the late 1990s, policies in both the jobs and the pricing sectors were reoriented in an effort to restore profits. In land use, clearly measurable goals along with strong and persistent public support and environmental group pressure promoted significant change. In forest practices, strong environmental pressures for change confronted the critical timber supply subsector head on, and change was carefully limited to a particular percentage reduction in allowable cut. With Aboriginal title, significant change may be in the offing as a result of the enhanced legal status of First Nations.

The policy regime framework introduced in Chapter 1 has helped illuminate the complex relations between change and stability. Actors typically make the best use of their resources in pursuit of their interests. But not all actors are equally endowed. Business groups have a special advantage because of their position in the economy. Activist governments can use their powers – with control over land, quite formidable in the BC forest sector – to advance their political interests, but they are constrained by the need to maintain business confidence. In this sector, significant policy change has not been possible without significant changes in background conditions. In all the cases of change, important shifts in background conditions were a crucial factor: greener public opinion in the land use and forest practices subsectors along with favourable market conditions in these two as well as the jobs and pricing subsectors. It is possible to create change without a burst of salience in public opinion if there is a government dedicated to change and favourable market conditions.

This framework also helps demonstrate the limitations on change created by the weight of history and the structural power of the forest industry. Significant change is possible. More wilderness can be preserved, and forest practices can be more environmentally sensitive, so long as they do not exceed a threshold that threatens the investment climate. More fundamental change is only possible if the fundamental dynamics of the system are transformed. Two such possibilities loom on the horizon. Because they directly alter the incentives of businesses and the government that depends on those businesses, the international market campaigns, if escalated, could promote a sharper departure from the status quo. Changes in Aboriginal title, brought about by externally produced changes in property rights, also portend more profound change.

The Future

As the decade of the 1990s came to a close, forest policy in BC was in turmoil. The government's search for sustainability, whether measured by environmental, social, economic, or political indicators, had failed. Squeezed between the rising floor of costs and the declining ceiling of product prices and weak markets overseas, the BC forest sector lost $1 billion in 1998. Just as business began to improve, the threat of major conflict with First Nations was pushed into the limelight by the decisions of several bands to begin logging their traditional lands without government approval. Meanwhile, the certification movement picked up steam as more major industrial customers – including Home Depot, Ikea, Lowes, and 84 Lumber – pledged to purchase only certified products. And in a move that rocked the province, American forest giant Weyerhaeuser took over MacMillan Bloedel, a major institution in BC forestry that had moved beyond traditional industrial forestry to a more ecologically sensitive set of harvesting practices. The softwood lumber dispute with the United States also continues to fester, increasing the pressures to make timber pricing more sensitive to market signals.

The widespread sense of crisis present at the dawn of the new century clearly reveals that BC forest policy has not achieved any semblance of equilibrium. Given what we have learned in our review of the 1990s, we see four major dynamics shaping the future. First, the issue of Aboriginal title is likely to be an even more dominant issue in the coming decade. Because it involves direct control over who owns the land, the question of title is fundamental to forest policy. It is conceivable that a model can be developed that will not bring about major changes in ownership and policy, but it seems more likely that significant changes are in the offing.

Second, the goal of ending the "war in the woods" is likely to continue to be elusive because of the deep gulf between the main protagonists. The development coalition has acceded to modest policy changes in some areas. But by the very nature of the conflict, responses to these pressures in the forms of forest practices codes and land use delineations never get to the heart of the critique of large-scale industrial forestry that many environmental groups fundamentally hold. The more profound policy change favoured by environmentalists would be met by fierce opposition from the development coalition.

The depth of this discord was revealed again by the most recent attempt to forge some sort of policy consensus out of the noisy conflicts of the late 1990s. In July 1999, the Clark government ordered the Forest Policy Review, appointing underemployed Jobs and Timber Accord advocate Garry Wouters to lead the review.[18] The review process began by issuing a "vision paper" supported by eight more detailed discussion papers. The vision paper proposed a compromise: in exchange for granting industry more secure tenure rights on a fixed area of land, industry would voluntarily give back harvesting rights that could then be distributed to create new conservation areas or new community forests, or to address Aboriginal land claims. During the final months of 1999, Wouters conducted a number of invitation-only consultation workshops throughout the province, followed by more public meetings.

The final report was issued in April 2000. Its recommendations included:

- establishing a competitive log market within six months and replacing administered stumpage pricing on the coast
- eliminating the rules requiring timber licences to build and maintain local sawmills
- establishing new forms of tenure and expanding interim agreements to allow First Nations and forest communities a more active role in forest management
- dismantling the Ministry of Forests by creating two agencies, one dealing with commercial timber issues and the other with forest stewardship
- creating clear rules regarding compensation to tenure holders
- dismantling FRBC and replacing it with regional funding agencies
- establishing a comprehensive independent commission to review how the AAC is set.[19]

As expected, a chasm divided the responses of environmental groups and forest industry officials. Ron MacDonald of the Council of Forest Industries was quick to argue that the report "doesn't recognize the importance of the forest industry" because it failed to call for an immediate designation of the working forest and an increase in the AAC.[20] David Emerson, president and CEO of Canfor, similarly argued that not enough attention was paid to achieving industry success. The IWA complained that the report's recommendations "pile more burdens on an already overburdened industry." COFI was trumpeting its own policy review. Its *Blueprint for Competitiveness* stressed the familiar themes of greater security of tenure and a more "flexible" approach to regulating forest practices. Underlining the connection among timber, jobs, and government revenues, the industry offered to "kickstart" the slumping provincial economy if the AAC could be raised from 70 million to 100 million cubic metres.[21]

On the other side, environmental groups criticized the report for not going far enough, with the David Suzuki Foundation's Cheri Burda noting that, although the recommendation to review the AAC was welcome, the review failed to offer any fundamental departure from existing forestry practices.[22] Environmentalists presented a package of reforms calling for a move toward ecosystem management as the basis for forest operations, a dramatic expansion of community-based tenures, and a significant decrease in the allowable annual cut.[23]

Given the wide gap between the principal protagonists, it is hard to see how Wouters could ever have found a compromise. Several prescriptions that "expand the pie" of total benefits available have the potential to attract support from both sides. Value-added production promises to create more income and jobs with less impact on the forest resource. Intensive zoning promises to reduce the amount of land required to produce a given level of harvest. But finding a middle ground even in these more promising areas is proving to be a challenge due to the fundamentally conflicting goals and objectives of the two competing coalitions of interests. The industry coalition wants to have a globally competitive industry, with the protection of environmental values as a secondary goal. The environmental coalition makes ecosystem integrity the central goal with timber harvesting secondary. Both sides claim that the two goals can be achieved simultaneously, but a comparison of the allowable harvest levels being recommended by the two sides highlights the gulf separating them. Peace in the woods will remain an elusive goal.

The third fundamental dynamic comes through the electoral arena, in which partisan politics is likely to reverse directions on many of the green initiatives pushed by the NDP. The BC Liberal Party, the presumptive government-in-waiting, has been clear that its focus will be primarily on economic development and refining the environmentally oriented regulatory initiatives of the NDP. Liberal leader Gordon Campbell has outlined a forest policy platform that echoes the main criticisms of forest industry officials documented throughout this book. Campbell has committed a Liberal government to increasing the allowable cut, giving serious consideration to dismantling Forest Renewal BC, and compensating forest companies for NDP initiatives. Campbell's modest environmental concerns are found in his call to reduce regulations under the Forest Practices Code and, borrowing a phrase from industry, to move toward a "results-based" code. He echoes Wouters' call for a market-based stumpage pricing system. Given deep-seated public attachment to the forest resource, Campbell's platform stops short of calling for the privatization of Crown land and the removal of raw log export restrictions.[24] Jim Fulton of the David Suzuki Foundation labels this platform "a blast from the past," but it is probably the direction in which the next ten years of public policy is headed.[25]

As our analysis of the 1990s shows, the party in power is central to any explanation of policy outcomes. The election and reelection of the NDP was essential to moving forest policy in a greener direction and enlarging the resources of the state to help workers and communities affected by these changes. A Liberal government would no doubt try to change the direction of many of these policy initiatives to reflect its support for business and a smaller role for government.

A Liberal government in pursuit of this agenda would certainly face obstacles. It goes without saying that these would not include business resistance or many of the forces of friction that we have linked to path dependence. On the other hand, those opposing the reversal of NDP policies would have some strong cards to play. For example, if a Liberal government is not sufficiently accommodating to Aboriginal claims, it is likely to confront a string of court injunctions on resource development. The public strongly opposes privatization of Crown forest land as well as any significant increase in corporate management authority on public land. There is also robust support within the BC public for new protected areas and maintaining strong environmental standards in the Forest Practices Code.[26] Land use decisions and the regulatory framework of the Code are now part of the policy legacy constraining change. Although there is public support for reducing regulatory red tape, if environmentalists could frame those changes as threatening environmental standards in the Code, their opposition would have powerful backing from the BC public. Perhaps more important than these domestic constraints are the newly emerging realities of the international marketplace.

The final major dynamic affecting the future of the forest sector is perhaps the most profound: debates over sustainable forestry policy in BC will increasingly be fought in the private sphere – through market and firm-level arenas, where environmental groups have so far enjoyed considerable success. Many environmental groups are increasingly engaging in strategies that focus directly on firm actions rather than using government to force change. The two most prominent examples in recent years are the threats of boycotts that have brought industry groups into direct negotiations with environmentalists in a major wilderness dispute and the drive to certify forest products.

After the success of the Clayoquot Sound campaign, environmentalists refocused on the Central Coast. Greenpeace and affiliated groups launched a major campaign in 1997, which has picked up considerable steam as an increasing number of major customers of BC forest products have expressed concern. Despite the ongoing stakeholder process under the auspices of the Central Coast Land and Resource Management Plan, the six major companies active in the region began secret negotiations with leading environmentalists to explore the possibility of an agreement on contested areas. When the negotiations were revealed in press accounts, however, bitter opposition emerged from actors who were not at the table, including local communities, First Nations, and even government.[27] Two companies defected from the negotiations, and environmentalists began a new round of escalating threats.[28] Despite these setbacks, in July 2000 the parties agreed to a truce, under which companies would defer logging in contested areas, and environmentalists would refrain from targeting the companies that signed on to the agreement in their international campaigns.[29]

The other major avenue for the privatization of forest policy is the certification movement. Several different certification schemes are now in competition, with the most influential being the Forest Stewardship Council. As described in Chapter 3, the FSC is an environmentally oriented organization that develops regionally applicable standards for forest management and then accredits organizations to certify whether a company meets those standards. A coalition of environmental groups has lined up commitments to only purchase FSC-certified products from an impressive number of major buyers.[30]

Although the FSC is new and regional standards for BC haven't even been developed yet, the organization is already having an immense impact on corporate behaviour. Because the FSC directly targets the financial incentives of firms, it has the potential to bring about changes in the behaviour of BC forest companies that government regulators were unable to achieve. And because it shifts the venue of forest practices regulation from government to private standard-setting bodies, it has the potential to revolutionize the governance of the forest sector. We have quoted Forest Minister Zirnhelt's famous remark "Don't forget the government can do anything." We have also quoted Merran Smith of the Sierra Club of BC, who in late 1999 told a *New York Times* reporter, "The government is irrelevant; it is the marketplace [that matters]."[31]

Of course, both statements are exaggerations, but we certainly can see the momentum shifting.

BC forest policy now appears highly unstable and potentially unsustainable. Whether the stresses and strains apparent in the current crisis are sufficient to overcome the powerful constraints on fundamental change revealed throughout this book, however, remains to be seen. The war in the woods is likely to persist as the battle rages between the defenders of the status quo and the advocates of change. Meanwhile the province's search for the elusive holy grail of sustainability will continue.

Notes

Chapter 1: Policy Cycles and Policy Regimes

1 Charles O. Jones, *An Introduction to the Study of Public Policy,* 3rd ed. (Monterey, CA: Brooks-Cole, 1997).
2 Michael Howlett and M. Ramesh, *Studying Public Policy: Policy Cycles and Policy Subsystems* (Toronto: Oxford University Press, 1996). It has also been used by the two major American texts on forest policy. See Frederick Cubbage et al., *Forest Resource Policy* (New York: John Wiley and Sons, 1993); Paul Ellefson, *Forest Resource Policy: Process, Participants, and Programs* (New York: McGraw-Hill, 1992).
3 Jones, *An Introduction*, 28-9; Howlett and Ramesh, *Studying Public Policy*, 11
4 Paul Sabatier, "Toward Better Theories of the Policy Process," *PS: Political Science and Politics* 24 (1991): 449-76; Paul Sabatier and Hank C. Jenkins-Smith, eds., *Policy Change and Learning: An Advocacy Coalition Approach* (Boulder, CO: Westview Press, 1993). For a recent defence of the cycle approach, see Peter deLeon, "The Stages Approach to the Policy Process: What Has It Done? Where Is It Going?" in Paul Sabatier, ed., *Theories of the Policy Process* (Boulder, CO: Westview Press, 1999), 19-34.
5 John Kingdon, *Agendas, Alternatives, and Public Policies*, 2nd ed. (New York: HarperCollins College Publishers, 1995).
6 Anne Schneider and Helen Ingram, "Social Construction of Target Populations – Implications for Politics and Policy," *American Journal of Political Science* 87,2 (1993): 334-47; Leslie Pal, *Beyond Policy Analysis: Public Issue Management in Turbulent Times* (Scarborough, ON: ITP Nelson, 1997); Deborah Stone, "Causal Stories and the Formation of Policy Agendas," *Political Science Quarterly* 104 (1989): 281-300.
7 Howlett and Ramesh, *Studying Public Policy.*
8 Kingdon, *Agendas, Alternatives, and Public Policies.*
9 Howlett and Ramesh, *Studying Public Policy.*
10 Charles Lindblom, "The Science of Muddling Through," *Public Administration Review* 19 (1959): 79-88.
11 John Forester, "Bounded Rationality and the Politics of Muddling Through," *Public Administration Review* 44,1 (January/February 1984): 23-31; Howlett and Ramesh, *Studying Public Policy,* chap. 7.
12 There is a mammoth literature on instrument choice. For recent overviews, see Howlett and Ramesh, *Studying Public Policy*, chap. 4; and Pal, *Beyond Policy Analysis*, chap. 4. In the context of BC forestry, see W.T. Stanbury and Ilan Vertinsky, "Governing Instruments for Forest Policy in British Columbia: A Positive and Normative Analysis," in Chris Tollefson, ed., *The Wealth of Forests: Markets, Regulation, and Sustainable Forestry* (Vancouver: UBC Press, 1998).
13 Howlett and Ramesh, *Studying Public Policy*, focus their implementation chapter (8) on instrument choice (see also Anderson 1994). Unquestionably, appropriately designed instruments are crucial to implementation success, but most other scholars consider instrument choice part of the decision-making stage (e.g., Pal, *Beyond Policy Analysis*).

14 Daniel Mazmanian and Paul Sabatier, *Implementation and Public Policy* (Glenview, IL: Scott, Foresman, 1983).
15 Brian W. Hogwood and B. Guy Peters, *Policy Dynamics* (Brighton, Sussex: Wheatsheaf, 1983).
16 For the Canadian literature, see Paul A. Pross, *Group Politics and Public Policy,* 2nd ed. (Toronto: Oxford University Press, 1992); William Coleman and Grace Skogstad, eds., *Policy Communities and Public Policy in Canada: A Structural Approach* (Mississauga, ON: Copp Clark Pitman, 1990); and Michael Atkinson and William D. Coleman, *The State, Business, and Industrial Change in Canada* (Toronto: University of Toronto Press, 1989). For the British literature, see R.A.W. Rhodes, "Policy Networks: A British Perspective," *Journal of Theoretical Politics* 2,3 (1990): 293-317; David Marsh and R.A.W. Rhodes, eds., *Policy Networks in British Government* (Oxford: Clarendon, 1992); David Marsh, *Comparing Policy Networks* (Buckingham: Open University Press, 1998).
17 William D. Coleman and Grace Skogstad, "Neo-Liberalism, Policy Networks, and Policy Change: Agricultural Policy Reform in Australia and Canada," *Australian Journal of Political Science* 30 (1995): 242-63; Michael Howlett and Jeremy Rayner, "Do Ideas Matter? Policy Network Configurations and Resistance to Policy Change in the Canadian Forest Sector," *Canadian Public Administration* 38 (1995): 382-410.
18 There is a vast literature on institutionalism. Two of the most prominent works are Peter Evans, Dietrich Rueschemeyer, and Theda Skocpol, eds., *Bringing the State Back In* (Cambridge: Cambridge University Press, 1985); Sven Steinmo, Kathleen Thelen, and Frank Longstreth, eds., *Structuring Politics: Historical Institutionalism in Comparative Analysis* (Cambridge: Cambridge University Press 1992). The concept of social learning comes from Hugh Heclo, *Modern Social Policies in Britain and Sweden: From Relief to Income Maintenance* (New Haven, CT: Yale University Press, 1974), and has been developed more recently by Peter Hall, "Policy Paradigms, Social Learning and the State," *Comparative Politics* 25,3 (April 1993): 275-96.
19 Terry Moe, "The Politics of Bureaucratic Structure," in J.E. Chubb and P.E. Peterson, eds., *Can the Government Govern*? (Washington, DC: Brookings Institution, 1989); Matthew McCubbins, Roger Noll, and Barry Weingast, "Structure and Process, Politics and Policy: Administrative Arrangements and the Political Control of Agencies," *Virginia Law Review* 75 (1989): 431-82.
20 Geoffrey Garrett and Barry Weingast, "Ideas, Interests, and Institutions: Constructing the European Community's Internal Market," in Judith Goldstein and Robert O. Keohane, eds., *Ideas and Foreign Policy* (Ithaca: Cornell University Press, 1993).
21 See the chapters by Phillips and Torgeson in Laurent Dobuzinskis, Michael Howlett, and David Laycock, eds., *Policy Studies in Canada: The State of the Art* (Toronto: University of Toronto Press, 1996).
22 Frank R. Baumgartner and Bryan D. Jones, *Agendas and Instability in American Politics* (Chicago: University of Chicago Press, 1993).
23 Sabatier and Jenkins-Smith, eds., *Policy Change and Learning;* Hank Jenkins-Smith and Paul Sabatier, "Evaluating the Advocacy Coalition Framework," *Journal of Public Policy* 14,2 (1994): 175-203; Ken Lertzman, Jeremy Rayner, and Jeremy Wilson, "Learning and Change in the BC Forest Policy Sector," *Canadian Journal of Political Science* 29 (March 1996): 111-33.
24 Jenkins-Smith and Sabatier, "Evaluating the Advocacy Coalition Framework," 182.
25 Ibid., 180.
26 For an elaboration on the differences between the ACF and the "policy regime" approach, see George Hoberg, "Distinguishing Learning from Other Sources of Policy Change: The Case of Forestry in the Pacific Northwest," prepared for delivery at the Annual Meeting of the American Political Science Association, Boston, MA, 3-6 September 1998.
27 This separation of government and private actors is supported by the findings of the empirical studies by Sabatier and colleagues that show government agencies tend to fall on the extremes of the belief clusters of advocacy coalitions. See Jenkins-Smith and Sabatier, "Evaluating the Advocacy Coalition Framework."
28 Coleman and Skogstad, eds., *Policy Communities*; Atkinson and Coleman, *The State,*

Business, and Industrial Change; Marsh and Rhodes, eds., *Policy Networks in British Government*; Marsh, *Comparing Policy Networks*.

29 For an insightful overview, see Peter Hall and Rosemary Taylor, "Political Science and the Three New Institutionalisms," *Political Studies* 44,5 (1996): 936-57.

30 Sven Steinmo and Jon Watts, "It's the Institutions, Stupid! Why Comprehensive National Health Insurance Always Fails in America," *Journal of Health Politics, Policy and Law* 20 (Summer 1995): 329-72.

31 Judith Goldstein and Robert O. Keohane, eds., *Ideas and Foreign Policy* (Ithaca: Cornell University Press, 1993).

32 Jenkins-Smith and Sabatier, "Evaluating the Advocacy Coalition Framework."

33 Wyn Grant and Anne MacNamara "When Policy Communities Intersect: The Cases of Agriculture and Banking," *Political Studies* 43 (1995): 509-15; George Hoberg and Edward Morawski, "Policy Change through Sector Intersection: Forest and Aboriginal Policy in Clayoquot Sound," *Canadian Public Administration* 40,3 (1997): 387-414.

34 Theodore Lowi, "American Business, Public Policy, Case Studies, and Political Theory," *World Politics* 16 (1964): 677-93; James Q. Wilson, ed., *The Politics of Regulation* (New York: Basic Books, 1980).

35 Mazmanian and Sabatier, *Implementation and Public Policy*.

36 Jenkins-Smith and Sabatier, "Evaluating the Advocacy Coalition Framework," 182. See also Peter Knoepfel and Ingrid Kissling-Naef, "Social Learning in Policy Networks," *Policy and Politics* 26,3 (1998): 343-67.

37 Lindblom, "The Science of Muddling Through"; Douglas Cater, *Power in Washington* (New York: Random House, 1964); Theodore Lowi, *The End of Liberalism* (New York: Basic Books, 1969); Grant McConnell, *Private Power and American Democracy* (New York: Alfred Knopf, 1969).

38 See Stephen Krasner, "Approaches to the State: Alternative Conceptions and Historical Dynamics," *Comparative Politics* 16,2 (1994): 223-46; Paul Pierson, "When Effect Becomes Cause: Policy Feedback and Policy Change," *World Politics* 45 (1993): 595-628; Paul Pierson, "Increasing Returns, Path Dependence, and the Study of Politics," *American Political Science Review* 94,2 (2000): 251-68.

39 Keith Banting, *Federalism and the Canadian Welfare State* (Kingston: McGill-Queen's University Press, 1987); Keith Krehbeil, *Pivotal Politics: A Theory of U.S. Lawmaking* (Chicago: University of Chicago Press, 1998).

40 The notable exception being Garrett and Weingast, "Ideas, Interests, and Institutions: Constructing the European Community's Internal Market."

41 Frank R. Baumgartner and Bryan D. Jones, "Agenda Dynamics and Policy Subsystems," *Journal of Politics* 53 (1991): 1044-74; Baumgartner and Jones, *Agendas and Instability*.

42 Baumgartner and Jones, *Agendas and Instability*.

43 Ibid.

44 Paul Sabatier, "Policy Change over a Decade or More," in Sabatier and Jenkins-Smith, eds., *Policy Change and Learning*, 34.

45 Charles Lindblom, *Politics and Markets* (New York: Basic Book, 1977); Fred Block, "The Ruling Class Does Not Rule: Notes on the Market Theory of the State," in J.E. Roemer, ed., *Foundations of Analytical Marxism* (Aldershot: International Library of Critical Writings in Economics, 1994), 2: 93-115.

46 For an approach to intergovernmental relations that emphasizes the importance of periodic fluctuations in salience, see Kathryn Harrison, *Passing the Buck: Federalism and Canadian Environmental Policy* (Vancouver: UBC Press, 1996).

47 Baumgartner and Jones, *Agendas and Instability*, 20.

48 Their policy formulation chapter is used to develop their ideas of policy subsystems and networks, which are not applied explicitly in other parts of the cycle. Hints at linkages can be found in the chapters on the other stages but they are not drawn as explicitly (Howlett and Ramesh, *Studying Public Policy*).

49 Actors not included here that might need to be included in a more exhaustive portrayal are communities, researchers (academics and consultants) who are involved in knowledge generation and policy evaluation, and professional foresters.

50 There is a significant amount of scholarship on the structural basis of business power. For two influential views originating from different ideological traditions, see Lindblom, *Politics and Markets;* and Fred Block, "The Ruling Class Does Not Rule."
51 Here we distinguish *authority* as formal policy-making power granted by law from *power* as a more general measure of influence. Thus, industry groups have considerable power but no authority.
52 Compare William Niskanen, *Bureaucracy and Representative Government* (Chicago: Aldine-Atherton, 1971), with James Q. Wilson, *Bureaucracy* (New York: Basic Books, 1989).
53 See Chapter 8 below, and Hoberg and Morawski, "Policy Change through Sector Intersection."
54 Gordon Hamilton, "Jack Munro finally calling it quits," *Vancouver Sun* (17 June 2000).
55 Lertzman, Rayner, and Wilson, "Learning and Change"; Jeremy Wilson, *Talk and Log: Wilderness Politics in British Columbia* (Vancouver: UBC Press, 1998).
56 For overviews on the history of BC forestry, see Ken Drushka, *In the Bight: The BC Forest Industry Today* (Madeira Park, BC: Harbour Publishing, 1999); Ken Drushka, *H.R.: A Biography of H.R. MacMillan* (Madeira Park, BC: Harbour Publishing, 1995); Wilson, *Talk and Log;* Richard A. Rajala, *Clearcutting the Pacific Rain Forest: Production, Science, and Regulation* (Vancouver: UBC Press, 1998); M. Patricia Marchak, Scott L. Aycock, and Deborah M. Herbert, *Falldown: Forest Policy in British Columbia* (Vancouver: David Suzuki Foundation and Ecotrust Canada, 1999), chap. 3
57 Jeremy Wilson, "Wilderness Politics in BC," in W.D. Coleman and G. Skogstad, eds., *Policy Communities in Canada* (Mississauga, ON: Copp Clark Pitman, 1990).
58 Wilson, *Talk and Log;* Benjamin Cashore, "Governing Forestry: Environmental Group Influence in British Columbia and the US Pacific Northwest," PhD dissertation, University of Toronto, 1997; George Hoberg, "Environmental Policy: Alternative Styles," in Michael Atkinson, ed., *Governing Canada: State Institutions and Public Policy* (Toronto: HBJ-Holt Canada, 1993), 307-42.
59 Of the six major cases involving environmental challenges to BC forest policies in the 1990s, only two could be classified as successful, one of which was subsequently overturned by another court, and the other was rendered moot by administrative response. The one major decision by the courts to constrain the discretion of the Ministry of Forests was actually a lawsuit brought by a major forest company against a decision by the chief forester to reduce their allowable cut level (George Hoberg, "How the Way We Make Policy Governs the Policy We Make," in Debra J. Salazar and Donald K. Alper, eds., *Sustaining the Forests of the Pacific Coast: Forging Truces in the War in the Woods* (Vancouver: UBC Press, 2000).
60 While far more formalistic in approach, the Forest Practices Code Act contains a similar provision. Section 165 authorizes the minister or a minister's delegate to "extend a time required to do anything under this Act, the regulations for the standards other than a review or appeal of a determination or the time to commence a proceeding."
61 Jim Beatty and Gordon Hamilton, "NDP grabs hint deficit near $1 billion," *Vancouver Sun* (13 September 1997): A1.
62 Jeremy Wilson, "Forest Conservation in British Columbia 1935-85: Reflections on a Barren Political Debate," *BC Studies* 76 (1987/8), 3-32; Wilson, *Talk and Log;* Lertzman, Rayner, and Wilson, "Learning and Change."
63 For earlier discussions of factors listed in this section, see George Hoberg, "The Politics of Sustainability: Forest Policy in British Columbia," in R.K. Carty, ed., *Politics, Policy, and Government in British Columbia* (Vancouver: UBC Press, 1996); Lertzman, Rayner, and Wilson, "Learning and Change"; George Hoberg, "Putting Ideas in Their Place: A Response to 'Learning and Change in the British Columbia Forest Policy Sector,'" *Canadian Journal of Political Science* 29 (March 1996): 135-44.
64 Cashore, "Governing Forestry"; Wilson, *Talk and Log.*
65 Steven Bernstein and Benjamin Cashore, "Globalization, Four Paths of Internationalization and Domestic Policy Change: The Case of Ecoforestry in British Columbia, Canada," *Canadian Journal of Political Science* 33,1 (March 2000): 67-99.
66 Riley E. Dunlap, George H. Gallup, Jr., Alec M. Gallup, *Health of the Planet: Results of a*

1992 International Environmental Opinion Survey of Citizens in 24 Nations (Princeton, NJ: The George H. Gallup International Institute, 1993); Donald Blake, Neil Guppy, and Peter Urmetzer, "Canadian Public Opinion and Environmental Action in Canada: Evidence from British Columbia," *Canadian Journal of Political Science* 30,3 (1997): 451-72.

67 Donald E. Blake, R.K. Carty, and Lynda Erickson, *Grassroots Politicians* (Vancouver: UBC Press, 1993), 63-4.

68 Federal jurisdiction in American forest policy allowed environmentalists there to use a nationalization strategy – moving from the regional area that dominated policy in the past to involve national interest groups and legislators – to reshape the distribution of power to their advantage. See Steven Lewis Yaffee, *The Wisdom of the Spotted Owl: Policy Lessons for a New Century* (Washington, DC: Island Press, 1994); George Hoberg, "From Localism to Legalism: The Transformation of Federal Forest Policy," in C. Davis, ed., *Western Public Lands and Environmental Politics* (Boulder, CO: Westview Press, 1997).

69 W.T. Stanbury and Ilan Vertinsky, "Boycotts in Conflicts over Forestry Issues: The Case of Clayoquot Sound," *Commonwealth Forestry Review* 76,1 (1997): 18-24; Bernstein and Cashore, "Globalization, Four Paths of Internationalization and Domestic Policy Change."

70 See the report of the Pearse Commission, British Columbia, "Timber Rights and Forest Policy in British Columbia: Report of the Royal Commission on Forest Resources," 2 vols. (Victoria: Queen's Printer, 1976). The 1912 Fulton Commission focused largely on tenure issues but did contain some concern for water quality. By the time of the 1945 and 1956 Sloan Commissions, the central concerns were pricing, tenure, and timber supply, but again there was some interest in water quality and salmon protection.

71 Several issues that have received a great deal of attention in previous decades are not treated here as distinct subsectors: silviculture; protecting forests from fire, pests, and disease; and trade. We address silvicultural issues in both the forest practices and jobs subsectors. While undeniably crucial to the forest sector, forest protection has not been a major issue on the policy agenda in the 1990s. Note, for example, the absence of discussion of the issue in the reviews beginning and ending the decade (Forest Resources Commission, "The Future of Our Forests" [Victoria: Queen's Printer, 1991]; British Columbia, "Shaping Our Future: BC Forest Policy Review" [Victoria: Queen's Printer, 2000]). While the trade issue could justifiably be considered a distinctive subsector, we choose to cover the relevant issues in the pricing sector.

72 For an analysis that stresses the need to focus on the subsectoral level, see Grant Jordan, William A. Maloney, and Andrew M. McLaughlin, "Characterizing Agricultural Policy-Making," *Public Administration* 72 (1994): 505-26; for a critique, see Michael Cavanagh, David Marsh, and Martin Smith, "The Relationship between Policy Networks at the Sectoral and Sub-Sectoral Levels: A Response to Jordan, Maloney and McLaughlin," *Public Administration*. 73 (Winter 1995): 627-33.

73 In an influential article, Peter Hall creates a hierarchical order, from low to high, between changes in settings, instruments, and objectives. We don't agree that the relationship needs to be hierarchical (Peter Hall, "Policy Paradigms").

Chapter 2: Experimentation on a Leash

1 On the institutional history of the Environment and Parks agencies, see Jeremy Wilson, *Talk and Log: Wilderness Politics in British Columbia* (Vancouver: UBC Press, 1998), 72-7.

2 Ibid., chap. 3.

3 Stephan Fuller, "Wilderness: A Heritage Resource," WAC 383, PABC (GR 1601), 20-1.

4 For accounts of the origins of the 12 percent target, see Larry Pynn, "12 percent preservation goal plucked from thin air," *Vancouver Sun* (30 November 1994); Harold Eidsvik, "Canada in a Global Context," in M. Hummel, ed., *Endangered Spaces: The Future for Canada's Wilderness* (Toronto: Key Porter Books, 1989), 44; and M.A. Sanjayan and M.E. Soule, *Moving beyond Brundtland: The Conservation Value of British Columbia's 12 Percent Protected Area Strategy* (Vancouver: Greenpeace, 1997), 3-4.

5 See Hummel, *Endangered Spaces*.

6 Keith Moore, *Coastal Watersheds: An Inventory of Watersheds in the Coastal Temperate Forests*

of British Columbia (Earthlife Canada Foundation and Ecotrust/Conservation International, 1991).

7 See Wilson, *Talk and Log*, 252-3.

8 See Ministry of Forests and Ministry of Environment, Lands and Parks, "Summary of Public Response to Parks and Wilderness for the 90s" (Victoria: 1991).

9 Wilson, *Talk and Log*, 262-4.

10 John W. Kingdon, *Agendas, Alternatives, and Public Policies* (Boston: Little, Brown and Company, 1984), 16-22, and 89-94.

11 Wilson, *Talk and Log*, 273-5.

12 See ibid., chaps. 6, 8, and 9.

13 British Columbia, Legislative Assembly, *Commissioner on Resources and Environment Act* (Victoria: 1992).

14 CORE (Commission on Resources and the Environment, "1992-93 Annual Report to the Legislative Assembly" (Victoria: June 1993), 19.

15 British Columbia, "Towards a Protected Areas Strategy for BC Parks and Wilderness for the 90s."

16 British Columbia, Land Use Coordination Office, "A Protected Areas Strategy for British Columbia: The Protected Areas Component of BC's Land Use Strategy" (Victoria: 1996), 6.

17 As quoted in Keith Baldrey, "Deal aims to bring peace to war in woods," *Vancouver Sun* (28 March 1994). In the same issue, also see Justine Hunter, "Rapid Transit heading north of the Fraser," and Vaughn Palmer, "NDP goes out on limb for forest workers."

18 Richard Schwindt (Commissioner), *Report of the Commission of Inquiry into Compensation for the Taking of Resource Interests*, 21 August 1992.

19 Ibid., 108.

20 For LUCO's perception of the differences between the CORE and LRMP processes, see Land Use Coordination Office, "Special Management Zone Project: Information Report," 12 June 1998, 11.

21 Mike Harcourt, interview with the author and Ben Cashore, 19 April 1996; Murray Rankin, interview with the author and Ben Cashore, 15 June 1995.

22 For a fuller account, see Wilson, *Talk and Log*, 278-89.

23 Stephen Owen, interview with the author and Ben Cashore, 13 December 1994.

24 Rankin interview.

25 These zones were referred to in various ways in the CORE reports and government approved land use plans. Consistent use of the Special Management Zone term came after 1996. See Wilson, *Talk and Log*, 279-80, and Land Use Coordination Office, "Special Management Zone Project: Information Report," 3-4.

26 British Columbia, Land Use Coordination Office, "Special Management Zone Project: Information Report," 7. Totals include SMZs set aside in the four CORE regions, Clayoquot Sound, and the Kamloops LRMP area.

27 Kaaren Lewis and Susan Westmacott, "Provincial Overview and Status Report" (Victoria: Land Use Coordination Office, 1996), 5.

28 Early in its term, the Clark government accepted recommendations of the Lower Mainland Regional Public Advisory Committee that called for preservation of about 50,000 hectares of the Stoltmann area, including the Upper Lillooet and Clendenning watersheds, which had received considerable attention in the WCWC campaign. The WCWC continued to campaign for the full proposal, protesting the decision to allow logging in important areas such as Sims Creek and the Upper Elaho.

29 Merran Smith of the Sierra Club of BC, quoted in James Brooke, "Loggers Find Canada Rain Forest Flush with Foes," *New York Times* (22 October 1999).

30 See Gordon Hamilton, "Eco-action hurt forest firms' sales to Germany," *Vancouver Sun* (17 March 2000); Gordon Hamilton, "BC's green tag revolution," *Vancouver Sun* (5 May 2000); Gordon Hamilton, "Harvesting a green dream," *Vancouver Sun* (6 May 2000); and Barrie McKenna, "U.S. environmentalists swing axe at Canadian forest industry," *Globe and Mail* (22 January 2000).

31 LUCO, "Status Report on Land and Resource Management Plans: April 2000." See the LUCO website (www.luco.gov.bc.ca/lrmp/lrmpstat.htm) for updates.

32 LUCO, "Land-use Planning in British Columbia: Making a World of Difference," at www.luco.gov.bc.ca/slupinbc/wrldiff.htm, last updated 1 June 2000.
33 See British Columbia, Office of the Premier, "Clark Announces Protection of Northern Rockies," 8 October 1997.
34 LUCO, "BC's New Protected Areas, Park Additions and Upgrades" (last updated 6 August 1999), at www.luco.gov.bc.ca/pas/newpas/ag99pas.html. Updates appear regularly.
35 Darcy Riddell, *Sierra Report* (Spring 1999), 6.
36 See Gordon Hamilton, "Coastal loggers seek eco-truce," *Vancouver Sun* (16 March 2000); Gordon Hamilton, "Logging-environmental negotiations break down," *Vancouver Sun* (30 May 2000); and Ann Gibbon, "Talks seek truce in BC forestry battle," *Globe and Mail* (20 March 2000).
37 Wayne Sawchuk, "Northern Rockies Victory," *BC Environmental Report* 8,4 (Winter 1997): 21.
38 "Fraser River Headwaters at Risk," *BC Environmental Report* 9,1 (Spring 1998): 13.
39 Ric Careless, foreword to *Keeping the Special in Special Management Zones,* by Jim Cooperman (Gibsons, BC: BC Spaces for Nature, 1998).
40 British Columbia, Land Use Coordination Office, Special Management Zone Working Group Project, "Status of Special Management Zones: Technical Supplement," introduction. See www.luco.gov.bc.ca/smz/statusintro.htm. Accessed 1 June 2000.
41 Cooperman, *Keeping the Special in Special Management Zones*, 78
42 Ibid., 78 and executive summary.
43 See, for example, Paul Senez, "Timber Grab on Vancouver Island," *Watershed Sentinel* 10,3 (June/July 2000), 2-5; and Sierra Club of BC, "Government Withholds Key Report from Public," Press Release, 8 May 2000.
44 Lewis and Westmacott, *Provincial Overview and Status Report,* and updated figures supplied by LUCO, June 1999.
45 Lewis and Westmacott, "Provincial Overview," 14.
46 Figures from an analysis by Baden Cross, Inland Rainforest Working Group. See also Terje Vold and Andy MacKinnon, "Old-growth Forests Inventory for British Columbia, Canada," *Natural Areas Journal* 18,4 (1998): 308-18. MacKinnon and Vold conclude that about 9 percent of remaining low elevation old-growth forests are protected, compared to about 16 percent of subalpine old growth. MacKinnon and Vold begin with an age class definition of old growth that varies by region and forest type. Old growth consists of coastal forests older than 250 years, and Interior forests older than 120 years in the case of lodgepole pine or deciduous species, or older than 140 years in the case of other forest types. They acknowledge that this definition includes areas that would be excluded under an ecological (structural characteristics) definition or under one that excludes "shorter" old-growth forests.
47 For one elaboration of the latter argument, see Sanjayan and Soule, *Moving Beyond Brundtland.*
48 See Patricia Lush, "MacBlo pursues BC compensation," *Globe and Mail* (25 September 1997); and the documents and affidavit accompanying MacMillan Bloedel's petition to the Supreme Court of British Columbia, "In the Supreme Court of British Columbia, In the Matter of the Forest Act, R.S.B.C. 1996 . . . ," filed 23 September 1997. See also Justine Hunter, "Claim for lost timber rights could cost BC $200 million," *Vancouver Sun* (24 September 1997).
49 British Columbia, Ministry of Forests, "Parks Settlement Agreement – MacMillan Bloedel" (Victoria: 1999), at www.for.gov.bc.ca/PAB/News.mb/overview.htm (accessed 19 May 1999). On the Forest Act provisions, see Wilson, *Talk and Log*, 159, 275-7.
50 Justine Hunter and Kelly Sinoski, "Privatize BC's forest lands, Miller says in call for overhaul," *Vancouver Sun* (10 April 1999).
51 See, for example, David Boyd, "The Folly of Privatizing BC's Crown Jewels"; Sierra Club Press Release, "More Forest Privatization in the Works," 6 May 1999. Similar concerns were expressed by some government officials. See "Bureaucrats forbidden to talk about MB land deal," *Vancouver Sun* (14 May 1999).
52 "Forest firms eye compensation," *Vancouver Sun* (8 May 1999).
53 Wilson, *Talk and Log*, 276 and 403n76.

54 David Perry, quoted in Ann Gibbon, "BC public wants to axe forest swap," *Globe and Mail* (16 August 1999).
55 "Zirnhelt 'gets message' over MacBlo land compensation," *Vancouver Sun* (16 August 1999).
56 For one exception, see Ben Parfitt, "Province can give MacBlo squat," *Georgia Straight* (6-13 May 1999).
57 British Columbia, Land Use Coordination Office, "Special Management Zone Project: Information Report," 21.

Chapter 3: The 6 Percent Solution

1 The roles and incentives of various actors in forest practices are outlined in W.T. Stanbury and Ilan Vertinsky, "Governing Instruments for Forest Policy in British Columbia: A Positive and Normative Analysis," in Chris Tollefson, ed., *The Wealth of Forests: Markets, Regulation, and Sustainable Forestry* (Vancouver: UBC Press, 1998).
2 In contrast to virtually all other areas of environmental regulation of forestry in British Columbia, the federal government plays a significant role in fisheries protection. This results from the federal government's constitutional authority over inland fisheries and anadromous fish. The Fisheries Act, first enacted in 1868, is perhaps the strongest federal tool for environmental protection. Section 32 requires that "No person shall destroy fish by any means other than fishing except as authorized by the minister or under regulations made by the minister." Section 35 states that "No person shall carry on any work or undertaking that results in the harmful alteration, disruption, or destruction of fish habitat," and Section 36 says "no person shall deposit or permit the deposition of a deleterious substance of any type in water frequented by fish or in any place under any conditions where such deleterious substance or any other deleterious substance that results from the deposit of such deleterious substance may enter any such water."
3 W.T. Stanbury and Ilan Vertinsky, "Boycotts in Conflicts over Forestry Issues: The Case of Clayoquot Sound," *Commonwealth Forestry Review* 76,1 (1997): 18-24; Steven Bernstein and Benjamin Cashore, "Globalization, Four Paths of Internationalization and Domestic Policy Change: The Case of Ecoforestry in British Columbia, Canada," *Canadian Journal of Political Science* 33,1 (March 2000): 67-99.
4 Jeremy Wilson, *Talk and Log* (Vancouver: UBC Press, 1998), chap. 12.
5 See, for example, Clark S. Binkley, "Preserving Nature through Intensive Plantation Forestry: The Case for Forestland Allocation with Illustrations from British Columbia," *Forestry Chronicle* 73,5 (September/October 1997): 553-9; and Sivaguru Sahajananthan, David Haley, and John Nelson, "Planning for Sustainable Forests in British Columbia through Land Use Zoning," *Canadian Public Policy* 24 (Supplement 2), 1998: S73-S81. For a critical view, see Jeremy Rayner, "Priority-Use Zoning: Sustainable Solution or Symbolic Politics?" in Chris Tollefson, ed., *The Wealth of Forests: Markets, Regulation, and Sustainable Forestry* (Vancouver: UBC Press, 1998), 232-54.
6 See, for example, M. Patricia Marchak, Scott L. Aycock, and Deborah M. Herbert, *Falldown: Forest Policy in British Columbia* (Vancouver: David Suzuki Foundation and Ecotrust Canada, 1999).
7 The descriptive material in this section, as well as the following section on the development and implementation of the Code, draws heavily on George Hoberg, "The British Columbia Forest Practices Code: Formalization and Its Effects," in M. Howlett, ed., *Canadian Forest Policy: Regimes, Policy Dynamics, and Institutional Adaptations* (Toronto: University of Toronto Press, forthcoming).
8 For a comprehensive treatment of this issue up through 1965, see Richard A. Rajala, *Clearcutting the Pacific Rain Forest: Production, Science, and Regulation* (Vancouver: UBC Press, 1998), 5.
9 Jeremy Wilson, "Wilderness Politics in BC," in W.D. Coleman and G. Skogstad, eds., *Policy Communities and Public Policy in Canada* (Mississauga, ON: Copp Clark Pitman, 1990), 155.
10 The full text of the section follows. Note that (a), (b), (d), and (e) mostly address economic values:
(a) encourage maximum productivity of the forest and range resources in the Province;
(b) manage, protect and conserve the forest and range resources of the Crown, having

regard to the immediate and long term economic and social benefits they may confer on the Province;

(c) plan the use of the forest and range resources of the Crown, so that the production of timber and forage, the harvesting of timber, the grazing of livestock and the realization of fisheries, wildlife, water, outdoor recreation and other natural resource values are coordinated and integrated, in consultation and cooperation with other ministries and agencies of the Crown and with the private sector;

(d) encourage a vigorous, efficient and world competitive processing industry in the Province; and

(e) assert the financial interest of the Crown in its forest and range resources in a systematic and equitable manner.

British Columbia *Ministry of Forests Act*, chap. 272, sec. 4, 1979 R.S.C.

11 British Columbia, Ministry of Forests, *Okanagan Timber Supply Area: Integrated Resource Management Harvesting Guideline* (Victoria: Ministry of Forest, 1992), 10.

12 British Columbia, The Scientific Panel for Sustainable Forest Practices in Clayoquot Sound, *Sustainable Ecosystem Management in Clayoquot Sound, Report 5*, 57.

13 Ibid., 59.

14 British Columbia, Ministry of Forests, *Okanagan Timber Supply Area.*

15 British Columbia, Ministry of Forests, "A Forest Practices Code: A Public Discussion Paper" (Victoria: 1991).

16 John Kingdon, *Agendas, Alternatives, and Public Policies*, 2nd ed. (New York: HarperCollins, 1995).

17 Wilson, *Talk and Log,* chap. 11.

18 New Democratic Party, "A Better Way for British Columbia" (Victoria: New Democratic Election Platform, 1991).

19 Aaron Doyle, Brian Elliott, and David Tindall, "Framing the Forests: Corporations, The BC Forest Alliance, and the Media," in W. Carroll, ed., *Organizing Dissent: Contemporary Social Movements in Theory and Practice*, 2nd ed. (Toronto: Garamond Press, 1997), 250.

20 British Columbia, Ministry of Forests, *News Release*, 30 July 1992.

21 Derek Tripp, A. Nixon, and R. Dunlop, *The Application and Effectiveness of the Coastal Fisheries Forestry Guidelines in Selected Cut Blocks on Vancouver Island,* prepared for the Ministry of Environment, Lands and Parks (Nanaimo, BC: Tripp Biologica Consultants Ltd., 1992), iii.

22 Derek Tripp, *The Use and Effectiveness of the Coastal Fisheries Forestry Guidelines in Selected Forest Districts of Coastal British Columbia* (Nanaimo, BC: Tripp Biologica Consultants Ltd., 1994).

23 Forest Resources Commission, "Final Report: The Future of Our Forests" (Victoria: Queen's Printer, 1991), 88.

24 Previous documents used by the NDP while in opposition did not contain reference to a Forest Practices Act. See New Democratic Party, "Sustainable Development."

25 This element of the problem is poorly defined and confused two aspects: the clarity of legal powers and the complexity of the regulatory framework. The clarity issue is dealt with more directly in the second element below.

26 British Columbia, Ministry of Forests, "British Columbia Forest Practices Code Discussion Paper" (Victoria: 1993).

27 On the concept of formalization, see Hoberg, "The British Columbia Forest Practices Code."

28 Forest Resources Commission, *Providing the Framework: A Forest Practices Code* (Victoria: Forest Resources Commission, 1992).

29 Wilson, *Talk and Log,* chap. 12.

30 Stanbury and Vertinsky, "Governing Instruments."

31 G.L. Baskerville, *Forest Practices Code: Summary of Presentations by Stakeholder Groups – A Report to the Ministry of Forests*. (Victoria: Queen's Printer, 1993), 21.

32 Gordon Hamilton and Keith Baldrey, "Forest critics seek meat in NDP's plan," *Vancouver Sun* (17 May 1994): A1.

33 British Columbia, Ministry of Forests and Ministry of Environment, Lands and Parks, "New Mandatory Standards to Improve BC's Forest Practices," News Release, 30 May 1994.

34 A seventh plan, the Range Use Plan, was also required but is not addressed in this analysis.
35 British Columbia, Ministry of Forests and Ministry of Environment, Lands and Parks, "Forest Practices Code of British Columbia – Riparian Management Area Guidebook" (Victoria: Queen's Printer, December 1995).
36 Under the Forest Act, licensees had the right to appeal but specially constituted boards had to be established for each appeal.
37 Gordon Hamilton, "Jobs, stability at risk if forest code altered, NDP warned," *Vancouver Sun* (25 November 1994): E3.
38 Vaughn Palmer, "NDP cuts red tape to make sure enough wood is cut," *Vancouver Sun* (6 February 1995): A8; Gordon Hamilton, "Forestry code logs support of protectionists, industry," *Vancouver Sun* (13 April 1995): A1.
39 A leaked Ministry of Forests Communication Plan submitted to cabinet states: "In November 1994, as key elements of the code neared completion, a team of specialists from the Ministries of Forests and Environment, Lands, and Parks, were directed to assess the impacts of these elements on timber supply, and to ensure that the impacts on the code on timber supply did not exceed six percent." BC Ministry of Forests Public Affairs Branch, *Communication Plan. Subject: Cabinet Submission on Biodiversity Letter,* Internal document, 19 August 1997.
40 British Columbia, Ministry of Forests and Ministry of Environment, Lands and Parks, "Forest Practices Code Timber Supply Analysis" (Victoria: Queen's Printer, February 1996).
41 Wilson, *Talk and Log*, chap. 12.
42 Ibid., 6.
43 Personal interview, Larry Pedersen, Chief Forester, 11 June 1998.
44 See Gerry Armstrong and Tom Gunton, "Implementation of Biodiversity Guidebook," Memorandum to regional managers, Ministry of Forests, and regional directors, Ministry of Environment, Lands and Parks, 15 August 1995.
45 Charles Lindblom, "The Science of Muddling Through," *Public Administration Review* 19 (1959): 79-88.
46 Pedersen interview.
47 Personal communication, Kerrie Brownie, Local and Landscape Unit Planning Specialist, Forest Practices Branch, Ministry of Forests, 13 July 1999.
48 British Columbia, Ministry of Forests and Ministry of Environment, Lands and Parks, "Forest Practices Code – Biodiversity Guidebook" (Victoria: Queen's Printers, 1995), 8.
49 British Columbia, Ministry of Forests and Ministry of Environment, Lands and Parks, "Release and Implementation of the Landscape Unit Planning Guide," Memorandum, 17 March 1999.
50 British Columbia, Ministry of Forests, "Just the Facts – A Review of Silviculture and Other Forestry Statistics, 1997." (Available at www.for.gov.bc.ca/hfp/forsite/jtfacts/cutblock. htm.)
51 Sierra Legal Defence Fund, "Stream Protection Under the Code: The Destruction Continues" (Vancouver: February 1997).
52 British Columbia, Forest Practices Board, *Forest Planning and Practices in Coastal Areas with Streams* (Victoria: Forest Practices Board, 1998), 4.
53 British Columbia, Ministry of Forests, Compliance and Enforcement Branch, "The Annual Report of Compliance and Enforcement Statistics for the Forest Practices Code," for the years 1996, 1997, 1998, available at www.for.gov.bc.ca/tasb/legsregs/fpc/fpc.htm.
54 KPMG, "Financial State of the Forest Industry and Delivered Wood Cost Drivers," prepared for the British Columbia Ministry of Forests, April 1997.
55 Of the $12.22 per cubic metre code costs, $4.52 was credited to planning and administration and $7.70 was credited to forest practices. The significance of forest practices, in order, were cutblock size, road and landing requirements, soil conservation requirements, riparian management requirements, greenup and adjacency, and visual quality objectives (ibid.).
56 PricewaterhouseCoopers, *The BC Forest Industry: Unrealized Potential* (Vancouver: PricewaterhouseCoopers, 2000).
57 Daniel Mazmanian and Paul Sabatier, *Implementation and Public Policy* (Glenview, IL: Scott, Foresman, 1983).

58 This point is emphasized even by the Clayoquot Sound Scientific Panel that environmentalists so frequently trumpet as state of the art ecosystem management.
59 Clark S. Binkley, "Preserving Nature through Intensive Plantation Forestry"; Fred Bunnell, "Next Time Try Data: A Plea for Variety in Forest Practices," working paper, UBC Centre for Conservation Biology, 1999.
60 Sierra Legal Defence Fund, "Stream Protection Under the Code" (Vancouver: 1997).
61 Sierra Legal Defence Fund, "British Columbia's Clear Cut Code" (Vancouver: November 1996).
62 Sierra Legal Defence Fund, "Wildlife at Risk" (Vancouver: April 1999).
63 Sierra Legal Defence Fund, "Going Downhill Fast: Landslides and the Forest Practices Code" (Vancouver: June 1997).
64 Sierra Legal Defence Fund, "British Columbia's Forestry Report Card" (Vancouver: 1997-8).
65 Greenpeace Canada, Greepeace International, and Greenpeace San Francisco, "Broken Promises: The Truth about What's Happening to British Columbia's Forests" (Vancouver: Greenpeace Canada et al., April 1997).
66 Council of Forest Industries, "COFI Submits 30-Day List," Press Release, 30 November 1998.
67 Council of Forest Industries, "A Blueprint for Competitiveness" (Vancouver: COFI, October 1999).
68 British Columbia, Forest Practices Board, "Forest Practices Board – 1997 Annual Report" (Victoria: 1998), 5.
69 See also British Columbia, Forest Practices Board, "An Audit of the Government of British Columbia's Framework for Enforcement of the Forest Practices Code," FPB/ARE/O1 (Victoria: 1999).
70 Keith Moore, Chair, Forest Practices Board, "Presentation to House of Commons Standing Committee on Natural Resources and Government Operations," 14 May 1999. At www.fpd.gov.bc.ca/background/MPs.htm.
71 British Columbia, Ministry of Environment, Lands and Parks and Ministry of Forests, "Revisions Make Forest Practices Code More Efficient, Effective," Press Release, 9 June 1997 (emphasis added).
72 For a revealing look at the government's assessment of the opinion climate at the time, see British Columbia, Ministry of Forests, Public Affairs Branch, "Communication Plan – Subject: Cabinet Submission on Biodiversity Letter," internal document, 19 August 1997.
73 British Columbia, Ministry of Forests and Ministry of Environment, Lands and Parks, "Forest Practices Code Amendments Reduce Costs."
74 Larry Pedersen, Chief Forester, "The Big Picture Policy Framework," address to Southern Interior Silviculture Committee Winter Workshop, Penticton, BC, 10 March 1999, 9.
75 The Five-Year Silviculture Plan was eliminated and replaced by policy requirements, relevant provisions of the Access Management Plan were incorporated into the Forest Development Plan, and the Logging Plan was incorporated into the Silviculture Prescription.
76 British Columbia, Ministry of Forests and Ministry of Environment, Lands and Parks, "Forest Practices Code Amendments Reduce Costs."
77 Forest Alliance of BC, "Code Changes a Progressive Step," News Release, 2 April 1998.
78 Gordon Hamilton, "BC chops red tape to save forestry firms $300 million," *Vancouver Sun* (3 April 1998): A1.
79 Sierra Legal Defence Fund, "Forest Practices Code Rollbacks Confirmed," News Release, 2 April 1998.
80 The letter was eventually published on the Board website: www.fpb.gov.bc.ca/reports/brd.htm.
81 British Columbia, Ministry of Forests, "Moving towards a Results-Based, Incentives-Driven Code," Forest Practices Code discussion paper – Streamlining Planning Initiative – internal document, November 1998, 1.
82 Ibid., 3.
83 Larry Pedersen, Chief Forester, "The Big Picture Policy Framework," 12.

84 British Columbia, Ministry of Forests, "Forest Land Commission to Oversee Private Land Logging," News Release, 9 June 1999. The basis for this agreement had previously been announced in January.
85 British Columbia, "Forest Practices on Private Forest Land, Summary of Standards and Administration," June 1999, accessible at www.for.gov.bc.ca/PAB/News/privateland/pflregulations.htm.
86 Ibid.
87 Ibid., Appendix II, sec. II (1.3).
88 A revised version of this section appeared as George Hoberg, "The Coming Revolution in Regulating Our Forests," *Policy Options* 20 (December 1999): 53-6.
89 MacMillan Bloedel, "MacMillan Bloedel to Phase Out Clearcutting," Press Release, 10 June 1998. Available at www.macblo.com/enviro/forestproject/tfp_press.htm.
90 Kathryn Harrison, "Talking with the Donkey: Cooperative Approaches to Environmental Protection," *Journal of Industrial Ecology* 2,3: 51-72.
91 David Humphreys, "The Global Politics of Forest Conservation since the UNCED," *Environmental Politics* 5,2 (1996): 231-56, at 247.
92 Errol Meidinger, "Look Who's Making the Rules: International Environmental Standard Setting by Non-Governmental Organizations," *Human Ecology Review* 4,1 (1997): 52-4.
93 These are accessible through www.web.net/fscca/.
94 The original text stated: "Primary forests ... shall be conserved. Such areas shall not be replaced by tree plantations or other land uses." The new version states: "Management Activities in high conservation value forests shall maintain or enhance the attributes which define such forests. Decisions regarding high conservation value forests shall always be considered in the context of a precautionary approach." See www.fscoax.org/principal.htm.
95 Forest Alliance Director Patrick Moore stated that the move could be used to "further marginalize extremists who are trying to use the FSC for their own purposes." See Gordon Hamilton, "Forestry Alliance pledges to join environmentalists in greening effort," *Vancouver Sun* (5 June 1999): D1. The Alliance was accepted as a member in June 2000.
96 For an early comparison of the three schemes, see Martin von Mirbach, "Demanding Good Wood," *Alternatives* 23 (Summer 1997): 10-17.
97 George Hoberg and Edward Morawski, "Policy Change through Sector Intersection: Forest and Aboriginal Policy in Clayoquot Sound," *Canadian Public Administration* 40,3 (Fall 1997): 387-414; Marchak et al., *Falldown*.
98 See Hoberg, "The British Columbia Forest Practices Code."

Chapter 4: The Politics of Long-Term Policy Stability

1 See British Columbia, Ministry of Forests, "B.C's First Four Community Forest Pilot Sites Announced" (Victoria: 1999). WWW document at (as of 9 June 1999): www.for.bc.ca/pscripts/pab/newrel/mofnews.asp; and Gordon Hamilton, "Pilot Project Planned to Change Forest Tenure," *Vancouver Sun* (24 October 1997): C4.
2 See Ministry of Forests, "Harrop-Proctor Society Gets Community Forest," News Release, 16 July 1999. Available at www.for.gov.bc.ca/pscripts/pab/newsrel/mofnews.asp.
3 See Michael Sasges, "Tree Farm Licences Put on Hold: Minister Seeks Public Opinion," *Vancouver Sun* (19 January 1989): B5; and Ken Bernsohn, "Call It a Giant Turf War: BC's Debate over Public v. Private control of Forests Is Renewed," *Western Report* 4,2 (January 1989): 42. See also Salasan Associates Ltd., *Summary of Public Response to the Proposed Forest Licence Conversion Policy* (Victoria: Ministry of Forests, 1990). Microlog Document #90-00752.
4 British Columbia, Ministry of Forests, "Proposed Policy and Procedures for the Replacement of Major Volume Based Tenures with Tree Farm Licences" (Victoria: 21 July 1988).
5 Jeremy Wilson, *Talk and Log: Wilderness Politics in British Columbia* (Vancouver: UBC Press, 1998), esp. p. 264; and Jeremy Wilson, "Implementing Forest Policy Change in British Columbia: Comparing the Experiences of the NDP Governments of 1972-75 and 1991-?" in T.J. Barnes and R. Hayter, eds., *Troubles in the Rainforest: British Columbia's Forest Economy in Transition* (Victoria: Western Geographical Press, 1997), 76-97.

6 Interview with Gerry Armstrong, Deputy Minister, Ministry of Forests, 1 March 1995, conducted by Jeremy Wilson and Ben Cashore.
7 On the impact of institutions on political choice, see Margaret Weir, "Ideas and the Politics of Bounded Innovation," in S. Steinmo, K. Thelen, and F. Longstreth, eds., *Structuring Politics: Historical Institutionalism in Comparative Analysis* (Cambridge: Cambridge University Press, 1992), 188-216.
8 On path dependency see David Wilsford, "The *Conjoncture* of Ideas and Interests," *Comparative Political Studies* 18, no. 3 (1985): 357-72, and "Path Dependency, or Why History Makes It Difficult but Not Impossible to Reform Health Care Systems in a Big Way," *Journal of Public Policy* 14,3 (1994): 251-84.
9 See British Columbia, Ministry of Forests, *Annual Report 1995/96* (Victoria: 1997), and David Haley and Martin K. Luckert, *Forest Tenures in Canada: A Framework for Policy Analysis* (Ottawa: Forestry Canada, 1990). For a short summary of major existing tenure types in BC and their relative weight in provincial harvests, see Appendices 4.1 and 4.2.
10 British Columbia, "Timber Rights and Forest Policy in British Columbia: Report of the Royal Commission on Forest Resources" (Victoria: Queen's Printer, 1976); and British Columbia, Forest Resources Commission, "The Future of Our Forests, Executive Summary" (Victoria: April 1991).
11 See David Haley, "The Forest Tenure System as a Constraint on Efficient Timber Management: Problems and Solutions," *Canadian Public Policy* 11 (1985, supplement): 315-20; and Ken Drushka, "Forest Tenure: Forest Ownership and the Case for Diversification," in K. Drushka, B. Nixon, and R. Travers, eds., *Touch Wood: BC Forests at the Crossroads* (Madeira Park, BC: Harbour Publishing, 1993), 1-22.
12 A position adopted by the CCF and NDP in opposition. See Jeremy Wilson, "Forest Conservation in British Columbia 1935-85: Reflections on a Barren Political Debate," *BC Studies* 76 (Winter 1987/8): 3-32.
13 See Michael M'Gonigle and Ben Parfitt, *Forestopia: A Practical Guide to the New Forest Economy* (Madeira Park, BC: Harbour Publishing, 1994). See also Herb Hammond, *Public Forests or Private Timber Supplies? ... The Need for Community Control of British Columbia's Forests* (Winlaw, BC: Silva Ecosystem Consultants Ltd, 1989); Herb Hammond, "Community Control of Forests," *Forest Planning Canada* 6,6 (November/December 1990): 43-6; Herb Hammond, *Wholistic Forest Use* (Winlaw, BC: Silva Ecosystem Consultants Ltd, 1991); Herb Hammond, *Seeing the Forest among the Trees: The Case for Wholistic Forest Use* (Vancouver: Polestar Press, 1991); Herb Hammond and Susan Hammond, "Sustainable Forest Planning and Use," *Forest Planning Canada* 1,4 (September-October 1985): 8-10.
14 Many different suggestions have been made about exactly how to structure and mandate local boards and the relationships that they would have with provincial authorities. For example, see Duncan Taylor and Jeremy Wilson, "Ending the Watershed Battles: BC Forest Communities Seek Peace through Local Control," *Environments* 22,3 (1994): 93-102; M'Gonigle and Parfitt, *Forestopia*; Cheri Burda, Deborah Curran, Fred Gale, and Michael M'Gonigle, *Forests in Trust: Reforming British Columbia's Forest Tenure System for Ecosystem and Community Health* (Victoria: University of Victoria Eco-Research Chair in Environmental Law and Policy 1997), report series R97-2; Cheri Burda, "It's Time to Set a New Table, Invite New Diners to BC's Forestry Feast," *Vancouver Sun* (2 April 1998): A19; and Cheri Burda, Fred Gale, and Michael M'Gonigle, "Eco-Forestry versus the State(us) Quo: Or, Why Innovative Forestry Is Neither Contemplated nor Permitted within the State Structure of British Columbia." *BC Studies* 119 (Autumn 1998): 45-72.
15 Interview with Gerry Armstrong, Deputy Minister, Ministry of Forests, 1 March 1995, conducted by Jeremy Wilson and Ben Cashore.
16 Wilson, "Forest Conservation in British Columbia," 30. See also Sheldon Kamieniecki, "Testing Alternative Theories of Agenda Setting: Forest Policy Change in British Columbia, Canada," paper presented at the Annual Meeting of the American Political Science Association, Boston, 3-6 September 1998.
17 For a review of the literature on ecosystem management, see Jessica Clogg, "Tenure Reform for Ecologically and Socially Responsible Forest Use in British Columbia" (MES thesis, York University, 1997). See also Michael M'Gonigle, "From the Ground Up:

Lessons from the Stein River Valley," in W. Magnusson, et al., eds., *After Bennett: A New Politics for British Columbia* (Vancouver: New Star Books, 1986), 169-91; Michael M'Gonigle, "Local Economies Solve Global Problems," *The New Catalyst* (Spring 1987), 17-21; Michael M'Gonigle, "Developing Sustainability: A Native/Environmentalist Prescription for Third-Level Government," *BC Studies* 84 (1989/90): 65-99; Tin Wis Coalition, "Community Control, Developing Sustainability, Social Solidarity" (Vancouver: Tin Wis Coalition, 1991); Village of Hazelton, "Framework for Watershed Management (formerly the Forest Industry Charter of Rights)" (Hazelton, BC: The Corporation of the Village of Hazelton, 1991); and Alice Maitland, "Forest Industry Charter of Rights," *Forest Planning Canada* 6,2 (March/April 1990): 5-9.

18 See Holly Nathan, "Aboriginal Forestry: The Role of First Nations," in K. Drushka, B. Nixon and R. Travers, eds., *Touch Wood: BC Forests at the Crossroads* (Madeira Park, BC: Harbour Publishing, 1993), 137-70; Chris McKee, *Treaty Talks in British Columbia: Negotiating a Mutually Beneficial Future* (Vancouver: UBC Press, 1997); and Christopher A. Lee and Phil Symington, "Land Claims Process and Its Potential Impact on Wood Supply," *Forestry Chronicle* 73,3 (1997): 349-52.

19 On path dependency, see Margaret Weir, "Ideas and the Politics of Bounded Innovation," 188-216; Paul A. David, "Clio and the Economics of QWERTY," *American Economic Review* 75,2 (1985): 332-7; and Richard Rose, "Inheritance before Choice in Public Policy," *Journal of Theoretical Politics* 2,3 (1990): 263-91.

20 See David S. Cohen and Brian Radnoff, "Regulation, Takings, Compensation, and the Environment: An Economic Perspective," in Chris Tollefson, ed., *The Wealth of Forests: Markets, Regulation, and Sustainable Forestry* (Vancouver: UBC Press, 1998), 299-341.

21 Clogg, "Tenure Reform." See also Ken Drushka, "Time bomb ticks in ruling on royalty case," *Vancouver Sun* (30 June 1999): D2; and Gordon Hamilton, "MB getting cash instead of land for timber loss," *Vancouver Sun* (4 September 1999): H1.

22 Heather Scofield, "NAFTA trio warned of corporate lawsuits: Plug loophole to avoid claims, think-tank says," *Globe and Mail* (23 June 1999): A7. More generally see Alan M. Rugman, John Kirton, and Julie Soloway, *Environmental Regulations and Corporate Strategy: A NAFTA Perspective* (Oxford: Oxford University Press, 1999). Recent jurisprudence in this area appears to be limiting the effect of this section. See the interim dispute panel ruling on the Pope and Talbot case at: www.dfait-maeci.bc.ca/TNA-NAC/dispute-e.asp#chapter%2011.

23 Richard Schwindt and Steven Globerman, "Takings of Private Rights to Public Natural Resources: A Policy Analysis," *Canadian Public Policy* 22,3 (1996): 216.

24 See Gordon Hamilton, "Forest firms sue NDP for $240 million," *Vancouver Sun* (29 December 1995): A1; Justine Hunter, "Forest giant seeks $200 million from BC for lost timber rights," *Vancouver Sun* (24 September 1997): A1.

25 See Ministry of Forests, "MacBlo Will Not Get Crown Land for Park Compensation," News Release, 3 September 1999. Available at www.for.gov.bc.ca/pscripts/pab/newsrel/mofnews.asp?refnum=1999%3A103. The report, by lawyer David Perry, is British Columbia, *MacMillan Bloedel Parks Settlement Agreement Decision* (Victoria: Ministry of Forests, 1999), available at www.for.gov.bc.ca/pab/news/mb/decision/index.htm.

26 J.A.K. Reid, *Significant Events and Developments in the Evolution of Timberland and Forestry Legislation in British Columbia* (Victoria: Ministry of Forests, 1985), 2; and Robert E. Cail, *Land, Man, and the Law: The Disposal of Crown Lands in British Columbia 1871-1913* (Vancouver: UBC Press, 1974).

27 Cail, *Land, Man, and the Law*, 125-68.

28 On the history of the forest industry during and immediately after construction of the transcontinental railway, see Joseph Collins Lawrence, "Markets and Capital: A History of the Lumber Industry of British Columbia (1778-1952)" (MA thesis, University of British Columbia, 1957), 38-100; and G.W. Taylor, *Timber: History of the Forest Industry in British Columbia* (Vancouver: J.J. Douglas, 1975).

29 "An Act to Amend the 'Timber Act,'" 49 Vict. chap. 22, 1886.

30 Reid, *Significant Events*, 3.

31 "The Land Act," 51 Vict. chap. 16, 1888. According to the terms of the new act, the dura-

tion of timber leases was fixed at thirty years, subject to a ground rent payment and payment of a volume-based stumpage levy. A variety of other temporary and special licences for hand-loggers, and covering tracts of land smaller than 400 hectares, were also included in the 1888 legislation. Gilbert Piché, "Forestry Notes in Connection with Stumpage Dues in Quebec," in Quebec, Report of the Minister of Lands and Forests of the Province of Quebec for the Twelve Months Ending 30th June 1910 (Quebec: King's Printer, 1911), 115.

32 G. Carrothers, "Forest Industries in British Columbia," in A.R.M. Lower, ed., *The North American Assault on the Canadian Forest* (New York: Greenwood, 1968), 233-4; and Reid, *Significant Events,* 3.

33 This is an opinion shared by Marris who argues that, during this period, a "short-term desire for Crown Revenue was paramount." See Robert Howard Marris, "Pretty Sleek and Fat: The Genesis of Forest Policy in British Columbia, 1903-1914" (MA Thesis, University of British Columbia, 1979), 16.

34 Between 1901 and 1903 the government entered into long-term pulpwood agreements with five companies, although pulp mills were not constructed in the province until after US tariffs on newsprint were removed beginning in 1909. Reid, *Significant Events,* 4; and R.E. Gosnell, "The Year Book of British Columbia and Manual of Provincial Information" (Victoria: Legislative Assembly, 1903), 247. US tariffs restricted the demand for pulpwood leases, but sawmill construction continued to increase as manufacturers in the eastern and Midwest US faced both reduced supplies and US conservationist legislation that had withdrawn large forest reserves from the current harvest.

35 "An Act to Amend the 'Land Act,'" 5 Edw. VII c. 33, 1905.

36 The government did, however, attempt to keep a general provision in place forcing provincial timber to be manufactured within the province. In 1901 the province joined with Ontario in establishing a "manufacturing condition" in its timber regulations, forcing the manufacture of timber in the province on threat of forfeiture of timber and cancellation of timber leases. See Carrothers, "Forest Industries in British Columbia," 236. Legislation implementing the condition was only passed in 1906. See "An Act Respecting the Use and Manufacture, within British Columbia, of Timber Cut on Lands of the Crown," 6 Edw. VII c. 42, 1906. In 1903-4 the government attempted to extend the manufacturing condition to alienated lands by instituting a schedule of taxes on timber not subject to royalty, with the provision that the taxes would be rebated if the timber was manufactured within the province.

37 Carrothers, "Forest Industries in British Columbia," 237.

38 As Quebec chief forester Piché and Ontario chief forester Judson Clark noted, the 1905 legislation "had the effect of giving all the best forests of British Columbia to the lumber merchants." See Piché, "Forestry Notes," 115. Between the passage of the licence legislation in 1905 and the cancellation of the system in late 1907, over 14,000 licences were issued by the provincial government covering over 36,230 square kilometres of timber land. In association with fee increases for other small licences and increases in stumpage rates and ground rents in the burgeoning sawmill industry, the funds generated by the speculation in timber licences resulted in the proportion of total provincial revenues supplied by the forest sector rising from 7 percent in 1901 to over 41 percent by 1908. Martin Robin, *The Rush for Spoils: The Company Province, 1871-1933* (Toronto: McClelland and Stewart, 1972), 17, 92-3.

39 It met with US chief forester Gifford Pinchot on two occasions and with Aubrey White and Bernhard Fernow in Toronto. See British Columbia, "Final Report of the Royal Commission of Inquiry on Timber and Forestry 1909-1910" (Victoria: King's Printer 1910), 9.

40 Ibid., 46-56.

41 H.N. Whitford and R.D. Craig, "Forests of British Columbia" (Ottawa: Commission of Conservation, 1918), 90-1. See also A.C. Flumerfelt, "Forest Resources," in Adam Shortt and Arthur G. Doughty, eds., *Canada and Its Provinces: A History of the Canadian People and Their Institutions by One Hundred Associates.* Vol. 22, *The Pacific Province Part II* (Glasgow: Brook and Company, 1914), 496-7.

42 Whitford and Craig, "Forests of British Columbia," 95. As Pearse pointed out in 1974, this was the origin of the present system of calculation of Crown timber charges in the province. See Peter H. Pearse, *Crown Charges for Early Timber Rights: Royalties and Other Levies for Harvesting Rights on Timber Leases, Licences and Berths in British Columbia – First Report of the Task Force on Crown Timber Disposal* (Victoria: Ministry of Forests, 1974), 61-2.
43 The entire tenure situation became more complex in 1930 when, after years of negotiation, the federal government in transferring control over natural resources to Manitoba, Alberta, and Saskatchewan also transferred control over the remaining unsold lands in the Railway Belt and Peace River District back to the province under the terms of the Railway Belt Retransfer Agreement Act: "An Act Respecting the Transfer of the Railway Belt and the Peace River Block," 20-1 George V 1930 (Canada). Part of the act obliged the provincial government to honour all of the commitments made by the federal government in granting timber licences, known as Timber Berths, under federal regulations administered by the Federal Department of the Interior.
44 On the complexity of the largely ad hoc system of rules and amendments put into place by this time, see E.C. Manning, "The Administration of Crown Lands in British Columbia," *Journal of Forestry* 36,10 (1938): 314-25. Generally, on the situation in BC between 1890 and 1925, see chap. 6 of R. Peter Gillis and Thomas R. Roach, *Lost Initiatives: Canada's Forest Industries, Forest Policy and Forest Conservation* (New York: Greenwood, 1986).
45 This consolidation effort began in the mid-1920s as moves were made in the direction of enhancing forest conservation. In the early 1920s the first provincial forest reserves were created under the terms of the Forest Act, and in 1927 the provincial Forest Service began a new inventory of provincial forest resources. By 1937 the results of this inventory were published in the Mulholland Report on Forest Resources. They demonstrated the rapid depletion of coastal forests that had occurred since the turn of the century. The report argued that the era of "devastation," or pure exploitation of provincial forests, was long past and that the restriction of the government role to that of protection only was insufficient to guarantee the continued long-term viability of the provincial forest industry, government revenues, and resource-dependent communities. The report urged the adoption of a regime of sustained yield regulation based on the establishment of large "working circles" that would allow natural regeneration to take place after logging and in which harvesting rates would match the rate of growth of timber. See F.D. Mulholland, *The Forest Resources of British Columbia* (Victoria: King's Printer, 1937).
46 British Columbia, "Report of the Commissioner Relating to the Forest Resources of British Columbia 1945," 2 vols. (Victoria: King's Printer, 1945), 51.
47 Ibid., 143.
48 Ibid., 146.
49 Ibid., 147.
50 W. Young, *Timber Supply Management in British Columbia: Past, Present and Future* (Victoria: Ministry of Forests, 1981).
51 British Columbia, "Report of the Commissioner Relating to the Forest Resources of British Columbia 1945," 127.
52 British Columbia, "Report of the Commissioner Relating to the Forest Resources of British Columbia 1956," 2 vols. (Victoria: Queen's Printer 1956).
53 Ibid., 93.
54 Ibid., 96-7.
55 "An Act to Amend the "Forest Act,'" 6-7 Eliz. II c. 18, 1958.
56 On the Sommers affair, see Paddy Sherman, *Bennett* (Toronto: McClelland and Stewart, 1966), 147-78.
57 The need to rationalize the existing system in order to acknowledge the decline of competitive conditions in resource allocation and pricing is discussed in G.S. Nagel, "Economic and Public Policy in the Forestry Sector of British Columbia" (PhD diss., Yale University, 1970).
58 British Columbia, "Timber Rights," 19-21.
59 On agenda setting and policy windows see John Kingdon, *Agendas, Alternatives, and Public*

Policies (Boston: Little Brown, 1984). On the Barrett record see Jeremy Wilson, "Implementing Forest Policy Change in British Columbia: Comparing the Experiences of the NDP Governments of 1972-75 and 1991-?" in T.J. Barnes and R. Hayter, eds., *Troubles in the Rainforest: British Columbia's Forest Economy in Transition* (Victoria: Western Geographical Press, 1997), 76-97; and Jeremy Wilson, *Talk and Log.*

60 Pearse, *Timber Rights,* 51-2.

61 Pearse had earlier joined other UBC forest economists in criticizing pure sustained yield forestry as uneconomic. See Peter H. Pearse, "Conflicting Objectives in Forest Policy: The Case of British Columbia," *Forestry Chronicle* 46 (1970): 281-7.

62 Reid, *Significant Events,* 18-19.

63 The forest industry disliked both the second Task Force report and the Forest Products Stabilization Act, which together, according to MacMillan Bloedel chief forester Grant Ainscough, relegated the logging industry "to the status of a public utility, limited to earning the rate prescribed in the allowance for profit and risk." Ainscough objected to this since "the industry has neither the low-risk nor a fixed price for its products enjoyed by a utility." In addition, the powers given to the Forest Products Board to regulate the industry were considered to be much too broad, permitting the board "to enter into or influence virtually any phase of the forest industry." G.L. Ainscough, "Comment," in CCPA, *Recent Forest Policy Developments in Manitoba, Saskatchewan, Alberta and British Columbia,* Proceedings of the National Forest Management Group Meeting held at Edmonton, Alberta (Montreal: Canadian Pulp and Paper Association, 1975), 70-1. A useful discussion of the situation in 1974-5 can also be found in Alan K. Wilkinson, "British Columbia," in David A. Wilson, ed., *Report of Proceedings of the Twenty-Eighth Tax Conference* (Toronto: Canadian Tax Foundation, 1976), 559-66.

64 The reasons behind the formation of the royal commission and excerpts from interviews with Pearse and Minister of Lands, Forests, and Water Resources Bob Williams are contained in Ken Bernsohn, *Cutting Up the North: History of the Forest Industry in the Northern Interior* (Vancouver: Hancock House, 1981), 154-63. At the time of the commission's establishment, the Task Force Terms of Reference cited above also served as the basic statement of provincial forest policy. See the province's submission to Canadian Council of Resource and Environment Ministers Task Force on Forest Policy, *Forest Policies in Canada: Major Objectives, Problems and Trends,* vol. 1 (Montreal: Canadian Council of Resource and Environment Ministers, 1976), 26-7.

65 British Columbia, "Timber Rights." The second volume has useful chapters on the concentration of timber holdings of major forest companies as well as on the provincial taxation system and its effects on the forest industry. Useful short discussions of the Pearse Royal Commission are contained in J.A. Gray, "Royal Commission and Forest Policy in British Columbia: A Review of the Pearse Report," *Canadian Public Policy* 3,2 (1977): 219-23; and Richard Schwindt, "The Pearse Commission and the Industrial Organization of the British Columbia Forest Industry," *BC Studies* 41 (Fall 1979): 3-35.

66 The historical significance of the 1976 royal commission is debatable, even though many changes in legislation and administrative practice followed its recommendations. Pearse has argued that the report marked the beginning of a new period in BC forest policy akin to those ushered in by the Fulton report in 1912 and the 1947 Sloan Commission. See Peter H. Pearse, *Forest Policy in Canada* (Victoria: UBC Forest Economics and Policy Analysis Project, 1985), BC-13. Richard Schwindt, on the other hand, has argued that the 1976 royal commission did not call for a major redirection of forest policy but acted "to legitimize and thereby to entrench the concentration of harvesting rights (and, therefore, concentration throughout the sector) in large part induced by previous policy." See Schwindt, "The Pearse Commission," 3-35.

67 "The Forest Act," 27 Eliz. II c. 23, 1978 and "The Ministry of Forests Act," 27 Eliz. II c. 27, 1978.

68 For a list of changes made in the acts, see Reid, *Significant Events,* 19-21.

69 Pearse, *Forest Policy in Canada,* BC-25.

70 See, for example, Jack Toovey, "Forest Renewal: British Columbia," in Canadian Forestry Association, *Tomorrow's Forests ... Today's Challenge? Proceedings of the National Forest Regeneration Conference* (Quebec City: Canadian Forestry Association, 1977), 117-25; and

F.L.C. Reed and Associates, *Forest Management in Canada,* vol. 1 (Ottawa: Canadian Forestry Service, 1978), 31-7. Both discussions argue much work had been done to date in the province but that much more work was required, especially concerning artificial regeneration of cut-over lands.

71 On the technical reports, analysis, and program components of the Forest and Range Resource process, see British Columbia, Ministry of Forests, "Meeting the Challenge: Guide to the Ministry of Forests Organization and Programs" (Victoria: 1999), 10-12; British Columbia, "Forest Planning Framework" (Victoria: 1983); and British Columbia, "Ministry of Forests Management System" (Victoria: 1983).

72 See British Columbia, *The Small Business Program: A White Paper for Discussion Purposes* (Victoria: Ministry of Forests, 1979); British Columbia, *Alternatives for Crown Timber Pricing: A White Paper for Discussion Purposes* (Victoria: Ministry of Forests, 1979); and British Columbia, *Incentives for Intensive Forest Management: A White Paper for Discussion Purposes* (Victoria: Ministry of Forests, 1979).

73 British Columbia, *Forest Management Review: British Columbia* (Victoria: Ministry of Forests and Lands, 1987), 3, 9, 11.

74 British Columbia, Ministry of Forests and Lands, "News Release: Major Shift in Forest Policy for British Columbia" (Victoria: 15 September 1987).

75 The question of reducing government stumpage credits under Section 88 arose due to the depressed final product prices. Given the links between the provincial appraisal system and these prices, stumpage was greatly reduced, resulting in some firms generating excess credits against stumpage. This meant that not only did the provincial government forest revenues suffer, but its forest expenditures rose. See British Columbia, "Five Year Forest and Range Resource Program 1985-1990" (Victoria: Ministry of Forests, 1985), 9; British Columbia, "New Directions for Forest Policy in British Columbia" (Victoria: Ministry of Forests and Lands, 1987); and British Columbia, "Forest Policy Review: A Summary of Major Decisions" (Victoria: Ministry of Forests and Lands, 1987).

76 On the industry clamour that anticipated and followed the new rates, see Jennifer Hunter, "BC dramatically changes its forest tax," *Globe and Mail* (16 September 1987): B1-B2; Jennifer Hunter, "BC forest company executives stymied in move to head off stumpage charges," *Globe and Mail* (2 September 1987): B8; John Schreiner, "BC hikes forest tax bite," *Financial Post* (21 September 1987): 6; and "BC sawmill owners say new timber fees will put axe to the logging business," *Globe and Mail* (4 December 1987): B13.

77 British Columbia, Ministry of Forests, "Proposed Policy and Procedures." See also Bernsohn, "Call It a Giant Turf War," 42.

78 British Columbia, Ministry of Forests, "Proposed Policy and Procedures."

79 Wilson, *Talk and Log.*

80 Michael Sasges, "Two more groups demand probe into use of forests," *Vancouver Sun* (1 February 1989).

81 Lyle Stewart, "Shifting winds in the forest," *Monday Magazine* (9-15 February 1989).

82 Ben Parfitt, "Heated response delays TFL move," *Vancouver Sun* (22 March 1989); British Columbia, Ministry of Forests, "Summary of Public Input" (Victoria: 1986); British Columbia, "Written Submissions Received in Support of Presentations Made at Public Information Session on the Proposed Policy and Procedures for the Replacement of Major Volume-Based Tenures with Tree Farm Licences: Parksville, British Columbia March 10, 1989," vol. 1 (Victoria: Ministry of Forests, 1989).

83 See Michael Sasges, "Official defends tree licence policy," *Vancouver Sun* (12 January 1989): B5; British Columbia, Ministry of Forests, "Tree Farm Licences in British Columbia" (Victoria: 1988).

84 See Michael Sasges, "Tree Farm Licences put on hold: Minister seeks public opinion," *Vancouver Sun* (19 January 1989): B5. See Salasan Associates Ltd., *Summary of Public Response*; British Columbia, Ministry of Forests, "Tree Farm Licences."

85 British Columbia, Ministry of Forests, "News Release, Permanent Forest Resources Commission Established," 29 June 1989; Ben Parfitt, "New body to monitor forests," *Vancouver Sun* (29 June 1989); and Lyle Stewart, "Business as usual," *Monday Magazine* (6-12 July 1989).

86 See Forest Resources Commission, *The Future of Our Forests: Executive Summary* (Victoria: Forest Resources Commission, 1991); and Sterling Wood Group, "Review of Forest Tenures in British Columbia," in *Forest Resources Commission, Background Papers* vol. 3 (Victoria: Forest Resources Commission, 1991).

87 NDP, "A Better Way for British Columbia: New Democrat Election Platform," resolutions 17-21, and 32.

88 Wilson, *Talk and Log*. See also M'Gonigle and Parfitt, *Forestopia*, 97-8.

89 British Columbia, Ministry of Aboriginal Affairs, "The Benefits and Costs of Treaty Settlements in British Columbia: A Summary of the KPMG Report" (Victoria: December 1996). Available at www.aaf.gov.bc.ca/aaf/pubs/kpmgsum.htm. See also Lee and Symington, "Land Claims Process," 349-52.

90 For example, Deputy Premier Dan Miller in a speech to the Northern Forest Products Association in April 1999. See Justine Hunter and Kelly Sinoski. "Privatize BC's forest lands, Miller says in call for overhaul," *Vancouver Sun* (10 April 1999): A1.

91 See Steve Weatherbe, "Tenure reform: Thinking the unthinkable," *Business Examiner*, (Victoria, 18 May 1999): 1; and Barrie McKenna, "Privatizing BC forests not viable: Union," *Globe and Mail* (16 June 1999): B12.

92 See Drushka, "Forest Tenure."

93 Truck Loggers Association, "Options for the Forest Resources Commission: Review, Reconsideration, Recommendations" (Vancouver: TLA, 1990), 29; and Truck Loggers Association, "BC Forests: A Vision for Tomorrow – An Overview" (Vancouver: TLA, 1990), 10-11. See also Truck Loggers Association, "BC Forests – A Vision for Tomorrow (Working Papers)" (Vancouver: TLA, 1990).

94 Truck Loggers Association, "BC Forests: A Vision for Tomorrow," 19-21.

95 Truck Loggers Association, "Options for the Forest Resources Commission," 29-33.

96 Forest Resources Commission, "Final Report," 43-9.

97 Ibid., 45-6.

98 Clark S. Binkley, "A Crossroad in the Forest: The Path to a Sustainable Forest Sector in BC," *BC Studies* 113 (Spring 1997): 39-68.

99 The analysis rests on the assumption that larger silvicultural incentives would flow to timber maximization units since old-growth forests would be liquidated more quickly in these regions. See Sivaguru Sahajananthan, David Haley, and John Nelson, "Planning for Sustainable Forests in British Columbia through Land Use Zoning," *Canadian Public Policy* 24, Supplement 2 (1998): S73-S81.

100 Tom Stephens, "White Paper for Discussion: Stumpage and Tenure Reform in BC" (Vancouver: MacMillan Bloedel, June 1998).

101 David Haley and Martin K. Luckert, "Tenures as Economic Instruments for Achieving Objectives of Public Forest Policy in British Columbia," in Chris Tollefson, ed., *The Wealth of Forests: Markets, Regulation, and Sustainable Forestry* (Vancouver: UBC Press, 1998), 148.

102 See Gordon Hamilton, "Public meetings planned to chart course for BC's forest industry," *Vancouver Sun* (16 July 1999): F1.

103 See Ministry of Forests, "Tenure and Pricing Issues and Directions." Available from ftp.for.gov.bc.ca/pab/review/index.htm.

104 See ABCPF, "Do you care about the future of your forests?" *Vancouver Sun* (9 September 1999): A13; and Gordon Hamilton, "Policy review won't fit time frame, foresters say," *Vancouver Sun* (9 September 1999): F1.

105 See Gordon Hamilton, "COFI launches public support bid," *Vancouver Sun* (9 October 1999): D4.

106 See Ministry of Forests, "Wouters Calls for Changes in Forest Policy," News Release, 5 April 2000. Available at www.for.gov.bc.ca/pscripts/pab/newsrel/mofnews.asp? refnum=2000%3A036; and Garry Wouters, "Shaping Our Future: BC Forest Policy Review" (Victoria: Office of the Jobs and Timber Accord Advocate, 5 April 2000). Available at www.for.gov.bc.ca/pab/review/index.htm.

107 See Gordon Hamilton, "Pilot project planned to change forest tenure," *Vancouver Sun* (24 October 1997): C4.

108 Darrell Frank and Kim Allan, "Community Forests in British Columbia: Models that Work," *Forestry Chronicle* 70 (1994): 720-4.
109 On the need for such distinctions, see Morris Zelditch Jr., William Harris, George M. Thomas, and Henry A. Walker, "Decisions, Nondecisions and Metadecisions," *Research in Social Movements, Conflict and Change* 5 (1983): 1-32; and Richard A. Smith, "Decision Making and Non-Decision Making in Cities: Some Implications for Community Structural Research," *American Sociological Review* 44,1 (1979): 147-61.
110 See M. Howlett and M. Ramesh, *Studying Public Policy: Policy Cycles and Policy Subsystems* (Toronto: Oxford University Press, 1998); and Bryan D. Jones, *Reconceiving Decision-Making in Democratic Politics: Attention, Choice and Public Policy* (Chicago: University of Chicago Press, 1994).
111 On nondecisions, see Peter Bachrach and Morton S. Baratz, "Decisions and Nondecisions: An Analytical Framework," *American Political Science Review* 57,2 (1963): 632-42; and Peter Bachrach and Morton S. Baratz, *Power and Poverty: Theory and Practice* (New York: Oxford University Press, 1970), esp. chap. 3. For a critique of the notion, see Geoffrey Debnam, "Nondecisions and Power: The Two Faces of Bachrach and Baratz," *American Political Science Review* 69,3 (1975): 889-900; and Peter Bachrach and Morton S. Baratz, "Power and Its Two Faces Revisited: A Reply to Geoffrey Debnam," *American Political Science Review* 69,3 (1975): 900-7.
112 See, for example, Morris Zelditch Jr. and Joan Butler Ford, "Uncertainty, Potential Power and Nondecisions," *Social Psychology Quarterly* 57,1 (1994): 64-74; Johanna H. Kordes-de Vaal, "Intention and the Omission Bias: Omissions Perceived as Nondecisions," *Acta Psychologica* 93 (1996): 161-72; and Mark Spranca, Elisa Minsk, and Jonathan Baron, "Omission and Commission in Judgment and Choice," *Journal of Experimental Social Psychology* 27 (1991): 76-105.
113 See Susan Phillips, "Discourse, Identity, and Voice: Feminist Contributions to Policy Studies," in L. Dobuzinskis, M. Howlett, and D.H. Laycock, eds., *Policy Studies in Canada: The State of the Art* (Toronto: University of Toronto Press, 1998), 242-65.
114 See Michael Howlett, "Acts of Commission and Acts of Omission: Legal-Historical Research and the Intentions of Government in a Federal State," *Canadian Journal of Political Science* 19,2 (1986): 363-70. On arrested policy cycles, see Roger W. Cobb and Marc Howard Ross, "Denying Agenda Access: Strategic Considerations," in Cobb and Ross, eds., *Cultural Strategies of Agenda Denial: Avoidance, Attack and Redefinition* (Lawrence: University Press of Kansas, 1997).
115 See Matthew A. Crenson, *The Un-Politics of Air Pollution: A Study of Non-Decisionmaking in the Cities* (Baltimore: Johns Hopkins Press, 1971); Raymond E. Wolfinger, "Nondecisions and the Study of Local Politics," *American Political Science Review* 65 (1971): 1063-80; and Frederick W. Frey, "Comment: On Issues and Nonissues in the Study of Power," *American Political Science Review* 65 (1971): 1081-101.
116 See Weir, "Ideas and the Politics of Bounded Innovation."
117 Ibid., 189.
118 See Gordon Hamilton, "Forest firms eye compensation," *Vancouver Sun* (8 May 1999); and Gordon Hamilton, "MB volunteers to trade lands for tenure reform," *Vancouver Sun* (11 June 1999).
119 On this notion of policy making as an unstable dynamic equilibrium process, see Michael Howlett, "Understanding Policy Change and Policy Stability: Elements of a Vector Theory of Policy Dynamics," paper presented to the Annual Meeting of the British Columbia Political Studies Association, Vancouver, 1999. Baumgartner and Jones have suggested that a key related factor in explaining policy change, or the lack of it, is also the stability or instability of the relevant subsystem. When a stable subsystem – a policy monopoly – exists, it is expected that policy will only change very slowly, if at all, over time. See Frank R. Baumgartner and Bryan D. Jones, *Agendas and Instability in American Politics* (Chicago: University of Chicago Press, 1993).
120 Michael Cavanagh, "Offshore Health and Safety Policy in the North Sea: Policy Networks and Policy Outcomes in Britain and Norway," in D. Marsh, ed., *Comparing Policy Networks* (Buckingham: Open University Press, 1998), 90-109; Wyn Grant, William Paterson, and

Colin Whitson, *Government and the Chemical Industry: A Comparative Study of Britain and West Germany* (Oxford: Clarendon, 1988); Andrew Jordan and John Greenaway, "Shifting Agendas, Changing Regulatory Structures and the 'New' Politics of Environmental Pollution: British Coastal Water Policy, 1955-1995," *Public Administration* 76 (1998): 669-94; and Michael Cavanagh, David Marsh, and Martin Smith, "The Relationship between Policy Networks at the Sectoral and Sub-Sectoral Levels: A Response to Jordan, Maloney and McLaughlin," *Public Administration* 73 (Winter 1995): 627-33.

Chapter 5: Policy Venues, Policy Spillovers, and Policy Change

1 See Melody Hessing and Michael Howlett, *Canadian Natural Resource and Environmental Policy: Political Economy and Public Policy* (Vancouver: UBC Press, 1997). On policy spillovers, see David Dery, "Policy by the Way: When Policy Is Incidental to Making Other Policies," *Journal of Public Policy* 18,2 (1999): 163-76; and Paul A. Sabatier, "Knowledge, Policy-Oriented Learning, and Policy Change," *Knowledge: Creation, Diffusion, Utilization* 8,4 (1987): 649-92. On spillovers in the forest sector specifically, see George Hoberg and Edward Morawski, "Policy Change through Sector Intersection: Forest and Aboriginal Policy in Clayoquot Sound," *Canadian Public Administration* 40,3 (Fall 1997): 387-414.

2 This is the case with many other aspects of Canadian-First Nations relations. See Paul Tennant, "Delgamuuk'w and Diplomacy: First Nations and Municipalities in British Columbia," paper presented at the Fraser Institute Conference co-hosted by the Canadian Property Rights Research Institute, Ottawa, 26-7 May 1999.

3 Claudia Notzke, *Aboriginal Peoples and Natural Resources in Canada* (Toronto: Centre for Aboriginal Management, Education and Training, 1994).

4 Michael Howlett, "Policy Paradigms and Policy Change: Lessons from the Old and New Canadian Policies towards Aboriginal Peoples," *Policy Studies Journal* 22,4 (1994): 631-52.

5 Douglas Sanders, "The Supreme Court of Canada and the 'Legal and Political Struggle' over Indigenous Rights," *Canadian Ethnic Studies* 22 (1990): 122-9.

6 K.T. Miller and G. Lerchs, *The Historical Development of the Indian Act* (Ottawa: Treaties and Historical Research Centre, Indian and Northern Affairs Canada, 1978), 128-9.

7 J.E. Hodgetts, *The Canadian Public Service: A Physiology of Government, 1867-1970* (Toronto: University of Toronto Press 1973), 105. In the first major government-commissioned anthropological studies into the state of Canada's Aboriginal populations in the late 1950s and early 1960s, prominent social scientists condemned the paternalism and bureaucratization of central federal control over Native affairs that characterized the existing assimilationist paradigm. See H.B. Hawthorne, ed., *A Survey of Contemporary Indians of Canada* (Ottawa: Queen's Printer, 1967); and R.W. Dunning, "Some Aspects of Governmental Indian Policy and Administration," *Anthropologica* 4,2 (1962): 209-31.

8 See I. Johnson, *Helping Indians to Help Themselves: The 1951 Indian Act Consultation Process* (Ottawa: Indian and Northern Affairs Canada, 1984); and J.F. Leslie, "Vision versus Revision: Native People, Government Officials and the Joint Senate/House of Commons Committee on Indian Affairs, 1946-1948 and 1959-1961," in *Papers Presented to the Canadian Historical Association Annual Meeting* (Victoria: Canadian Historical Association, 1984).

9 See J. Taylor, *Canadian Indian Policy During the Interwar Years, 1918-1939* (Ottawa: Indian Affairs and Northern Development 1984); and Brian Titley, *A Narrow Vision: Duncan Campbell Scott and the Administration of Indian Affairs in Canada* (Vancouver: UBC Press, 1986).

10 See J. Rick Ponting and Roger Gibbins, *Out of Irrelevance* (Toronto: Butterworths, 1980); J. Rick Ponting, ed., *Arduous Journey* (Toronto: McClelland and Stewart, 1986); J.A. Long, "Political Revitalization in Canadian Native Indian Society," *Canadian Journal of Political Science* 23 (1990): 751-74; J.A. Long and Menno Boldt, "Leadership Selection in Canadian Native Communities: Reforming the Present and Incorporating the Past," *Great Plains Quarterly* 7 (1987): 103-15; T. Hall, "Self-Government or Self-Delusion? Brian Mulroney and Aboriginal Rights," *Canadian Journal of Native Studies* 6 (1986): 77-89; and S. Weaver, "The Joint Cabinet/National Indian Brotherhood Committee: A Unique

Experiment in Pressure Group Relations," *Canadian Public Administration* 25 (1982): 211-39.

11 See introduction in J. Anthony Long and Menno Boldt, eds., *Governments in Conflict? Provinces and Indian Nations in Canada* (Toronto: University of Toronto Press, 1988), 3-20.

12 On the history of the treaties, see L. Upton, "The Origins of Canadian Indian Policy," *Journal of Canadian Studies* 8 (1973), 51-61; Titley, *A Narrow Vision;* and D. Madill and W.E. Daugherty, *Indian Government Under Indian Act Legislation, 1868-1951* (Ottawa: Indian Affairs and Northern Development, 1980).

13 See Roger Townsend, "Specific Claims Policy: Too Little Too Late," and Murray Angus, "Comprehensive Claims: One Step Forward, Two Steps Back," in Diane Englestad and John Bird, eds., *Nation to Nation: Aboriginal Sovereignty and the Future of Canada* (Toronto: Anansi, 1992), 60-6 and 67-77.

14 See *Guerin et al.* v. *Her Majesty the Queen,* [1984] 2 S.C.R. 335.

15 See Ian G. Scott and J.T.S. McCabe, "The Role of the Provinces in the Elucidation of Aboriginal Rights in Canada," in Long and Boldt, *Governments in Conflict?,* 59-71.

16 On the general issues of Aboriginal rights and title, see N.H. Mickenberg and Peter Cumming, eds., *Native Rights in Canada* (Toronto: General Publishing 1972); P.J. Usher, F.J. Tough, and R.M. Galois, "Reclaiming the Land: Aboriginal Title, Treaty Rights and Land Claims in Canada," *Applied Geography* 12 (1992): 109-32. See also B. Slattery, "The Constitutional Guarantee of Aboriginal and Treaty Rights," *Queen's Law Journal* 8 (1983): 232-73. On the earlier constitutional status of Native rights, see N. Bankes, "Indian Resource Rights and Constitutional Enactments in Western Canada, 1871-1930," in L. Knafla, ed., *Law and Justice in a New Land: Essays in Western Canadian Legal History* (Toronto: Carswell, 1986), 129-64.

17 Canada, *Indian Treaties and Surrenders from 1680 to 1890,* vols. 1-3. (Ottawa: Queen's Printer, 1905); W.E. Daugherty, *Maritime Indian Treaties in Historical Perspective* (Ottawa: Indian and Northern Affairs Canada, 1983); W.J. Eccles, "Sovereignty-Association, 1500-1783," *Canadian Historical Review* 65,4 (1984): 475-510; and Richard H. Bartlett, *Indian Reserves in Quebec,* no. 8 (Saskatoon: University of Saskatchewan Native Law Centre, Studies in Aboriginal Rights, 1984).

18 See Frank Tough, *"As Their Natural Resources Fail": Native Peoples and the Economic History of Northern Manitoba, 1870-1930* (Vancouver: UBC Press, 1996); and Jean Friesen and Kerry Abel, eds., *Aboriginal Resource Use in Canada: Historical and Legal Aspects* (Winnipeg: University of Manitoba Press, 1991).

19 Alexander Morris, *The Treaties of Canada with the Indians of Manitoba, the North-West Territories, and Kee-wa-tin* (Toronto: Belfords, Clarke and Co., 1880); and Norman K. Zlotkin, "Post-Confederation Treaties," in B. Morse, ed., *Aboriginal Peoples and the Law* (Ottawa: Carleton University Press, 1985).

20 See Royal Commission on Aboriginal Peoples, *Treaty Making in the Spirit of Co-Existence: An Alternative to Extinguishment* (Ottawa: Minister of Supply and Services, 1994); Howlett, "Policy Paradigms," 631-52; and S. Weaver, "A New Paradigm in Canadian Indian Policy for the 1990s," *Canadian Ethnic Studies* 22 (1990): 8-18.

21 Quebec, *The James Bay and Northern Quebec Agreement: Agreement between the Government of Quebec ...* (Quebec: Éditeur Officiel du Quebec, 1976); Canada, *Northeastern Quebec Agreement* (Ottawa: Information Canada, 1978); Canada, Indian and Northern Affairs, *The Western Arctic Claim: The Inuvialuit Final Agreement* (Ottawa: Indian and Northern Affairs Canada, 1984); Canada, *Comprehensive Land Claim Umbrella Final Agreement between the Government of Canada, the Council for Yukon Indians and the Government of the Yukon* (Ottawa: Indian and Northern Affairs Canada, 1990). Canada, Indian and Northern Affairs, *Agreement in Principle for the Nunavut Settlement Area* (Ottawa: Indian and Northern Affairs Canada, 1992); Canada, Indian and Northern Affairs, *Gwich'in Comprehensive Land Claim Agreement* (Ottawa: Indian and Northern Affairs Canada, 1992); Canada, Indian and Northern Affairs, *Comprehensive Land Claim Agreement ...* (Ottawa: Indian and Northern Affairs Canada, 1993); and Canada, British Columbia, Nisga'a Nation, *Nisga'a Final Agreement* (Ottawa: Federal Treaty Negotiation Office, 4 August 1998).

22 See British Columbia, Ministry of Aboriginal Affairs, "Present Status of the BCTC Process,

December 1997" (Victoria: 1997),available at www.aaf.gov.bc.ca/aaf/treaty/fnprocess.htm; and Canada, "Background Paper: The Comprehensive Land Claim Negotiations of the Labrador Inuit, The Government of Newfoundland and Labrador, and the Government of Canada" (Ottawa: Indian and Northern Affairs Canada, 1990).

23 See *R.* v. *Marshall* [1999] 3 S.C.R. 533 for a discussion of the recent case interpreting a 1760 British treaty as providing the Mi'kmaq nation in the Atlantic provinces with resource rights, including a right to access to the fisheries of the area. The old military "protectionist" paradigm ended in most of the settled part of the country after the War of 1812. At that time a period of experimentation with a reserve system began that culminated in multiple inquiries into "bettering the state of natives" held in all of the provinces of British North America between 1815 and 1860. See Bartlett, *Indian Reserves in Quebec*; Great Britain, "Report on the Indians of Upper Canada: 1839" (Toronto: Canadiana House 1968); J. Hodgetts, *Pioneer Public Service: An Administrative History of the United Canadas, 1841-1867* (Toronto: University of Toronto Press 1955); J.F. Leslie, *Commissions of Inquiry into Indian Affairs in the Canadas: 1828-1858* (Ottawa: Indian and Northern Affairs Canada, 1985); H.F. McGee, *The Native Peoples of Atlantic Canada* (Toronto: McClelland and Stewart 1974); and D.C. Scott, "Indian Affairs 1840-1867," in A. Shortt and A. Doughty, eds., *Canada and Its Provinces* (Toronto: Publishers Association of Canada, 1914), 331-62. The new "assimilationist" paradigm was institutionalized following the transfer of control over Indians to the colonial governments in 1860, and it received its organizational form in the reserve and band systems contained in the various Indian Acts promulgated by the colonial governments and replicated by the federal government following Confederation in 1867. See Miller and Lerchs, *Historical Development*; and Upton, "The Origins of Canadian Indian Policy."

24 Notzke, *Aboriginal Peoples*.

25 See K. Lysyk, "Approaches to Settlement of Indian Title Claims: The Alaska Model," *UBC Law Review* 8 (1973): 321-42; and A.M. Ervin, "Contrasts between the Resolution of Native Land Claims in the United States and Canada Based on Observations of the Alaska Native Land Claims Movement," *Canadian Journal of Native Studies* 1 (1981): 123-40. See also Constance Hunt, "Approaches to Native Land Settlements and Implications for Northern Land Use and Resource Management Policies," in R.F. Keith and J.B. Wright, eds., *Northern Transitions* (Ottawa: Canadian Arctic Resources Committee 1978), 5-41.

26 M.W. Wagner, "Footsteps along the Road: Indian Land Claims and Access to Natural Resources," *Alternatives* 18 (1991): 22-8. See also Kevin R. Gray, "The Nunavut Land Claims Agreement and the Future of the Eastern Arctic: The Uncharted Path to Effective Self-Government," *University of Toronto Faculty of Law Review* 52,2 (1994): 300-45.

27 On water rights in general, which are also a major issue on the Prairies, see Notzke, *Aboriginal Peoples*.

28 On the British Columbia process, see Chris McKee, *Treaty Talks in British Columbia: Negotiating a Mutually Beneficial Future* (Vancouver: UBC Press, 1997).

29 Notzke, *Aboriginal Peoples*.

30 André Légaré, "The Process Leading to a Land Claims Agreement and Its Implementation: The Case of the Nunavut Land Claims Settlement," *Canadian Journal of Native Studies* 16,1 (1996): 139-63; and Frank Duerden, Sean Black, and Richard G. Kuhn, "An Evaluation of the Effectiveness of First Nations Participation in the Development of Land-Use Plans in the Yukon," *Canadian Journal of Native Studies* 16,1 (1996): 105-24.

31 J. Keeping, *The Inuvialuit Final Agreement* (Calgary: Canadian Institute of Resources Law, 1989); L. MacLachlan, "The Gwich'in Final Agreement," *Resources* 36 (1991): 6-11; E.J. Peters, *Existing Aboriginal Self-Government Arrangements in Canada: An Overview* (Kingston: Queen's University Institute of Intergovernmental Relations, 1987); and Magdalena A.K. Muir, *Comprehensive Land Claims Agreements of the Northwest Territories: Implications for Land and Water Management* (Calgary: Canadian Institute of Resources Law, 1994).

32 See *Guerin* v. *R.*, [1984] 2 S.C.R. 335.

33 *R.* v. *St. Catherine's Milling and Lumber Co.* 14 App. Cas. 46 (1888).

34 See S. Barry Cottam, "The Twentieth Century Legacy of the St. Catherine's Case: Thoughts on Aboriginal Title in Common Law," in B.W. Hodgins, S. Heard, and J.S.

Milloy, eds., *Co-Existence? Studies in Ontario-First Nations Relations* (Peterborough: Frost Centre for Canadian Heritage and Development Studies, 1992), 118-27; and Kent McNeil, "The Meaning of Aboriginal Title," in Michael Asch, ed., *Aboriginal and Treaty Rights in Canada: Essays on Law, Equity, and Respect for Difference* (Vancouver: UBC Press, 1997), 135-54.

35 Taylor, *Canadian Indian Policy*.

36 Douglas Sanders, "The Queen's Promises," in L.A. Knafla, ed., *Law and Justice in a New Land: Essays in Western Canadian Legal History* (Toronto: Carswell 1986): 101-27.

37 Long, "Political Revitalization"; and Paul Tennant, *Aboriginal Peoples and Politics: The Indian Land Question in British Columbia, 1849-1989* (Vancouver: UBC Press, 1990).

38 At the First Ministers level, in the 1960s and 1970s Canada's Aboriginal organizations managed to link their own demands for collective rights to those of francophones in Quebec and secured a place at the constitutional bargaining table as the federal government grappled with the crisis provoked by threats of Quebec secessionism. The Aboriginal position was well-articulated throughout this process and received the support of prominent parliamentarians. However, First Nations were left out of the 11th-hour bargaining that resulted in the 1982 Canada Act and secured only several provisions in the new document entrenching existing treaties and a promise that the next round of constitutional discussions would focus entirely on their aspirations. Brian Slattery, "The Hidden Constitution: Aboriginal Rights in Canada," *American Journal of Comparative Law* 32 (1984): 361-76; and B.H. Wildsmith, *Aboriginal People and Section 25 of the Canadian Charter of Rights and Freedoms* (Saskatoon: University of Saskatchewan Native Law Centre, 1988). The promised Aboriginal round ended in failure in 1987, and later constitutional efforts to entrench self-government and Native rights ground to a halt with the failure of the Canadian public to endorse the Charlottetown Accord on the Constitution in October 1992. See Kathy L. Brock, "The Politics of Aboriginal Self-Government: A Canadian Paradox," *Canadian Public Administration* 34 (1991): 272-86; Douglas Sanders, "An Uncertain Path: The Aboriginal Constitutional Conferences," in J.M. Weiler and R.M. Elliot, eds., *Litigating the Values of a Nation* (Toronto: Carswell, 1986): 63-77; Douglas Sanders, "The Indian Lobby," in R. Simeon and K. Banting, eds., *And No One Cheered: Federalism, Democracy and the Constitution Act* (Toronto: Methuen, 1983); and Mary Ellen Turpel, "The Charlottetown Discord and Aboriginal Peoples' Struggle for Fundamental Political Change," in K. McRoberts and P. Monahan, eds., *The Charlottetown Accord, Referendum, and the Future of Canada* (Toronto: University of Toronto Press 1993): 117-51.

39 Paul Tennant, "Native Indian Political Organization in British Columbia, 1900-1969: A Response to Internal Colonialism," *BC Studies* 55 (1982): 3-49; Paul Tennant, "Native Indian Political Activity in British Columbia 1969-1983," *BC Studies* 57 (1983): 112-36. Land claims in particular were pursued vigorously by some B.C. First Nations over a long period of time. See D. Raunet, *Without Surrender, Without Consent: A History of the Nishga Land Claims* (Vancouver: Douglas and McIntyre, 1984); and E.P. Patterson, "A Decade of Change: Origins of the Nishga and Tsimshian Land Protests in the 1880s," *Journal of Canadian Studies* 18 (1983): 40-54.

40 See Paul Tennant, "Aboriginal Peoples and Aboriginal Title in British Columbia Politics," in R.K. Carty, ed., *Politics, Policy, and Government in British Columbia* (Vancouver: UBC Press, 1996): 45-66; and Tennant, *Aboriginal Peoples and Politics*.

41 See Frank Cassidy, "Aboriginal Land Claims in British Columbia," in Ken Coates, ed., *Aboriginal Land Claims in Canada: A Regional Perspective* (Toronto: Copp Clark Pitman, 1992), 11-44.

42 See McNeil, "The Meaning of Aboriginal Title."

43 See, for example, the Sechelt self-government agreement. Canada, *Sechelt Indian Band Self-Government Act* (Ottawa: Queen's Printer, 1986).

44 Sanders, "The Supreme Court of Canada." See also Frank Cassidy, ed., *Aboriginal Title in British Columbia: Delgamuuk'w* v. *The Queen* (Vancouver: Oolichan Books/Institute for Research on Public Policy 1992); and James S. Frideres, "Government Policy and Indian Natural Resource Development," *Canadian Journal of Native Studies* 4,1 (1984): 51-66.

45 See *Calder* v. *British Columbia* (A.G.), [1973] S.C.R. 313. See also K. Lysyk, "The Indian Title

Question in Canada: An Appraisal in the Light of Calder," *Canadian Bar Review* 51 (1973): 450-80; W.H. McConnell, "The Calder Case in Historical Perspective," *Saskatchewan Law Review* 38 (1974): 88-122; and Douglas Sanders, "The Nishga Case," *BC Studies* 19 (1973): 3-20.

46 On focusing events, see Thomas Birkland, *After Disaster: Agenda-Setting, Public Policy and Focusing Events* (Washington: Georgetown University Press, 1997); and John Kingdon, *Agendas, Alternatives, and Public Policies* (Boston: Little Brown, 1984).

47 On the injunction process in the James Bay, Baker Lake, Lubicon, and Meares Island cases, see J. O'Reilly, "The Courts and Community Values: Litigation Involving Native Peoples and Resource Development," *Alternatives* 15 (1988): 40-8; Evelyn Pinkerton, "Taking the Minister to Court: Changes in Public Opinion about Forest Management and Their Political Expression in Haida Land Claims," *BC Studies* 57 (1983): 68-85; and Darlene Abreu Ferreira, "Oil and Lubicons Don't Mix: A Land Claim in Northern Alberta in Historical Perspective," *Canadian Journal of Native Studies* 12,1 (1992): 1-35. On the negotiation of comprehensive land claims, see Thomas Berger, "Native History, Native Claims and Self-Determination," *BC Studies* 57 (1983): 10-23; Frank Cassidy, ed., *Reaching Just Settlements: Land Claims in British Columbia* (Vancouver: Oolichan Books/Institute for Research on Public Policy, 1991); Harvey A. Feit, "Negotiating Recognition of Aboriginal Rights: History, Strategies and Reactions to the James Bay and Northern Quebec Agreement," *Canadian Journal of Anthropology* 1 (1980): 159-70. On Métis lands and smaller so-called "specific claims" arising out of disputes over the implementation of existing treaties, see Coates, *Aboriginal Land Claims in Canada*.

48 See *Delgamuuk'w* v. *British Columbia* Dec. 11, 1997 S.C.C. file no. 23799.

49 *See R.* v. *Van der Peet,* [1996] S.C.R. 507; *R.* v. *Gladstone,* [1996] 2 S.C.R. 723; *R.* v. *N.T.C. Smokehouse Ltd.,* [1996] 2 S.C.R. 672; and *R.* v. *Sparrow,* [1990] 1 S.C.R. 1075. See also *Council of Haida Nations and Miles Richardson* v. *Ministry of Forests, Attorney General for British Columbia and MacMillan Bloedel Ltd.* BC Court of Appeal 7 Nov. 1997, docket no. CA021277. Also see Andrea Bowker, "Sparrow's Promise: Aboriginal Rights in the BC Court of Appeal," *University of Toronto Law Review* 53,1 (1997): 1-48; and S. Bradley Armstrong, "Defining the Boundaries of Aboriginal Title after *Delgamuuk'w*," *Resources* 62 (1998): 2-7.

50 *Sparrow* was a significant case in that it set out the parameters of existing Aboriginal rights and established the position of the Supreme Court that such rights are not "frozen" but should be interpreted as "living" entities. While precedent setting, *Sparrow* was difficult to operationalize. A subsequent case dealing with the question of how such rights could be determined – *Van der Peet* – was more significant vis-à-vis resource management. See Michael Asch and Patrick Macklem, "Aboriginal Rights and Canadian Sovereignty: An Essay on *R.* v. *Sparrow*," *Alberta Law Review* 29 (1991): 498-517; Peter Usher, "Some Implications of the Sparrow Judgment for Resource Conservation and Management," *Alternatives* 18 (1991): 20-2; Thomas Isaac, "Balancing Rights: The Supreme Court of Canada, *R.* v. *Sparrow,* and the Future of Aboriginal Rights," *Canadian Journal of Native Studies* 13,2 (1993): 199-219; and Bowker, "Sparrow's Promise."

51 See British Columbia, Claims Task Force, "Report" (Vancouver: 1991); and Tennant, "Aboriginal Peoples and Aboriginal Title."

52 National Aboriginal Forestry Association, "Provincial Forest Resource Access Policies," in *Forest Land and Resources for Aboriginal Peoples: An Intervention Submitted to the Royal Commission on Aboriginal Peoples by the National Aboriginal Forestry Association* (Ottawa, ON: National Aboriginal Forestry Association, 13 July 1993). Available at http://sae.ca/nafa/roycom3.htm#anchor1206203.

53 National Aboriginal Forestry Association, "Co-Management and Other Forms of Agreement in the Forest Sector [5 Part Strategy: Tools for Aboriginal Communities to Promote Good Forest Land Management Practices]." Available at http://sae.ca/nafa/comanage.htm.

54 British Columbia, Ministry of Aboriginal Affairs, "The Benefits and Costs of Treaty Settlements in British Columbia: A Summary of the KPMG Report" (Victoria: December 1996). Available at www.aaf.gov.bc.ca/aaf/pubs/kpmgsum.htm.

55 See Christopher A. Lee and Phil Symington, "Land Claims Process and Its Potential Impact on Wood Supply," *Forestry Chronicle* 73,3 (1997): 349-52. See also Stewart Bell, "Tribe ready to shed token role in woods," *Vancouver Sun* (18 July 1999): A9. Note that not all of the claims currently under negotiation involve timberlands, especially those covering urban areas in the Lower Mainland area of the province. See Gerry Bellett, "Land claims a dilemma for Delta," *Vancouver Sun* (13 July 1998): B1, and Larry Pynn, "Negotiators face challenge," *Vancouver Sun* (11 July 1998): A1.

56 "An Act to give effect to the Nisga'a Final Agreement," 48-9 Eliz. II, 1999-2000. This agreement was negotiated outside of the new treaty process as its commencement pre-dated its creation. See Canada, British Columbia, Nisga'a Nation, *Nisga'a Final Agreement.* The Sechelt Agreement-in-Principle (from which the Sechelt First Nation walked away in June 2000) provides for Aboriginal management at or above provincial standards on Sechelt treaty lands with fee simple title exercised over a maximum of 3,055 hectares to be selected over a twenty-four year period. See British Columbia, "Sechelt Agreement-in-Principle" (Victoria: Ministry of Aboriginal Affairs, 1999). Available at www.aaf.gov.bc.ca/aaf/nations/sechelt/secheltaip199.htm.

57 See British Columbia, Legislative Assembly, Select Standing Committee on Aboriginal Affairs, "Towards Reconciliation: Nisga'a Agreement-in-Principle and British Columbia Treaty Process – First Report – July 1997," Second Session, Thirty-sixth Parliament Legislative Assembly of British Columbia, July 1997. Available at www.legis.gov.bc.ca/cmt/cmt01/1997/1report/.

58 See Chapter 5 of Canada, British Columbia, Nisga'a Nation, "Nisga'a Final Agreement." Note the land areas covered in this treaty have come under fire from neighbouring First Nations. See Neil Sterritt, "The Nisga'a Treaty: Competing Claims Ignored," *BC Studies* 120 (Winter 1998/99): 73-98; and Robert Matas, "BC band seeking to halt treaty signing," *Globe and Mail* (8 July 1998): A4.

59 See Canadian Council of Forest Ministers, "National Forest Strategy (1998-2003): Sustainable Forests – A Canadian Commitment" (Ottawa: Canadian Council of Forest Ministers, May 1998), 31. See also National Aboriginal Forestry Association "Aboriginal and Treaty Rights to Renewable Resources," in *Forest Land and Resources for Aboriginal Peoples: An Intervention Submitted to the Royal Commission on Aboriginal Peoples by the National Aboriginal Forestry Association* (Ottawa, ON: July 1993). Available at sae.ca/nafa/roycom2.htm#anchor1135311. More generally, see Evelyn Pinkerton, ed., *Co-operative Management of Local Fisheries: New Directions for Improved Management and Community Development* (Vancouver: UBC Press, 1989). See also Debra Wright, "Self-Government, Land Claims and Silviculture: An Aboriginal Forest Strategy," *Forestry Chronicle* 70,3 (1994): 238-41; and Richard H. Bartlett, "Resource Development and Aboriginal Title in Canada," in Bartlett, ed., *Resource Development and Aboriginal Land Rights* (Calgary: Canadian Institute of Resources Law, 1991), 1-38.

60 See National Aboriginal Forestry Association, introduction to *Forest Land and Resources for Aboriginal Peoples.*

61 National Aboriginal Forestry Association, "Co-Management and Other Forms of Agreement in the Forest Sector."

62 National Aboriginal Forestry Association, introduction to *Forest Land and Resources for Aboriginal Peoples.* See also Geoff Quaile and Peggy Smith, "An Aboriginal Perspective on Canada's Progress toward Meeting Its National Commitments to Improve Aboriginal Participation in Sustainable Forest Management," paper presented to the Eleventh World Forestry Congress, Antalya, Turkey, 13-22 October 1993, vol. 5, topic 2. Available at http://193.43.36.7/waicent/faoinfo/forestry/wforcong/publi/v5/t29e/default.htm.

63 Immediately prior to the *Delgamuuk'w* decision, court actions claiming a lack of appropriate consultation had succeeded in overturning some cutting permits. See Gordon Hamilton, "Court halts permit on contested Kitkatla area," *Vancouver Sun* (22 October 1997): D24.

64 British Columbia, Ministry of Aboriginal Affairs, "Province of British Columbia Crown Land Activities and Aboriginal Rights Policy Framework" (Victoria: October 1997). Available at www.aaf.gov.bc.ca/aaf/pubs/crown.htm.

65 Westland Resource Group, "A Review of the Forest Practices Code of British Columbia and Fourteen Other Jurisdictions" (1995). Available at www.for.gov.bc.ca/pab/ publctns/westland/1-10.htm.
66 National Aboriginal Forestry Association, "Co-Management and Other Forms of Agreement."
67 See the lists of IMAs provided at the Ministry of Aboriginal Affairs website at www.aaf.gov.bc.ca/treaty/interim/interim.htm.
68 See British Columbia Chiefs of the Central Region, "Province of British Columbia and the Chiefs of the Central Region of the Nuu-chah-nulth Tribal Council Sign Interim Measures Agreement for Clayoquot Sound" (Victoria: Ministry of Aboriginal Affairs, 19 March 1994).
69 See Hoberg and Morawaski, "Policy Change through Sector Intersection."
70 See Larry Pynn, "MB, environmentalists agree to pact on Clayoquot logging," *Vancouver Sun* (19 June 1999): A3.
71 See Gordon Hamilton, "MacBlo turns over rights in Clayoquot," *Vancouver Sun* (5 October 1999): D4.
72 See Oral Reasons for Judgment, Supreme Court of British Columbia, 12 June 1998, *Chief Councillor Mathew Hill, also known as Tha-lathatk, on his own behalf and on behalf of all other members of the Kitkatla Band* v. *The Minister of Forests, et al. and International Forest Products,* docket no. 19980612; Oral Reasons for Judgement, Court of Appeal for British Columbia, 6 July 1998, *Chief Councillor Mathew Hill, also known as Tha-lathatk, on his own behalf and on behalf of all other members of the Kitkatla Band,* v. *The Minister of Forests, et al. and International Forest Products,* docket no, CA024761; and Jim Beatty, "Forest industry joins Interfor in logging fight," *Vancouver Sun* (19 June 1998): A1.
73 See Kim Pemberton, "Siska band blocks road to halt logging," *Vancouver Sun* (7 October 2000): B7.
74 See Kim Pemberton, "Westbank Indians keep logging," *Vancouver Sun* (10 September 1999): A6; and Kim Pemberton, "Court turns down Victoria bid to stop Westbank logging," *Vancouver Sun* (28 September 1999): A4.
75 Kim Pemberton, "Vancouver Island band joins logging action," *Vancouver Sun* (9 October 1999).
76 See Shawn Blore and Robert Matas, "BC Natives ask court to ban logging," *Globe and Mail* (20 June 1998): A5; Stewart Bell, "A fragile detente on land claims fades away," *Vancouver Sun* (27 June 1998): A1; "BC Indian chiefs press treaty talks," *Vancouver Sun* (27 June 1998): B6; and Marina Jimenez, "Final Nisga'a treaty close, Clark claims," *Vancouver Sun* (30 June 1998).
77 See Dianne Rinehart, "Prospect of land deal galvanizes talks," *Vancouver Sun* (9 July 1998): A1.
78 See Ministry of Forests, "Land and Resources Accord Signed with Wet'suwet'en," Press Release, 27 April 2000. Available at www.for.gov.bc.ca/pscripts/pab/newsrel/mofnews.asp?refnum=2000%3A043.
79 See Royal Commission on Aboriginal People, *Restructuring the Relationship,* vol. 2, part 2, chap. 4, at www.indigenous.bc.ca/v2/vol2ch4s1tos2.asp.
80 In 1999, Aboriginal affairs minister Dale Lovick suggested that signing ten treaties over the 1999-2004 period was the provincial government's expectation. See Dirk Meissner, "Minister sees 10 BC treaties over next five years," *Vancouver Sun* (4 January 1999): A1.
81 See Rick Ouston, "BC Indian chiefs lay claim to entire province, resources," *Vancouver Sun* (2 February 1998): A1; Stewart Bell, "Ruling on Native land rights puts treaty bid in disarray," *Vancouver Sun* (31 January 1998): B1; "BC Natives seek court remedies in Royal Oak fight," *Globe and Mail* (26 January 1998): B3; and Stephen Hume, "Crown land lawsuit looms," *Vancouver Sun* (17 January 1998): B3.
82 Pursuing both routes is also possible, as is occurring with the Haida Nation in the Queen Charlotte Islands. See Paul Willcocks, "Haida tired of slow treaty talks, going to court to claim islands," *Globe and Mail* (23 January 1999): A5.
83 "The Chiefs Speak:" We assert our Aboriginal title to all of BC," *Vancouver Sun* (5 February 1998): A19.

84 Several bands withdrew, leaving forty-three active negotiations in December 1997. See British Columbia, Ministry of Aboriginal Affairs, "Present Status of the BCTC Process, December." Available at www.aaf.gov.bc.ca/aaf/treaty/fnprocess.htm. See also Gordon Hamilton, "Haida ruling threatens Tree Farm Licences," *Vancouver Sun* (12 November 1997): A1; Stewart Bell, "Land claim ruling dispels old notion," *Vancouver Sun* (19 December 1997): A21; and Gordon Hamilton, "Deal or leave forest, Tsilhqot'in say," *Vancouver Sun* (19 February 1998): A10.

85 See Robert Matas, "Freeze demand ignored," *Globe and Mail* (16 March 1998): A3; Peter O'Neil, "Ottawa to revamp land claim process," *Vancouver Sun* (13 March 1998): A4.

86 Joseph Gosnell, "Let's get on with the land claim," *Vancouver Sun* (19 December 1997): E2.

87 As Chief Justice Lamer wrote: "What is required is that the government demonstrate 'both that the process by which it allocated the resource and the actual allocation of the resource which results from that process reflect the prior interest' of the holders of Aboriginal title in the land ... There is always a duty of consultation. Whether the Aboriginal group has been consulted is relevant to determining whether the infringement of Aboriginal title is justified, in the same way that the Crown's failure to consult an Aboriginal group with respect to the terms by which reserve land is leased may breach its fiduciary duty at common law." *Delgamuuk'w* v. *British Columbia,* 11 Dec. 1997 S.C.C. file no. 23799.

88 Ibid.

89 As Chief Justice Lamer argued in the 1990 *Sparrow* case, an infringement, such as a licence to cut trees, would be considered valid if it met three tests: "First, is there a valid legislative objective? ... If a valid legislative objective is found, the analysis proceeds to the second part of the justification issue ... The special trust relationship and the responsibility of the government vis-à-vis Aboriginals must be the first consideration in determining whether the legislation or action in question can be justified ... Within the analysis of justification, there are further questions to be addressed, depending on the circumstances of the inquiry. These include the questions of whether there has been as little infringement as possible in order to effect the desired result; whether, in a situation of expropriation, fair compensation is available; and, whether the aboriginal group in question has been consulted with respect to the conservation measures being implemented." *R.* v. *Sparrow,* [1990] 1 S.C.R. 1075.

90 *Delgamuuk'w* v. *British Columbia.*

91 Ibid.

92 Ibid.

93 Ibid.

94 Even if a claim to Aboriginal title was granted, Lamer argued "The content of aboriginal title contains an inherent limit in that lands so held cannot be used in a manner that is irreconcilable with the nature of the claimants' attachment to those lands. This inherent limit arises because the relationship of an aboriginal community with its land should not be prevented from continuing into the future. Occupancy is determined by reference to the activities that have taken place on the land and the uses to which the land has been put by the particular group. If lands are so occupied, there will exist a special bond between the group and the land in question such that the land will be part of the definition of the group's distinctive culture. Land held by virtue of aboriginal title may not be alienated because the land has an inherent and unique value in itself, which is enjoyed by the community with aboriginal title to it. The community cannot put the land to uses which would destroy that value." *Delgamuuk'w* v. *British Columbia.*

95 Ibid.

96 See Paul Willcocks, "Sechelt band to reject treaty that was hailed as 'Shining Light,'" *Vancouver Sun* (31 May 2000): A1; and Janet Steffenhagen and Paul Willcocks, "Treaty process on life support, chief says," *Vancouver Sun* (1 June 2000): A7.

97 Frank R. Baumgartner and Bryan D. Jones, *Agendas and Instability in American Politics* (Chicago: University of Chicago Press, 1993): 80.

98 Ibid., 26. See also 239-41.

99 Ibid., 32.

100 Ibid., 34.

101 On policy learning and the role of external crises, see Paul Sabatier and Hank Jenkins-Smith, "The Advocacy Coalition Framework: Assessment, Revisions, and Implications for Scholars and Practitioners," in Sabatier and Jenkins-Smith, eds., *Policy Change and Learning: An Advocacy Coalition Approach* (Boulder, CO: Westview 1993): 211-36.

102 On policy spillover, see Wyn Grant and Anne MacNamara, "When Policy Communities Intersect: The Cases of Agriculture and Banking," *Political Studies* 43 (1995): 509-15; and Hoberg and Morawski, "Policy Change through Sector Intersection."

103 On the sectoral and subsectoral linkage issue, see Michael Cavanagh, "Offshore Health and Safety Policy in the North Sea: Policy Networks and Policy Outcomes in Britain and Norway," in D. Marsh, ed., *Comparing Policy Networks* (Buckingham: Open University Press, 1998): 90-109; and Michael Cavanagh, David Marsh, and Martin Smith, "The Relationship between Policy Networks at the Sectoral and Sub-Sectoral Levels: A Response to Jordan, Maloney and McLaughlin," *Public Administration* 73 (Winter 1995): 627-33.

Chapter 6: Fine-Tuning the Settings

1 British Columbia, Ministry of Forests, "Tree Farm Licence 47: Rationale for Allowable Annual Cut Determination" (Victoria: Ministry of Forests, 1996), 59; an imperative to "consider" a "direction" is a minor masterpiece of bureaucratic obfuscation: where exactly does the buck stop here?

2 Ken Drushka, *In the Bight: The BC Forest Industry Today* (Madeira Park, BC: Harbour Publishing, 1999), 84-7.

3 Little-studied and shadowy bodies, TSA steering committees are described in Joan E. Vance, *Tree Planning: A Guide to Public Involvement in Forest Stewardship* (Vancouver: BC Public Interest Advocacy Centre, 1990), 36; their existence and importance was confirmed in personal interviews with MOF District Managers in 1990-1.

4 Jim Cooperman, personal communication, June 1999; David Peerla notes that the focus groups used by environmental organizations operate "two levels of abstraction" away from technical issues like timber supply planning (personal communication, June 1999).

5 *Evans Forest Products Ltd* v. *The Chief Forester of BC et al.* (1995) unreported (B.C.S.C.); *Western Canada Wilderness Committee* v. *Chief Forester* (1996), 62 A.C.W.S (3d) 779 (B.C.S.C.).

6 John Kingdon, *Agendas, Alternatives, and Public Policies*, 2nd edition (New York: HarperCollins, 1995), 71-7.

7 The phrase itself is Jeremy Wilson's in "Forest Conservation in British Columbia 1935-85: Reflections on a Barren Political Debate," *BC Studies* 76 (Winter 1987/8), 3-32; Jeremy Wilson, *Talk and Log* (Vancouver: UBC Press, 1998).

8 It could be argued that it takes us back a century or more to the debate over "timber famine" and the response of the US conservation movement, but here we will heed Kingdon's warning.

9 F.D. Mulholland, *The Forest Resources of British Columbia* (Victoria: King's Printer, 1937).

10 British Columbia, "Report of the Commissioner Relating to the Forest Resources of British Columbia, 1945," 2 vols. (Victoria: King's Printer, 1945).

11 The opposition of MacMillan and others is recorded in Ken Drushka, *H.R.: A Biography of H.R. MacMillan* (Madeira Park, BC: Harbour Publishing, 1995), 238-44 and 277-83.

12 British Columbia, "Report of the Commissioner Relating to the Forest Resources of British Columbia 1956," 2 vols. (Victoria: Queen's Printer 1956).

13 British Columbia, "Timber Rights and Forest Policy in British Columbia: Report of the Royal Commission on Forest Resources," 2 vols. (Victoria: Queen's Printer, 1976).

14 British Columbia, "Report of the Commissioner, 1956," vol. 1, 237.

15 British Columbia, "Timber Rights and Forest Policy," 85-7; Forest Resources Commission, "The Future of Our Forests" (Victoria: 1991), 39-40; Clark Binkley, "A Crossroad in the Forest: The Path to a Sustainable Forest Sector in BC," *BC Studies* 113 (1995): 39-61.

16 British Columbia, "Timber Rights and Forest Policy," 68, 260-2.

17 Sheldon Zakreski and Cory Waters, "Forestry Jobs and Timber" (Victoria: Sierra Club of BC, 1997).

18 Kingdon, *Agendas, Alternatives, and Public Policies*, 142.

19 O. Ray Travers, "Stewardship of Tree Farm Licenses 44 and 46 in the Proposed Management and Working Plans: An Evaluation," prepared for the Sierra Club of Western Canada, September 1991; Egan Ecological Services, "Environmental Impacts of Forest Management Practices in Tree Farm Licence 44 and Tree Farm Licence 46," prepared for the Sierra Club of Western Canada, September 1991.
20 Research in the US Pacific Northwest suggests that unmanaged forests may provide suitable habitat for most native species of forest-dwelling plants and animals in all three seral stages, but this is not necessarily true for the managed forests that are replacing them. See A.J. Hansen, T.A. Spies, F.J. Swanson, and J.L. Ohmann, "Conserving Biodiversity in Managed Forests: Lessons from Natural Forests," *BioScience* 41 (1991), 382-91; the management approach described therein found its way into the Coastal Biodiversity Guidelines (and later into the Biodiversity Guidebook), for example, in the objective of "planning harvesting activities to distribute a variety of seral stages and habitat patches across the landscape." See British Columbia, Ministry of Forests, "Guidelines to Maintain Biological Diversity in Coastal Forests" (Victoria: Ministry of Forests, 1992), 1.
21 Lois Dellert, "Sustained Yield Forestry in British Columbia: The Making and Breaking of a Policy (1900-1993)" (MES thesis, York University, 1994), 69-71.
22 *Reasons for Judgement of the Hon. Mr. Justice Smith between the Sierra Club and the Chief Forester, Appeal Board (A930623) and between the Province of British Columbia and David S. Cohen, Gary Bowden, Charles Gairns (A9223847)*, Supreme Court of BC, Vancouver, 22 December 1992.
23 HST Consortium, Forest Resources Commission, "A Summary of Technical Reviews of Forest Inventories and Allowable Annual Cut Determinations in BC," background papers, vol. 7 (Victoria: Forest Resources Commission, 1991); British Columbia, Ministry of Forests, "Review of the Timber Supply Analysis Process for BC Timber Supply Areas, Final Report," 2 vols. (Victoria: Ministry of Forests, 1991).
24 HST Consortium, "A Summary," 17.
25 British Columbia, Ministry of Forests, "Proposed Action Plan for the Implementation of Recommendations from the [Timber Supply Analysis] Report" (Victoria: Ministry of Forests, 1991).
26 R. Nixon, "Forestry and the End of Innocence," *Forest Planning Canada* 7,6 (November/December 1991): 21.
27 British Columbia, Ministry of Forests, "Timber Supply Review Backgrounder" available at www.for.gov.bc.ca/tsb/back/tsr/tsrbkg.htm.
28 David S. Cohen, "A Report on the Social and Economic Objectives of the Crown that the Minister of Forests Should Communicate to the Chief Forester in Relation to the Setting of Allowable Annual Cuts under Section 7 of the Forest Act" (N.p.: 28 July 1994).
29 Andrew Petter to John Cuthbert, Re: Economic and Social Objectives of the Crown, File 10100-01, 28 July 1994.
30 Trevor Jones, "A Critique of the Methods Used by the BC Government to Assess Socio-economic Implications of Forestry/Conservation Strategies," July 1994, available at www.imag.net/bcwild/jones.html; and compare the innovative social impact study carried out for the US Forest Ecosystem Management Assessment Team (FEMAT), "Forest Ecosystem Management: An Ecological, Economic and Social Assessment" (Washington, DC: US Government Printing Office, 1993-793-071).
31 For a succinct explanation of how this very large reduction came about, see George Hoberg, "From Localism to Legalism: The Transformation of Federal Forest Policy," in Charles Davis, ed., *Western Public Lands and Environmental Politics* (Boulder, CO: Westview Press, 1997), 55-60.
32 M. Patricia Marchak, Scott L. Aycock, Deborah M. Herbert, *Falldown: Forest Policy in British Columbia* (Vancouver: David Suzuki Foundation/Ecotrust Canada, 1999).
33 J.D. Nelson, "The Effect of Harvesting Guidelines on Timber Supply and Delivered Wood Costs," Canada-BC Partnership Agreement on Forest Resource Development (FRDA II), Economics and Social Analysis Program WP-6-001, February 1993; Andrew Petter to Larry Pederson, Re: The Crown's Economic and Social Objectives Regarding Visual Resources, File 16290-01, 26 February 1996.

34 Larry Pedersen, "The Truth Is Out There," address to the Annual Convention of the Northern Forest Products Association, Prince George, BC, 3 April 1997.
35 British Columbia, Ministry of Forests, "Kingcome Timber Supply Area: Rationale for Allowable Annual Cut Determination," 47. Effective 1 November 1996, available at www.for.gov.bc.ca/tsb/tsr1/ration/tsa/tsa33/httoc.htm.
36 For example, *Arrowsmith Timber Supply Area, Timber Supply Review Discussion Paper* (Victoria: Ministry of Forests, 1995), 8; where the timber supply projections provide more room for manoeuvre, mainly in TFLs, the target figure for decade over decade decline seems to be 10 percent.
37 TSR1 is given a failing grade in Sierra Legal Defence Fund, "British Columbia's Forestry Report Card, 1997-98" (Vancouver: 1998), and it is rejected as based on the wrong problem definition in Marchak et al., *Falldown.*
38 Forest Alliance of British Columbia, *Analysis of Recent British Columbia Government Forest Policy and Land Use Initiatives* (Vancouver: Price Waterhouse, 1995), 48. Note that this figure is strictly comparable with the 0.5 percent outcome as both include deciduous species and exclude the impact of new protected areas on AAC.
39 TSR 1 summary available at www.for.gov.bc.ca/tsb/tsr1.
40 This figure, based on the working assumptions used in TSR1, differs from the 40 percent figure cited in the Sierra Legal Defence Fund, "Report Card," where a later (and lower) MOF estimate of LTHL after withdrawal of new protected areas is used.
41 Memorandum from W. Dumont (Chief Forester, Western Forest Products) to Vancouver Island Resource Targets Working Group, 20 July 1995, file F36-Core HIA; the opposing point of view is found in Sierra Club of British Columbia, "Beyond Timber Targets: A Balanced Vision for Vancouver Island" (Victoria, June 1997).
42 The *locus classicus* of this argument in BC is Clark Binkley, "Preserving Nature through Intensive Plantation Forestry: The Case for Forest Land Allocation with Illustrations from British Columbia," *Forestry Chronicle* 73 (1997): 553-9; see also Jeremy Rayner, "Priority-Use Zoning: Sustainable Solution or Symbolic Politics?" in Chris Tollefson, ed., *The Wealth of Forests: Markets, Regulation, and Sustainable Forestry* (Vancouver: UBC Press, 1998), 232-54.
43 British Columbia, Ministry of Forests and Ministry of Environment, Lands and Parks, "Forest Practices Code Timber Supply Analysis" (Victoria: February 1996), 8. Available at www.for.gov.bc.ca/pab/publctns/fpctsa/tsa-toc.htm.
44 See the references to memoranda on Implementation of the Biodiversity Guidebook, 15 August 1995, and on Achieving Acceptable Biodiversity Timber Impacts, 25 August 1997, from the deputy ministers of the Ministry of Environment, Lands and Parks and the Ministry of Forests, in British Columbia, Ministry of Forests and Ministry of Environment, Land and Parks, "Release and Implementation of the Landscape Unit Planning Guidebook" (Victoria: 1999), ii.
45 www.for.gov.bc.ca/tsb/tsr2/tsr2.htm.
46 See above, note 29.
47 Drushka, *In the Bight*, 204-5.
48 Greg Utzig and Donna Macdonald, *Citizens' Guide to AAC Determinations: How to Make a Difference* (Vancouver: BCEN, 2000).
49 Interview with Glen Dunsworth, MacMillan Bloedel (now Weyerhaeuser), May 1994.
50 Larry Pedersen, personal communication, 14 July 2000.
51 On storylines in environmental policy, see John S. Dryzeck, *The Politics of the Earth: Environmental Discourses* (Oxford: Oxford University Press, 1997); and Maarten A. Hajer, *The Politics of Environmental Discourse: Ecological Modernization and the Policy Process* (Oxford: Clarendon Press, 1995). The link established between timber supply and zoning in order to break out of the zero sum game between timber and protected areas has many of the hallmarks of the ecological modernization story described in these works.
52 Peter Knoepfel and Ingrid Kissling-Naef, "Social Learning in Policy Networks," *Policy and Politics* 26,3 (1998), 355.
53 Ibid.
54 Ibid., 356.

55 The general implausibility of yield regulation as an instrument to achieve objectives other than timber targets is well described in Lois Dellert, "Sustained Yield: Why Has It Failed to Achieve Sustainability?" in Tollefson, *The Wealth of Forests.*
56 British Columbia, Ministry of Forests, "Timber Supply Analysis in British Columbia," available at www.for.gov.bc.ca/tsb/other/brochure/tsa_bro.htm.
57 Garry Wouters, *Shaping Our Future: BC Forest Policy Review* (Victoria: Office of the Jobs and Timber Accord Advocate, 2000), 31. Available at www.for.gov.bc.ca/pab/review/index.htm
58 Baumgartner and Jones, *Agendas and Instability*, 8.
59 British Columbia, Ministry of Forests, "Timber Supply Analysis in British Columbia," available at www.for.gov.bc.ca/tsb/other/brochure/tsa_bro.htm.

Chapter 7: Timber Pricing in British Columbia

1 We wish to thank Michael Whybrow, Graeme Auld, and Deanna Newsom for helpful comments on an earlier version of this chapter.
2 See Richard Schwindt and Terry Heaps, "Chopping Up the Money Tree: Distributing the Wealth from British Columbia's Forests" (Vancouver: David Suzuki Foundation, 1996), 31.
3 The stumpage system accounted for 94 percent of Ministry of Forest revenues in 1993. Other fees include royalty charges and annual rents (see Schwindt and Heaps, "Chopping Up the Money Tree": 33-4).
4 We use the term "timber pricing" to refer to resource rent policies, including stumpage. We use the term "market prices" to refer to what companies receive for timber in the marketplace.
5 Brian L. Scarfe, "Timber Pricing Policies and Sustainable Forestry," in Chris Tollefson, ed., *The Wealth of Forests: Markets, Regulation, and Sustainable Forestry* (Vancouver: UBC Press, 1998).
6 See Sierra Legal Defence Fund, "Profits or Plunder: Mismanagement of BC's Forests" (Vancouver: Sierra Legal Defence Fund, 1998).
7 Recent studies have posited a relationship between high stumpage and decreasing employment, and low stumpage and higher employment. See Janaki R.R. Alavalapati, Michael B. Percy, and Martin K. Luckert, "A Computable General Equilibrium Analysis of a Stumpage Price Increase Policy in British Columbia," *Journal of Forest Economics* 3,2 (1997): 143-70.
8 We also focus on other measures that influence price, such as raw log export restrictions. These restrictions play an indirect role as they deflate the market price of timber while, at the same time, encouraging further processing in the province. See Daowei Zhang, "Welfare Loss of Log Export Restrictions in British Columbia" (Auburn, AL: Forest Policy Center, School of Forestry and Wildlife Sciences Auburn University, 2000).
9 Changes made to the way government *calculates* stumpage that keeps the same policy instrument (e.g., administered pricing) we treat as setting changes.
10 A small-scale program that used competitive bidding was introduced in 1980 and expanded in 1987. However, even this program required bidders to meet employment and mill objectives, thus falling short of a purely competitive process.
11 A healthy forest industry is defined broadly to encompass Mackintosh's conception of the function of a staples economy that would support a number of economic development objectives, including profitability, employment, and revenues.
12 This term was first used by H.V. Nelles and is cited in Jeremy Wilson, *Talk and Log: Wilderness Politics in British Columbia* (Vancouver: UBC Press 1998), 91. As other chapters in this volume have also noted, this characterization mirrors Wilson's findings that, before the 1990s, forestry "discourse has been dominated by those most intensely involved in forest exploitation, by the interests of the development coalition – forest capital, forest labour and the government forest bureaucracy." See Jeremy Wilson, "Forest Conservation in British Columbia, 1935-85: Reflections on a Barren Political Debate," *BC Studies* 76 (Winter 1987/8): 3-32.
13 The high wages objective is treated narrowly to refer to decisions regarding resource rent instruments and settings that were directly designed to increase or address employment issues. As such, this objective does not refer to the belief/phenomenon that high profits will lead to high wages.

14 See also Figure 1.2 "BC Forest Industry Net Earnings."
15 Benjamin Cashore, "Flights of the Phoenix: Explaining the Durability of the Canada-US Softwood Lumber Dispute," *Canadian-American Public Policy* 32 (December 1997): 63.
16 See "Wilderness committee stumps for US industry," *Vancouver Sun* (18 October 1994): D3; Ron Wyden, Letter to Ambassador Burney Washington, DC (US Congress: 1992).
17 British Columbia, Ministry of Forests, "Stumpage and Royalty Changes" (Victoria: Ministry of Forests, Valuation Branch, 1994).
18 A competitive bidding/market-oriented instrument has difficulties coinciding with appurtenancy requirements for two reasons. First, these requirements necessarily limit the number of potential bidders, thus driving down the price. Second, and as a result, the US does not consider any competitive bidding with appurtenancy requirements as a measure of market value and would thus not address US complaints.
19 For a review and classification of economic development paradigms, see Melody Hessing and Michael Howlett, *Canadian Natural Resource and Environmental Policy: Political Economy and Public Policy* (Vancouver: UBC Press, 1996); and Stephen Brooks, *Public Policy in Canada: An Introduction,* 2nd ed. (Toronto: McClelland and Steward, 1993), 212.
20 Benjamin Cashore, "The Role of the Provincial State in Forest Policy: A Comparative Study of British Columbia and New Brunswick" (MA thesis, Carleton University, 1988).
21 Richard Schwindt, "The British Columbia Forest Sector: Pros and Cons of the Stumpage System," in Thomas Gunton and John Richards, eds., *Resource Rents and Public Policy in Western Canada* (Halifax: Institute for Research on Public Policy, 1987).
22 Harold A. Innis, *Problems of Staple Production in Canada* (Toronto: Ryerson Press, 1933); W.A. Mackintosh, "Economic Factors in Canadian History," in W.T. Easterbrook and M.H. Watkins, eds., *Approaches to Canadian Economic History* (Toronto: McClelland and Stewart, 1967).
23 A.R.M. Lower, ed., *The North American Assault on the Canadian Forest: A History of the Lumber Trade between Canada and the United States* (Toronto: The Ryerson Press, 1938).
24 British Columbia, "Report of the Fulton Royal Commission" (Victoria: Queen's Printer, 1910), 11.
25 Schwindt, "The British Columbia Forest Sector," 189.
26 Gordon Sloan, *Report of the Commissioner on the Forests and Resources of British Columbia* (Victoria: Charles F. Banfield, 1945).
27 See Cashore, "The Role of the Provincial State," 57, and Ian McAskill, "Public Charges and Private Values: A Study of British Columbia Timber Pricing" (MA thesis, Simon Fraser University, 1984).
28 Sloan, *Report,* 18 and 25.
29 The revenue shortfalls were largely owing to a population explosion in BC. The province witnessed a 700 percent increase between 1871 and 1911. See Patricia Marchak, *Green Gold: The Forest Industry in British Columbia* (Vancouver: UBC Press, 1983), 33.
30 Sloan, *Report,* 27.
31 Railway land grants were also given during this time as a way of promoting infrastructure development. Sloan says that, of the two forms of tenure, the lease was the more secure but necessitated a greater investment on account of the mill requirement (Sloan, *Report,* 29).
32 The public debt in 1904 was over $1.5 million, and the provincial government was forced to take out a loan from the Bank of Commerce. See Keith Reid and Don Weaver, "Aspects of the Political Economy of the BC Forest Industry," in P. Knox and P. Resnick, eds., *Essays in BC Political Economy* (Vancouver: New Star Books, 1974), 14-15. See also, W.A. Carrothers, "Forest Industries in British Columbia," in A.R.M. Lower, ed., *The North American Assault on the Canadian Forest: A History of the Lumber Trade between Canada and the United States* (Toronto: Ryerson Press, 1938), 29.
33 While satisfying the revenue objective, this policy instrument was clearly under the goal of maintaining a healthy forest industry. The province's Minister of Lands would note five years later that these temporary timber licences "let loose the flood of prosperity which the province has enjoyed ever since." Sloan, *Report,* 29.
34 British Columbia, "Timber Rights and Forest Policy in British Columbia: Report of the Royal Commission on Forest Resources," vol. 1 (Victoria: Queen's Printer, 1976), 28.

35 Cashore, "The Role of the Provincial State," 64.
36 The key distinguishing feature is that stumpage rates would be assessed to Crown lands licensed after 1912, but "royalty" charges would accrue to OTTs (see Sloan, *Report,* 30). These rate schedules are legislated and have remained unchanged over long periods. See McAskill, "Private Charges," 2.
37 British Columbia, "Timber Rights and Forest Policy," vol. 1, 28. An upset price was established and cutting rights would go to the firm with the highest bid. As Schwindt explains: "Forest Branch appraisers estimated both the value of the timber to be cut and the costs of the harvest; and by subtracting the latter from the former, they set a minimum or 'upset' price for the stand. The licences were allocated by bidding, with bids starting at the upset price" (see Schwindt, "The British Columbia Forest Sector," 191). As Pearse notes, the criteria used to determine this upset price were designed to yield a more discriminating estimate of the value of Crown timber than the royalties applicable to the old temporary tenures and some Grown grants. Pearse, *Timber Rights,* vol. 1, 28. Even before sustained yield came into effect, Schwindt notes upset price often became the actual price owing because economies of scale and transportation factors meant that there would often be only one bidder. See Schwindt, "The British Columbia Forest Sector," 192.
38 Old Temporary Tenure lands have gradually declined in importance. As of 1940, they still accounted for over one-quarter of the BC harvest. With the introduction of Tree Farm Licences, companies were given the incentive to merge their OTT lands into TFLs in exchange for increased cutting rights. As a result, about half of the OTTs were transformed into TFLs by 1966. As of the 1990s OTT royalty fees accounted for about 8 percent of the total volume harvested.
39 Pearse, *Timber Rights,* vol. 1, 28.
40 Cashore, "The Role of the State," 66.
41 This notion of sustained yield of timber supply comes from the German School of forestry and is different from a more holistic notion of forest sustainability, in which nontimber values are treated as important. See Benjamin Cashore, Ilan Vertinsky, and Rachana Raizada, "Firm Responses to External Pressures for Sustainable Forest Management in British Columbia and the US Pacific Northwest," in Debra J. Salazar and Donald K. Alper, eds., *Sustaining the Forests of the Pacific Coast: Forging Truces in the War in the Woods* (Vancouver: UBC Press, 2000); Greg Aplet, Nels Johnson, Jeffrey T. Olson, and V. Alaric Sample, eds., *Defining Sustainable Forestry* (Washington, DC: Island Press, 1993).
42 From Cashore, "The Role of the State," 66.
43 Schwindt, "The British Columbia Forest Sector," 194.
44 Royalties were still being assessed for the Old Temporary Tenures.
45 So strong were concerns that the existing licences gave too much security to firms that the second Sloan Commission was established in 1956. The (accepted) recommendation was that TFLs be limited to renewable twenty-one year terms. However, recommendations were not directed toward resource rent timber pricing issues.
46 Because the value of trees varies by species and region, the Forest Service was given the arduous task of deriving rates that corresponded to these differences, providing for fluctuations in stumpage across both time and space. In order to establish the market value (selling price) needed for stumpage calculations, timber harvest on the coast was calculated from the coastal log market while Interior calculations use the selling price of timber.
47 See David Haley "A Regional Comparison of Stumpage Values in British Columbia and the United States Pacific Northwest," *Forestry Chronicle* 56 (October 1980): 225-30; Marchak, *Green Gold,* 71; Laurence Copithorne, "Natural Resources anad Regional Disparities: A Skeptical View," *Canadian Public Policy* 5,2 (Spring 1979): 181-94.
48 See Wilson, *Talk and Log,* 125-9.
49 Peter Pearse, *Forest Tenures in British Columbia: Task Force on Crown Timber Disposal* (Victoria: British Columbia, Task Force on Crown Timber Disposal, 1974).
50 Pearse, *Timber Rights.*
51 Wilson, *Talk and Log,* 126.
52 Cashore, "Flights of the Phoenix," 63.

53 The argument was not that BC's prices were not low but that, even if they were, they were not giving "preferential" treatment to one industry. Under US trade law, "preferential" treatment must be proven before a subsidy dispute can proceed.
54 Kempf and Vander Zalm used this opportunity to announce that the government was reviewing BC's stumpage system. See James P. Groen, "British Columbia's International Relations: Consolidating a Coalition-Building Strategy," *BC Studies*, 102 (Summer 1994): 25-59. Vander Zalm said publicly that he hoped this would stall the countervail proceedings (ibid., 151), but it was clear that the government was not disappointed with the American countervail pressure.
55 Cashore, "The Role of the State," 140.
56 *Vancouver Sun* (5 February 1987).
57 There is some debate about the level of support Kempf had from the premier and cabinet in this regard, but there is evidence that the cash-strapped cabinet had deliberated over this strategy. Groen, "British Columbia's International Relations," quotes former Social Credit finance minister Mel Couvalier as saying "it wasn't like Jack Kempf pulled off a coup. In fact, shortly after taking office we worked out the strategy that we eventually used."
58 From Groen, "British Columbia's International Relations," interview with Jack Kempf, 29 August 1987.
59 After the passage of the new 1980 Forest Act, a Small Business Enterprise Program (SBEP) established competitive short-term timber sale licences in British Columbia. The SBEP was expanded in 1987 by taking a small percentage of the allowable annual cut away from major licence holders and was renamed the Small Business Forest Enterprise Program (SBFEP). Awards are made under two different categories: a "highest bid" category, where the timber sale goes to the highest monetary bidder, and a "bid proposal sale," where such issues as value added and other factors are taken into account. (Our thanks to Michael Whybrow for explaining this point.) See www.for.gov.bc.ca/RTE/woodlots/woodlot/ WOODLT1.HTM#sbfep.
60 An official from one large BC forest company notes that: "On the whole, BC has a pretty good [stumpage] system ... we like the stumpage system from the point of view that anybody can go and see where it is ... [We have had the] present system since the 1940s. [It has] gone through several royal commissions and a task force ... and ... the major thrust came from the United States [Countervail proceedings] in 1983. [I]t was examined with a fine tooth comb and it came through ... as clean as a whistle" (personal interview, Frank Howard, August 1987, Vancouver, BC).
61 Personal interviews, August 1987. Michael Percy and Christian Yoder, *The Softwood Lumber Dispute and Canada-US Trade in Natural Resources* (Halifax: Institute for Research on Public Policy, 1987), argued at this time that "it is not the formula that is at fault; it is the data used in the calculation that require attention" (73). Moreover, they note that historically industry profitability is poor compared with manufacturing industries in Canada and with US forest companies (74). Schwindt and Heaps, "Chopping Up the Money Tree," made a similar argument.
62 "If it is included in the stumpage system [it may end up affecting lumber not geared for the US]. We do not want to pay tariff to the US on stuff we're shipping to China." Personal interview, Frank Howard. August 1987.
63 In addition, the target rate system created what is known as the "waterbed effect." While Forest Service Valuation Branch officials would have discretion in lowering stumpage in a certain tract of land or region, rates would have to go up elsewhere to meet target objectives.
64 The target rates were initially set on 1 October 1987 as $10.59 per cubic metre on the coast and $8.59 per cubic metre in the Interior. As the official MOF explanation states, "Those target rates were set in consideration of a number of Provincial objectives, the most notable being the Government's desire to eliminate the federal tax on softwood lumber exported to the United States." See British Columbia, Ministry of Forests, "Stumpage Appraisal Information: Comparative Value Timber Pricing" (Victoria: 1991). Royalty fees (now covering 8 percent of the harvest) were also increased to about one-half of stumpage rates.
65 Pearse quoted in *Vancouver Sun* on 17 July 1987.

66 It must be noted that the high profits recorded in the forest sector are often owing to the cyclical nature of the industry. An extensive study by Richard Schwindt and Terry Heaps, "Chpping Up the Money Tree," found that the BC forest industry did not make excessive profits compared to other industries operating in the province.
67 A Price Waterhouse study found that the sawmill sector realized $1.2 billion in profits in 1993. It also found that the pulp and paper sector lost $700 million but appeared to be recovering.
68 See Steve Weatherbe, "Victoria says BC timber is underpriced," *Alberni Times* (12 April 1994).
69 Ibid.
70 Petter is reported to have said that "the government is selling public timber to the major forest companies for far less than what it is worth ... [L]umber prices and industry profits have skyrocketed in the last two years, without the BC public collecting its fair share of the rewards." See Weatherbe, ibid.
71 BC environmental groups were hindered somewhat by the US Commerce Department's 1991 ruling that BC's raw log export restrictions resulted in a 3.6% subsidy, more than half the alleged 6.51% subsidy finding. Stumpage prices were only found to account for a 2.91% subsidy. See Donald G. Balmer, "Escalating Politics of the US-Canadian Softwood Lumber Trade 1982-1992," *Northwest Environmental Journal* 9,1/2 (1993): 85-107. This provided a conundrum because environmental groups generally support raw log export restrictions as they believe this encourages value added production within the province. Yet removing raw log export restrictions had become a key goal of the US industry, making the environmental group/US timber industry alliance quite tenuous. Environmental groups were open to criticism that they were actually supporting removing raw log export restrictions, a charge levelled at them by BC's Minister of Forests. See "Protestors playing into hands of US, minister says," *Vancouver Sun* (20 October 1994): B10.
72 See Wilson, *Talk and Log,* 4.
73 The Ministry of Forests reports that "the concepts and magnitude of the stumpage and royalty increases were discussed at FSSC [Forest Sector Strategy Committee] ... with senior representatives of industry, labour and other groups ... but decisions were made by the government." See British Columbia, Ministry of Forests, "Stumpage and Royalty Changes" (Victoria: 1994), 3. The report also explained that "current and future cost increases, such as those expected because of the Forest Practices Code, were considered in arriving at these decisions" (ibid.).
74 Valuation Branch then implemented the government's timber pricing decisions. See British Columbia, Ministry of Forests, "Stumpage and Royalty Changes," 4.
75 Government documents explained that "the current structural change in the lumber market, driven by increased demand in the United States interacting with timber supply constraints (especially in the US PNW), is expected to persist. This has increased the value of the timber resource in British Columbia." See British Columbia, Office of the Premier, "British Columbia's Forest Renewal Plan: Working in Partnership" (Victoria: Government of British Columbia, 1994).
76 The Ministry of Forests explanation of the changes is clear: "If lumber prices fall below this range the average stumpage rate drops sharply, until at roughly US $250 per mbm and below there is no increase over pre-May 1, 1994 stumpage levels. Therefore, if lumber prices return to 1992 levels, the average stumpage rate will also return to 1992 levels." See British Columbia, Office of the Premier, "Working in Partnership," 5.
77 In fact, the government argued that the future of Forest Renewal was not related to fluctuating timber prices "because the fund is permanent, the amounts raised in good market periods should be sufficient to fund program expenditures through bad market periods, thereby protecting program expenditures from lumber market volatility." Ibid., 2.
78 The Ministry of Forests explained that "the increased costs due to the Forest Practices Code were considered in determining the magnitude of the increase in timber charges being implemented by the Province." Ibid.
79 See John Schreiner, "BC grabs $600 million from forest firms," *Financial Post* (15 April 1994): 1.

80 Miro Cernetig, "$2-billion BC forest tax to aid displaced workers: Companies accept levy for conservation, replanting," *Globe and Mail* (15 April 1994): A1, A5; Patricia Lush, "BC forest plan applauded: Unions, environmentalists, industry join to back proposal," *Globe and Mail* (16 April 1994): C1.
81 See Gordon Hamilton, "Foes close ranks for Timber Accord," *Vancouver Sun* (19 June 1997): D1, D6, who writes that the "accord had roots in falling profits."
82 Gordon Hamilton, "Desperate BC boosts overseas log shipments," *Vancouver Sun* (5 February 1998): D1, D18.
83 See Gordon Hamilton, "BC Stands to lose $1 billion in predicted forestry slump," *Vancouver Sun* (11 March 1998): A1, A2.
84 See "BC firm on stumpage cuts despite US concerns," *Canadian Press Wire Service* (19 May 1998). Indeed, "The minister said he will approve the cuts within weeks despite the fact that talks with the US agencies are ongoing without any sign of harmony."
85 See "Clark plans to defend stumpage cut: BC's premier is taking to Washington, DC," *Vancouver Sun* (7 July 1998): A3.
86 Government officials noted that the review was triggered by "unforeseen" cost increases owing to the Forest Practices Code and that reducing the stumpage rates were necessary to offset these costs, with the goal of increasing BC forest industry competitiveness and forest sector employment (personal interviews). The study referred to is KPMG, "Financial State of the Forest Industry and Delivered Wood Cost Drivers," prepared for the Ministry of Forests, April 1997.
87 Lush reports that companies called BC stumpage prices before the decreases the "highest costs in the world." See Lush, "BC said to slash stumpage fees," *Globe and Mail* (27 January 1998): B3.
88 The KPMG data was used as the basis for calculating that if the true impacts of the Forest Practices Code and other initiatives on companies' operational costs had been known, stumpage increases would have been $4.89 (Cdn) lower on average (personal interviews, government documents).
89 For the first time the new formula also included chip prices in determining the target rate's upward and downward moves as a result of the estimated Statistics Canada price.
90 The KPMG study was not asked to calculate differences between the Interior and coast in implementing the Code. However, the compromise deal between coastal and Interior operators over the level of stumpage decreases did make just such a distinction. An explanation for this compromise deal is traced not to differences in FPC impact but to different impacts on the coastal industry from the Asian collapse and the inability of coastal companies to redirect to the US market because of the quota system. Once the reduction was announced, companies were quick to point out its immediate effects in allowing them to increase their workforce. MB CEO Tom Stephens is reported to have said that the reductions resulted in the immediate rehiring of 1,000 employees. See Justine Hunter and David Hogben, "Stumpage cuts trigger rehiring in BC forests: The premier's initiative brings negative reaction from US lumber industry," *Vancouver Sun* (29 May 1998): A1-2.
91 It is uncertain whether changing to competitive bidding and eliminating raw log export restrictions would completely eliminate US countervail pressures. Some have argued that the US coalition may turn to other policy areas to assert alleged subsidies, such as environmental policy differences. See T.M. Apsey and J.C. Thomas, "Lessons of the Softwood Lumber Dispute: Politics, Protectionism and the Panel Process" (Vancouver: Council of Forest Industries and Thomas and Davis, 1997); and Cashore, "Flights of the Phoenix."
92 MacMillan Bloedel, "A White Paper for Discussion: Stumpage and Tenure Reform in BC" (Vancouver, BC: MacMillan Bloedel, 1998), 3.
93 See "It's time to cut BC logging costs says outgoing Fletcher Challenge president," *Financial Post* (10 December 1998). President Whitehead is quoted as saying, "Under the present stumpage system, there's no way the US would agree with cuts (to stumpage), so what I'm saying is you've got to cut the costs to make BC competitive ... The only way to do it is a transparent way: We've got to change the system to market bid. What I'm saying is to take some of this TFL and sell it off to be fee simple land."
94 COFI President Ron MacDonald argued that the Softwood Lumber Agreement should be

scrapped only after a new "free market" solution is in place. His comments were in response to Premier Clark's charge that, owing to constant scrutiny by US forest companies and difficulty with the current quota system, he might just scrap the deal. MacDonald said, "unless BC was to first adopt a free-market stumpage system [similar to the US, in which private landowners sell cutting rights to mills at market-sensitive rates] and address the issue of raw log export restrictions – two issues where the Americans claim BC is favouring its forest industry – the US government would likely respond with a countervailing duty on all Canadian lumber exports." Christopher Chipello, "MacMillan Bloedel's CEO pushes innovative programs: Asian Crisis has masked forest-product giant's restructuring efforts," *Wall Street Journal* (14 December 1998).

95 Gordon Hamilton, "Industry, business condemn Wouters' forestry report: Eco-groups say the long-awaited review takes the first cautious steps toward lower annual harvest," *Vancouver Sun* (5 April 2000): D1.

96 See Gerard Young, "Report stirs forest industry debate: Push to tie stumpage rates to market draws fire from both sides," *Times-Colonist* (6 April 2000): A3; and Gordon Hamilton, "Campbell's forestry solution ranges wide: More access to timber, less regulation, end to FRBC included as Liberal leader unveils plan for BC industry," *Vancouver Sun* (14 January 2000): D1.

97 Barry Gunn, "Business gives forest report a failing grade," *Nanaimo Daily News* (6 April 2000): A1.

98 See Gordon Hamilton, "'Team BC' softwood fight going to capitals," *Vancouver Sun* (17 June 1999): F1, F22; Peter Morton, "Top lumber executives want full free trade agreement: End years of fighting," *Financial Post* (13 July 1999). Frank Dottori, president of Montreal-based Tembec Inc., is quoted as saying, "We're trying to convince Canadian and US authorities that free trade in an open and competitive market is best all around."

99 Ken Drushka, "Sticking point for lumber deal: Canadian sawmill industry finally agrees on principles governing free trade with the US but the Americans are suspicious," *Vancouver Sun* (24 May 2000).

100 Drushka, "Sticking Point."

101 See Peter Morton, "US lobby group a front for lumber producers," *Financial Post* (6 June 2000).

102 See Gordon Hamilton, "Stumpage moves soon, MB says," *Vancouver Sun* (22 October 1998), from www.brucebressette.com/tenure13.htm. Stephens's apparent excitement led him to believe that instrument changes were imminent: "MacMillan Bloedel president Tom Stephens said Wednesday the forest industry and the BC government are working together on a new stumpage system that would satisfy American concerns that BC lumber is subsidized." "He said industry chief executive officers met with Premier Clark last week on the issue and were satisfied they have the government's commitment to make changes to the stumpage." Stephens might have been buoyed by Evan Lloyd, ADM of forest-sector initiatives 1998 statement that the government was considering more market-sensitive proposals for timber pricing settings that might address US concerns (ibid.).

103 British Columbia, Ministry of Forests, "Tenure and Pricing: Issues and Directions" (Victoria: Ministry of Forests, 1999).

104 Personal communication, Ministry of Forests officials, June 2000.

105 Gordon Hamilton, "Lumber deal backfired, report shows: The contentious softwood lumber agreement has led to a flood of Canadian logs – rather than lumber – being exported to the U.S., a new study on North American timber supplies shows," *Vancouver Sun* (8 June 2000) (from www.vancouversun.com/newsite/business/000608/4240843.html).

106 The IWA's response is quite thoughtful and offers several departures from the status quo, including the elimination of appurtenancy requirements from existing tenure agreements in order to promote efficient use of the forest resources. At the same time, the paper discusses a number of mechanisms that would maintain the employment objective within the timber pricing cycle and recommends a new type of residual stumpage system. IWA Canada, "Toward a Value-Added Lumber Industry: Tenure and Stumpage Reform in British Columbia" (Vancouver, BC: IWA Canada, 1999).

107 Sierra Legal Defence Fund, "Profits or Plunder," 12.
108 See Cashore, "Flights of the Phoenix," and Apsey and Thomas, "Lessons of the Softwood Lumber Dispute."
109 See Sierra Legal Defence Fund, "Profits or Plunder," 12.
110 See Gordon Hamilton, "Stumpage rules blamed for lack of alternatives to clearcuts: Interfor has developed a method of logging that leaves two-thirds of the forest intact," *Vancouver Sun* (17 October 1997): B4. "Provincial stumpage regulations are discouraging loggers from trying out alternatives to clearcutting, says of one of the province's leading foresters." The article quotes Ric Slaco, chief forester at International Forest Products, as saying, "The province collects the same stumpage revenues whether companies clearcut or try more costly but environmentally friendly logging methods ... The end result is that companies are shouldering all the burden of extra cost, a huge disincentive at a time when high operating costs are pushing forestry into the red."
111 For example, most environmental groups currently support the Suzuki Foundation's argument that "Lowering stumpage rates and giving companies a break on their Hydro bills is not going to restore this industry." See David Suzuki Foundation, report and Press Release, "Fundamental Change Needed for Forest Industry Survival" (Vancouver: David Suzuki Foundation, 1998). The report argues that "these types of concessions have been given in the past and look where we are today. Jobs are being cut, just like they were in the 1980s, and forestry-dependent communities are again in turmoil."

Chapter 8: "Don't Forget Government Can Do Anything"

1 British Columbia, "Forest Renewal Plan" (Victoria: 1994), 4.
2 An investment consultant's report shows benefits and wages for forest workers in Western Canada to be nearly $16 an hour, compared to $12 an hour for central Canada, $11 an hour in Sweden and the US west, and less than $8 an hour in the US south. See RBC Dominion Securities, "BC Forest Products Sector," 12 June 1998.
3 Personal communication, Garry Wouters, Jobs and Timber Accord Advocate, 13 June 2000; British Columbia, Office of the Jobs and Timber Accord Advocate, "Annual Report 1998/99" (Victoria: September 1999).
4 Jeremy Wilson, *Talk and Log: Wilderness Politics in British Columbia* (Vancouver: UBC Press, 1998), 270-3, 281.
5 British Columbia, Office of the Premier, "Forest Sector Committee to Help Guide Future of BC's Forest Industry," News Release, 2 April 1993.
6 Vaughn Palmer, "NDP goes out on a limb for forest workers," *Vancouver Sun* (28 March 1994): A10; Keith Baldrey, "Deal aims to bring peace to war in woods," *Vancouver Sun* (28 March 1994): B2.
7 British Columbia, "Forest Renewal Plan."
8 Forest Resources Commission, The Future of our Forests" (Victoria: April 1991), 28.
9 Jim Beatty and Gordon Hamilton, "NDP grab hints deficit near $1 billion," *Vancouver Sun* (13 September 1997): A1.
10 Patricia Lush, "BC Forest Plan Applauded," *Globe and Mail* (16 April 1994): B4.
11 British Columbia, "Forest Renewal Plan."
12 British Columbia, Ministry of Forests, "Annual Report, 1996/97" (Victoria: 1998).
13 British Columbia, "Forest Renewal Plan," 15.
14 Charles Lindbom, "The Science of Muddling Through," *Public Administration Review* 19 (1959): 79-88.
15 British Columbia, "Forest Renewal Plan," 15.
16 KPMG, "The Financial State of the Forest Industry."
17 Forest Renewal BC, "Business Plan, 2000-01" (Victoria: Forest Renewal BC, 2000).
18 Forest Renewal BC, "Five Year Report, 1994-1999" (Victoria: Forest Renewal BC, 1999).
19 Beatty and Hamilton, "NDP grab," *Vancouver Sun* (13 September 1997): A1.
20 Ibid.
21 The comment took on a semi-legendary status, even being dubbed by the province's leading political commentator as the quote of the year for 1996. Vaughn Palmer, "Can the winner 'Clark' his way out of the quote of the year?" *Vancouver Sun* (20 December 1996): A18.

22 Gordon Hamilton, "Bureaucracy eating up Forest Renewal funds," *Vancouver Sun* (17 January 1997): A1; Gordon Hamilton, "Zirnhelt aims to cut FRBC costs," *Vancouver Sun* (18 January 1997): B1.
23 Gordon Hamilton, "Clark backs off raiding forest fund," *Vancouver Sun* (25 January 1997): A1.
24 Gordon Hamilton, "$100 million siphoned from Forest Renewal fund," *Vancouver Sun* (26 March 1997): A5.
25 Forest Renewal BC, "Business Plan, 2000-01," table 7.
26 British Columbia, Forest Renewal BC, "Forest Renewal BC Restructuring Announced," News Release, 2 February 1999.
27 Compare Forest Renewal BC, "Business Plan 1998-99," 30, and "Business Plan 1999-2000," 27.
28 Gordon Hamilton, "Forest Renewal BC axing budget, 78 jobs," *Vancouver Sun* (3 February 1999): D1.
29 Justine Hunter, "Forest Renewal admits it will drain key funds," *Vancouver Sun* (27 May 1999): A1.
30 Justine Hunter, "FRBC 'rainy day' arrives; Liberals denounce plan," *Vancouver Sun* (28 May 1999): H1.
31 Justine Hunter, "Huge amount wasted by Forest Renewal BC," *Vancouver Sun* (27 May 1999): A8.
32 Ibid.
33 Gordon Hamilton, "Revamp stumpage, dump the FRBC, forest union urges," *Vancouver Sun* (16 June 1999): D1.
34 Gordon Hamilton, "FRBC, union accused of subverting program," *Vancouver Sun* (27 August 1999).
35 Justine Hunter, "Forest Renewal BC beset by confusion and waste," *Vancouver Sun* (17 July 1999): A1.
36 Office of the Auditor General of British Columbia, *Forest Renewal BC: Planning and Accountability in the Corporation: The Silviculture Programs* (Victoria: Queen's Printer, 1999), 37.
37 Auditor General, *Forest Renewal BC*, 116.
38 Duane E. Leigh, *Does Training Work for Displaced Workers* (Kalamazoo, MI: W.E. Upjohn Institute for Employment Research, 1990).
39 Justine Hunter, "Clark orders 21,000 forest jobs," *Vancouver Sun* (22 March 1996): A1.
40 British Columbia, "Jobs and Timber Accord: Summary," (Victoria: June 1996).
41 The study argued that the Oregon and Washington numbers should be discounted by the fact that 10 to 15 percent of their fibre was imported from other jurisdictions and that the Oregon and Washington figures should be compared only to BC coastal figures, not the provincial average. In BC, logging in the Interior is far more mechanized so there are fewer jobs per cubic metre harvested. See Gary Bowden, "A Preliminary Review of Estimates Used to Compare Employment in the British Columbia, Washington, and Oregon Forest Industries," prepared for the Council of Forest Industries, March 1996, available at www.cofi.org/employ.html.
42 Justine Hunter, "Faltering forest industry struggles to survive as job promise gathers dust," *Vancouver Sun* (10 March 1997): A1.
43 Justine Hunter and Gordon Hamilton, "Government forest firms near job deal, Clark confirms," *Vancouver Sun* (23 May 1997): A7.
44 Vaugh Palmer, "Clark's 'jobs accord' strong on rhetoric, shaky on numbers," *Vancouver Sun* (18 June 1997): A12.
45 British Columbia, Office of the Premier, "Jobs and Timber Accord to Create Thousands of New Jobs," News Release (Victoria: 19 June 1997).
46 British Columbia, "Jobs and Timber Accord: Summary," June 1996.
47 Ibid.
48 Justine Hunter, "Some forest firms balking at jobs deal, premier admits," Vancouver Sun (18 June 1997): A5.
49 Gordon Hamilton, "Clark pledges $1.5 million for 40,000 jobs," *Vancouver Sun* (20 June 1997): A1.

50 Gordon Hamilton, "Lignum takes renewal dollars," *Vancouver Sun* (21 June 1997): H3.
51 British Columbia, Office of the Premier, "Jobs and Timber Accord to Create Thousands of New Jobs," 5.
52 British Columbia, "Forest Jobs for BC (Update #2)," (Victoria: May 1998).
53 These job figures are estimates by the accord advocate, but the report does not fully explain how the estimates are derived.
54 Office of the Jobs and Timber Accord Advocate, "Annual Report, 1998-99."
55 British Columbia, Ministry of Forests, "Communities to Play Greater Role in Forest Management," News Release (3 December 1997). Information about this initiative is available at www.for.gov.bc.ca/pab/jobs/community/index.htm.
56 Gordon Hamilton, "Policy review won't fit time frame, foresters say," *Vancouver Sun* (9 September 1999). See www.for.gov.bc.ca/pab/review/index.htm.
57 Justine Hunter, "Huge amount wasted by Forest Renewal BC," *Vancouver Sun* (27 May 1999): A8.

Chapter 9: Conclusion

1 Jeremy Wilson, *Talk and Log: Wilderness Politics in British Columbia, 1965-96* (Vancouver: UBC Press, 1998); Benjamin Cashore, "Governing Forestry: Environmental Group Influence in British Columbia and the US Pacific Northwest" (PhD diss., University of Toronto, 1997).
2 Ken Lertzman, Jeremy Rayner, and Jeremy Wilson, "Learning and Change in the BC Forest Policy Sector," *Canadian Journal of Political Science* 29 (March 1996): 111-33.
3 On the links between public opinion and environmental group influence, see Kathryn Harrison, *Passing the Buck: Federalism and Canadian Environmental Policy* (Vancouver: UBC Press, 1996).
4 Council of Forest Industries, "British Columbia Forest Industry Fact Book: 1998" (Vancouver: Council of Forest Industries, 1999).
5 The concept of path dependence is developed in Margaret Weir, "Ideas and the Politics of Bounded Innovation," in S. Steinmo, K. Thelen, and F. Longstreth, eds., *Structuring Politics: Historical Institutionalism in Comparative Analysis* (Cambridge: Cambridge University Press, 1992).
6 On the structural power of business, see Charles Lindblom, *Politics and Market* (New York: Basic Books, 1977).
7 John Kingdon, *Agendas, Alternatives, and Public Policies*, 2nd ed. (New York: HarperCollins College Publishers, 1995).
8 Frank R. Baumgartner and Bryan D. Jones, "Agenda Dynamics and Policy Subsystems," *Journal of Politics* 53 (1991): 1044-74; Frank Baumgartner and Bryan Jones, *Agendas and Instability in American Politics* (Chicago: University of Chicago Press, 1993).
9 Ibid.
10 Lertzman et al., "Learning and Change."
11 Margaret Weir, "Ideas and the Politics of Bounded Innovation"; Paul Pierson, "When Effect Becomes Cause: Policy Feedback and Policy Change," *World Politics* 45,4 (1993): 595-628.
12 For a recent perspective of the forest industry on NDP policies and their consequences for the investment climate, see PricewaterhouseCoopers, *The BC Forest Industry: Unrealized Potential* (Vancouver, BC: PricewaterhouseCoopers, January 2000).
13 Philip H. Pollock III, Stuart A. Lilie, and M. Elliot Vittes, "Hard Issues, Core Values and Vertical Constraint: The Case of Nuclear Power," *British Journal of Political Science* 23,1 (1993): 29-50.
14 George Hoberg, "How the Way We Make Policy Governs the Policy We Make," in Debra J. Salazar and Donald K. Alper, eds., *Sustaining the Forests of the Pacific Coast: Forging Truces in the War in the Woods* (Vancouver: UBC Press, 2000).
15 See, for example, US Department of Agriculture, Forest Services, "Ecosystem Management: 1993 Annual Report of the Forest Service" (Washington, DC: US Department of Agriculture, Forest Service, 1994); US Department of Agriculture Committee of Scientists, "Sustaining the People's Lands: Recommendations for

Stewardship of the National Forests and Grasslands into the Next Century" (Washington, DC: US Department of Agriculture, 1999); US General Accounting Office, "Forest Service Priorities: Evolving Mission Favors Resource Protection Over Production," GAO/RCED-99-166 (Washington, DC: General Accounting Office, 1999).

16 Benjamin Cashore, "US Pacific Northwest," in Bill Wilson, G. Cornelis van Kooten, Ilan Vertinsky, Louise Arthur, *Forest Policy: International Case Studies* (Oxon, UK: CABI Publishing, 1999); George Hoberg, "Distinguishing Learning from Other Sources of Policy Change: The Case of Forestry in the Pacific Northwest," prepared for delivery at the Annual Meeting of the American Political Science Association, Boston, MA, 3-6 September 1998.

17 George Hoberg and Edward Morawski, "Policy Change through Sector Intersection: Forest and Aboriginal Policy in Clayoquot Sound," *Canadian Public Administration* 40,3 (1997): 387-414.

18 Just after being launched, the review process was denounced in an uncharacteristically harsh manner by the Association of British Columbia Professional Foresters. The ABCPF purchased space in the province's leading newspapers to ensure that its "open letter to the premier" criticizing the review would get wide exposure. It stated: "We fear it is too prescriptive, narrow in scope and time limited to achieve anything approaching public consensus. We are particularly concerned by the intention to hold invitation-only community workshops – an approach inconsistent with openness and inclusiveness." Nick Arkle, President, ABCPF, Open Letter to Premier Dan Miller, 8 September 1999.

19 British Columbia, "Shaping Our Future: BC Forest Policy Review" (Victoria: Queen's Printer, 2000).

20 See Gordon Hamilton, "Industry, business condemn Wouters' forestry report: Eco-groups say the long-awaited review takes the first cautious steps toward lower annual harvest," *Vancouver Sun* (5 April 2000): D1.

21 Council of Forest Industries, "Blueprint for Competitiveness," available at www.cofi.org (accessed 26 June 2000).

22 Cheri Burda, "Forest policy needs real change, not tinkering," *Times-Colonist* (9 April 2000).

23 Gordon Hamilton, "Enviro-briefs counter views of forest industry," *Vancouver Sun* (23 November 1999).

24 Address by Gordon Campbell, leader of the official opposition, Truck Loggers Convention, Vancouver, BC, 13 January 2000.

25 Gordon Hamilton, "Campbell's forestry solution ranges wide: More access to timber, less regulation, end to FRBC included as Liberal leader unveils plan for BC industry," *Vancouver Sun* (14 January 2000): D1.

26 Environics Research Group, "Forest Policy Review – 2000: Public Attitude Survey," unpublished.

27 Gordon Hamilton, "Coastal loggers seek eco-truce: 'Significant' industry proposal would halt much of the logging on the remote north and central coast, environmentalists say," *Vancouver Sun* (16 March 2000).

28 Gordon Hamilton, "Eco-groups restart campaign against forest producers," *Vancouver Sun* (2 June 2000); Gordon Hamilton, "All old-growth logging will be opposed, coalition warns," *Vancouver Sun* (19 June 2000).

29 Canada NewsWire, "Forest Companies and Environmental Groups Pursue Unprecedented Solutions Initiative," 28 July 2000, www.newswire.ca/releases/July2000/28/ c7136.html.

30 George Hoberg, "The Coming Revolution in Regulating Our Forests," *Policy Options* 20,10 (December 1999): 53-6; Errol Meidinger, "Look Who's Making the Rules: International Environmental Standard Setting by Non-Governmental Organizations," *Human Ecology Review* 4,1 (1997): 52-4; Benjamin Cashore, "Competing for Legitimacy: Globalization, Sustainability, and the Emergence of Private Governance (Certification Programs) in the US and Canadian Forest Sectors," Auburn University Forest Policy Center Internal Working Paper 102 (July 2000).

31 James Brooke, "Loggers find Canada rain forest flush with foes," *New York Times* (22 October 1999).

Bibliography

Abel, Kerry, and Jean Friesen, eds. *Aboriginal Resource Use in Canada: Historical and Legal Aspects*. Winnipeg: University of Manitoba Press 1991

Ainscough, G.L. "Comment." In *Recent Forest Policy Developments in Manitoba, Saskatchewan, Alberta and British Columbia.* Proceedings of the National Forest Management Group Meeting held at Edmonton, Alberta. Montreal: Canadian Pulp and Paper Association 1975

Alavalapati, Janaki R.R., Michael B. Percy, and Martin K. Luckert. "A Computable General Equilibrium Analysis of a Stumpage Price Increase Policy in British Columbia." *Journal of Forest Economics* 3,2 (1997): 143-70

Anderson, James E. *Public Policymaking: An Introduction.* Boston, MA: Houston Mifflin 1994

Angus, Murray. "Comprehensive Claims: One Step Forward, Two Steps Back." In *Nation to Nation: Aboriginal Sovereignty and the Future of Canada*, ed. D. Engelstad and J. Bird, 60-6. Toronto: Anansi 1992

Aplet, Greg, Nels Johnson, Jeffrey T. Olson, and V. Alaric Sample, eds. *Defining Sustainable Forestry*. Washington, DC: Island Press 1993

Apsey, T.M., and J.C. Thomas. "Lessons of the Softwood Lumber Dispute: Politics, Protectionism and the Panel Process." Vancouver: Council of Forest Industries and Thomas and Davis 1997

Armstrong, Gerry, and Tom Gunton. "Implementation of Biodiversity Guidebook." Memorandum to regional managers, Ministry of Forests, and regional directors, Ministry of Environment, Lands and Parks, British Columbia 1995

Armstrong, S. Bradley. "Defining the Boundaries of Aboriginal Title after Delgamuuk'w." *Resources* 62 (1998): 2-7

Atkinson, Michael M., and William D. Coleman. *The State, Business, and Industrial Change in Canada*. Toronto: University of Toronto Press 1989

Bachrach, Peter, and Morton S. Baratz. "Decisions and Nondecisions: An Analytical Framework." *American Political Science Review* 2 (1963): 632-42

–. "Power and Its Two Faces Revisited: A Reply to Geoffrey Debnam." *American Political Science Review* 69,3 (1975): 900-4

–. *Power and Poverty: Theory and Practice.* New York: Oxford University Press 1970

Balmer, Donald G. "Escalating Politics of the US-Canadian Softwood Lumber Trade 1982-1992." *Northwest Environmental Journal* 9, 1/2 (1993): 85-107

Bankes, N. "Indian Resource Rights and Constitutional Enactments in Western Canada, 1871-1930." In *Law and Justice in a New Land: Essays in Western Canadian Legal History*, ed. L. Knafla. Toronto: Carswell 1986

Banting, Keith. *The Welfare State and Canadian Federalism.* Kingston: McGill-Queen's University Press 1987

Banting, Keith G., and Richard Simeon. *And No One Cheered: Federalism, Democracy and the Constitution Act.* Toronto: Methuen 1983

Bartlett, R.H. *Indian Reserves in Quebec*. Studies in Aboriginal Rights, no. 8. Saskatoon: University of Saskatchewan Native law Centre 1984

–. "Resource Development and Aboriginal Title in Canada." In *Resource Development and Aboriginal Land Rights*, ed. Bartlett. Calgary: Canadian Institute of Resources Law 1991

Baskerville, G.L. *Forest Practices Code: Summary of Presentations by Stakeholder Groups – A Report to the Ministry of Forests*. Victoria: Queen's Printer 1993

Baumgartner, Frank R., and Bryan D. Jones. "Agenda Dynamics and Policy Subsystems." *Journal of Politics* 53 (1991): 1044-74

–. *Agendas and Instability in American Politics*. Chicago: University of Chicago Press 1993

Berger, Thomas. "Native History, Native Claims and Self-Determination." *BC Studies* 57 (1983): 10-23

Bernsohn, Ken. "Call It a Giant Turf War: BC's Debate over Public v. Private Control of Forests Is Renewed," *Western Report*, 4,2 (January 1989): 42

–. *Cutting Up the North: History of the Forest Industry in the Northern Interior*. Vancouver: Hancock House 1981

Bernstein, Steven, and Benjamin Cashore. "Globalization, Four Paths of Internationalization and Domestic Policy Change: The Case of Ecoforestry in British Columbia, Canada." *Canadian Journal of Political Science* 33,1 (2000): 67-99

Binkley, Clark S. "A Crossroad in the Forest: The Path to a Sustainable Forest Sector in BC." *BC Studies* 113 (1995): 39-68

–. "Preserving Nature through Intensive Plantation Forestry: The Case for Forestland Allocation with Illustrations from British Columbia," *Forestry Chronicle* 73,5 (September/October 1997): 553-59 Birkland, Thomas. *After Disaster: Agenda-Setting, Public Policy and Focusing Events*. Washington: Georgetown University Press 1997

Blake, Donald E., R. Kenneth Carty, and Lynda Erickson. *Grassroots Politicians*. Vancouver: UBC Press 1993

Blake, Donald E., Neil Guppy, and Peter Urmetzer. "Canadian Public Opinion and Environmental Action in Canada: Evidence from British Columbia." *Canadian Journal of Political Science* 30,3 (1997): 451-72

Block, Fred. "The Ruling Class Does Not Rule: Notes on the Market Theory of the State." In *Foundations of Analytical Marxism*, ed. J.E. Roemer. Aldershot: International Library of Critical Writings in Economics 1994

Bowden, Gary. "A Preliminary Review of Estimates Used to Compare Employment in the British Columbia, Washington, and Oregon Forest Industries." Vancouver: Council of Forest Industries of BC 1996

Bowker, Andrea. "Sparrow's Promise: Aboriginal Rights in the BC Court of Appeal." *University of Toronto Law Review* 53,1 (1997): 1-48

British Columbia. "Alternatives for Crown Timber Pricing: A White Paper for Discussion Purposes." Victoria: Ministry of Forests 1979

–. Claims Task Force. "Report." Vancouver: British Columbia Claims Task Force 1991

–. Commission on Resources and the Environment. "1992-93 Annual Report to the Legislative Assembly." Victoria: Commission on Resources and the Environment 1993

–. "Final Report of the Royal Commission of Inquiry on Timber and Forestry 1909-1910." Victoria: King's Printer 1910

–. "Five Year Forest and Range Resource Program 1985-1990." Victoria: Ministry of Forests 1985

–. "Forest Jobs for BC (Update #2)." Victoria: British Columbia 1998

–. *Forest Management Review: British Columbia*. Victoria: Ministry of Forests and Lands 1987

–. "Forest Planning Framework." Victoria: Ministry of Forests 1983

–. "Forest Policy Review: A Summary of Major Decisions." Victoria: Ministry of Forests and Lands 1987

–. Forest Practices Board. *Forest Planning and Practices in Coastal Areas with Stream*. Victoria: Forest Practices Board 1998

–. Forest Practices Board. "Forest Practices Board: 1997 Annual Report." Victoria: Forest Practices Board 1998

–. Forest Practices Board. "An Audit of the Government of British Columbia's Framework for Enforcement of the Forest Practices Code." Victoria: Forest Practices Board 1999
–. "Forest Practices on Private Forest Land, Summary of Standards and Administration." 1999. www.for.gov.bc.ca/PAB/News/privateland/pflregulations.htm
–. Forest Renewal BC. "Forest Renewal BC Restructuring Announced." Forest Renewal BC 1999
–. "Forest Renewal Plan." Victoria: British Columbia 1994
–. Forest Resources Commission. "Background Papers: Volume 3." Victoria: Forest Resources Commission 1991
–. Forest Resources Commission. "The Future of Our Forests, Executive Summary." Victoria: Queen's Printer 1991
–. *Incentives for Intensive Forest Management: A White Paper for Discussion Purposes*. Victoria: Ministry of Forests 1979
–. "Jobs and Timber Accord." Victoria: British Columbia, 1997
–. Land Use Coordination Office. "Land-use Planning in British Columbia: Making a World of Difference." Victoria: Land Use Coordination Office 2000
–. Land Use Coordination Office. "A Protected Areas Strategy for British Columbia: The Protected Areas Component of BC's Land Use Strategy." Victoria: Land Use Coordination Office 1996
–. Land Use Coordination Office. "Special Management Zone Project: Information Report." Victoria: Land Use Coordination Office 1998
–. Land Use Coordination Office. "Status Report on Land and Resource Management Plans." Victoria: Land Use Coordination Office 2000
–. Legislative Assembly. "Commissioner on Resources and Environment Act." Victoria: Legislative Assembly 1992
–. Legislative Assembly. Select Standing Committee on Aboriginal Affairs. "Towards Reconciliation: Nisga'a Agreement-in-Principle and British Columbia Treaty Process." Victoria: Legislative Assembly, Select Standing Committee on Aboriginal Affairs 1997
–. Ministry of Aboriginal Affairs. "The Benefits and Costs of Treaty Settlements in British Columbia: A Summary of the KPMG Report." Victoria: Ministry of Aboriginal Affairs 1996
–. Ministry of Aboriginal Affairs. "Present Status of the BCTC Process, December." Victoria: Ministry of Aboriginal Affairs 1997
–. Ministry of Aboriginal Affairs. "Province of British Columbia Crown Land Activities and Aboriginal Rights Policy Framework." Victoria: Ministry of Aboriginal Affairs 1997
–. Ministry of Environment, Lands and Parks, and Ministry of Forests. "Revisions Make Forest Practices Code More Efficient, Effective." Victoria: Ministry of Environment et al. 1997
–. Ministry of Forests. "Annual Report 1995/96." Victoria: Ministry of Forests 1997
–. Ministry of Forests. "Annual Report 1996/97." Victoria: Ministry of Forests 1998
–. Ministry of Forests. "British Columbia Forest Practices Code Discussion Paper." Victoria: Ministry of Forests 1993
–. Ministry of Forests. "British Columbia's First Four Community Forest Pilot Sites Announced." Victoria: Ministry of Forests 1999
–. Ministry of Forests. "Communities to Play Greater Role in Forest Management." Victoria: Ministry of Forests 1997
–. Ministry of Forests. "Forest Land Commission to Oversee Private Land Logging." Victoria: Ministry of Forests 1999
–. Ministry of Forests. "Guidelines to Maintain Biological Diversity in Coastal Forests." Victoria: Ministry of Forests 1992
–. Ministry of Forests. "Meeting the Challenge: Guide to the Ministry of Forests Organization and Programs." Victoria: Ministry of Forests 1999
–. Ministry of Forests. "Moving Towards a Results-Based, Incentives-Driven Code." Victoria: Ministry of Forests 1998
–. Ministry of Forests. "News Release, Permanent Forest Resources Commission Established." Victoria: Ministry of Forests 1989

–. Ministry of Forests. *Okanagan Timber Supply Area, Integrated Resource Management Harvesting Guideline*. Victoria: Ministry of Forests 1992
–. Ministry of Forests. "Parks Settlement Agreement: MacMillan Bloedel." Victoria: Ministry of Forests 1999
–. Ministry of Forests. "Proposed Action Plan for the Implementation of Recommendations from the [Timber Supply Analysis] Report." Victoria: Ministry of Forests 1991
–. Ministry of Forests. "Proposed Policy and Procedures for the Replacement of Major Volume Based Tenures with Tree Farm Licences." Victoria: Ministry of Forests 1988
–. Ministry of Forests. "Review of the Timber Supply Analysis Process for BC Timber Supply Areas, Final Report." Victoria: Ministry of Forests 1991
–. Ministry of Forests. "Stumpage Appraisal Information: Comparative Value Timber Pricing." Victoria: Ministry of Forests 1991
–. Ministry of Forests. "Stumpage and Royalty Changes." Victoria: Ministry of Forests, Valuation Branch 1994
–. Ministry of Forests. "Summary of Public Input." Review of the 1986 roll-over public discussion. Victoria: Ministry of Forests 1986
–. Ministry of Forests. "Tenure and Pricing: Issues and Directions." Victoria: Ministry of Forests 1999
–. Ministry of Forests. "Tree Farm Licence 47: Rationale for Allowable Annual Cut Determination." Victoria: Ministry of Forests 1996
–. Ministry of Forests. "Tree Farm Licences in British Columbia." Victoria: Ministry of Forests 1988
–. Ministry of Forests, Compliance and Enforcement Branch. "The Annual Report of Compliance and Enforcement Statistics for the Forest Practices Code." For the years 1996, 1997, and 1998. Victoria: Ministry of Forests 1998
–. Ministry of Forests and Lands. "News Release: Major Shift in Forest Policy for British Columbia." Victoria: Ministry of Forests and Lands 1987
–. "Ministry of Forests Management System." Victoria: Ministry of Forests 1983
–. Ministry of Forests and Ministry of Environment, Lands and Parks. "Forest Practices Code Amendments Reduce Costs, Maintain Environmental Standards." Victoria: Ministry of Forests and Ministry of Environment, et al. 1998
–. Ministry of Forests and Ministry of Environment, Lands and Parks. "Forest Practices Code of British Columbia: Biodiversity Guidebook." Victoria: Queen's Printer 1995
–. Ministry of Forests and Ministry of Environment, Lands and Parks. "Forest Practices Code of British Columbia: Riparian Management Area Guidebook." Victoria: Queen's Printer 1995
–. Ministry of Forests and Ministry of Environment, Lands and Parks. "Forest Practices Code Timber Supply Analysis." Victoria: Ministry of Forests and Ministry of Environment, et al. 1996
–. Ministry of Forests and Ministry of Environment, Lands and Parks. "New Mandatory Standards to Improve BC's Forest Practices." Victoria: Ministry of Forests and Ministry of Environment et al. 1994
–. Ministry of Forests and Ministry of Environment, Lands and Parks. "Release and Implementation of the Landscape Unit Planning Guide." Victoria: Ministry of Forests and Ministry of Environment et al. 1999
–. Ministry of Forests and Ministry of Environment, Lands and Parks. "Summary of Public Responses to Parks and Wilderness for the 90s." Victoria: Ministry of Forests and Ministry of Lands and Parks 1991
–. Ministry of Forests Public Affairs Branch. "Communication Plan – Subject: Cabinet Submission on Biodiversity Letter." Victoria: 1997
–. "New Directions for Forest Policy in British Columbia." Victoria: Ministry of Forests and Lands 1987
–. Office of the Auditor General. *Forest Renewal BC: Planning and Accountability in the Corporation: The Silviculture Programs*. Victoria: Queen's Printer 1999
–. Office of the Jobs and Timber Accord Advocate. "Annual Report 1998/99." Victoria: Office of the Jobs and Timber Accord Advocate 1999

–. Office of the Premier. "British Columbia's Forest Renewal Plan: Working in Partnership." Victoria: Government of British Columbia 1994

–. Office of the Premier. "Clark Announces Protection of Northern Rockies." Victoria: Government of British Columbia 1997

–. Office of the Premier. "Forest Sector Committee to Help Guide Future of BC's Forest Industry." Victoria: Government of British Columbia 1993

–. Office of the Premier. "Jobs and Timber Accord to Create Thousands of New Jobs." Victoria: Government of British Columbia 1997

–. "Report of the Commissioner Relating to the Forest Resources of British Columbia." Victoria: King's Printer 1945

–. "Report of the Commissioner Relating to the Forest Resources of British Columbia." Victoria: Queen's Printer 1956

–. "Final Report of the Royal Commission of Inquiry on Timber and Forestry, 1909-10." Royal Commission on Timber and Forestry. Victoria: R. Wolfenden n.d.

–. "Sechelt Agreement-in-Principle." Victoria: Ministry of Aboriginal Affairs 1999

–. "Shaping Our Future: BC Forest Policy Review." Victoria: Queen's Printer 2000

–. "The Small Business Program: A White Paper for Discussion Purposes." Victoria: Ministry of Forests 1979

–. "Timber Rights and Forest Policy in British Columbia: Report of the Royal Commission on Forest Resources." 2 vols. Victoria: Queen's Printer 1976

–. "Written Submissions Received in Support of Presentations Made at Public Information Session on the Proposed Policy and Procedures for the Replacement of Major Volume-Based Tenures with Tree Farm Licences – Parksville, British Columbia." Victoria: Ministry of Forests 1989

British Columbia Chiefs of the Central Region of the Nuu-chah-nulth Tribal Council. "Province of British Columbia and Nuu-chah-nulth Tribal Council Interim Measures Agreement for Clayoquot Sound (Brief)." Victoria: Government of British Columbia 1994

Brock, Kathy L. "The Politics of Aboriginal Self-Government: A Canadian Paradox," *Canadian Public Administration* 34 (1991): 272-86

Stephen Brooks. *Public Policy in Canada: An Introduction*. 2nd ed. Toronto: McClelland and Stewart 1993

Bunnell, Fred. "Next Time Try Data: A Plea for Variety in Forest Practices." Working Paper Vancouver: UBC Centre for Conservation Biology 1999

Burda, Cheri, Deborah Curran, Fred Gale, and Michael M'Gonigle. *Forests in Trust: Reforming British Columbia's Forest Tenure System for Ecosystem and Community Health*. Victoria: Eco-Research Chair in Environmental Law and Policy 1997

Burda, Cheri, Fred Gale, and Michael M'Gonigle. "Eco-Forestry versus the State(us) Quo: Or Why Innovative Forestry Is Neither Contemplated Nor Permitted within the State Structure of British Columbia," *BC Studies* 119 (1998): 45-72

Cail, Robert Edgar. *Land, Man and the Law: The Disposal of Crown Lands in British Columbia, 1871-1913*. Vancouver: UBC Press 1974

Canada. British Columbia, Nisga'a Nation. "Nisga'a Final Agreement." Ottawa: Federal Treaty Negotiation Office 1998

–. *Comprehensive Land Claim Umbrella Final Agreement between the Government of Canada, the Council for Yukon Indians and the Government of the Yukon*. Ottawa: Indian and Northern Affairs Canada 1990

–. Indian and Northern Affairs. *Agreement in Principle for the Nunavut Settlement Area*. Ottawa: Indian and Northern Affairs Canada 1992

–. Indian and Northern Affairs. "Background Paper: The Comprehensive Land Claim Negotiations of the Labrador Inuit, the Government of Newfoundland and Labrador, and the Government of Canada." Ottawa: Indian and Northern Affairs Canada 1990

–. Indian and Northern Affairs. *Comprehensive Land Claim Agreement between Her Majesty the Queen in Right of Canada and the Dene of Colville Lake, Déline, Fort Good Hope and Fort Norman and the* Métis *of Fort Good Hope, Fort Norman and Norman Wells in the Sahtu Region of the Mackenzie Valley as Represented by the Sahtu Tribal Council*. Ottawa: Indian and Northern Affairs Canada 1993

–. Indian and Northern Affairs. *Gwich'in Comprehensive Land Claim Agreement.* Ottawa: Indian and Northern Affairs Canada 1992
–. Indian and Northern Affairs. *The Western Arctic Claim: The Inuvialuit Final Agreement.* Ottawa: Indian and Northern Affairs Canada 1984
–. *Indian Treaties and Surrenders from 1680 to 1890.* 3 vols. Ottawa: Queen's Printer 1905
–. *Northeastern Quebec Agreement.* Ottawa: Information Canada 1978
–. *Sechelt Indian Band Self-Government Act.* Ottawa: Queen's Printer 1986
Canadian Council of Forest Ministers. "National Forest Strategy (1998-2003). Sustainable Forests: A Canadian Commitment." Ottawa: Canadian Council of Forest Ministers 1998
Canadian Council of Resource and Environment Ministers Task Force on Forest Policy. *Forest Policies in Canada: Major Objectives, Problems and Trends.* Vol. 1. Montreal: Canadian Council of Resource and Environment Ministers 1976
Careless, Ric. Foreword to *Keeping the Special in Special Management Zones,* by Jim Cooperman. Gibsons, BC: BC Spaces for Nature 1998
Carroll, William. *Organizing Dissent: Contemporary Social Movements in Theory and Practice.* Toronto: Garamond Press 1997
Carrothers, W.A. "Forest Industries in British Columbia." In *The North American Assault on the Canadian Forest: A History of the Lumber Trade between Canada and the United States,* ed. A.R.M. Lower, 227-343. Toronto: Ryerson Press, 1938
Cashore, Benjamin. "Competing for Legitimacy: Globalization, Sustainability, and the Emergence of Private Governance (Certification Programs) in the US and Canadian Forest Sectors." Auburn University Forest Policy Center Internal Working Paper 102 (July 2000)
–. "Flights of the Phoenix: Explaining the Durability of the Canada-US Softwood Lumber Dispute," *Canadian-American Public Policy* 32 (December 1997): 1-63
–. "Governing Forestry: Environmental Group Influence in British Columbia and the US Pacific Northwest." PhD diss., University of Toronto 1997
–. "The Role of the Provincial State in Forest Policy: A Comparative Study of British Columbia and New Brunswick." MA thesis, Carleton University 1988
–. "US Pacific Northwest." In *Forest Policy: International Case Studies,* ed. Bill Wilson, G. Cornelis van Kooten, Ilan Vertinsky, and Louise Arthur. Oxon, UK: CABI Publishing 1999
Cashore, Benjamin, Ilan Vertinsky, and Rachana Raizada. "Firm's Responses to External Pressures for Sustainable Forest Management in British Columbia and the US Pacific Northwest." In *Sustaining the Pacific Coast Forests: Forging Truces in the War in the Woods,* ed. Debra J. Salazar and Donald K. Alper. Vancouver: UBC Press 2000
Cassidy, Frank. "Aboriginal Land Claims in British Columbia." In *Aboriginal Land Claims in Canada: A Regional Perspective,* ed. K. Coates. Toronto: Copp Clark Pitman 1992
–, ed. *Aboriginal Title in British Columbia:* Delgamuuk'w *v.* The Queen. Vancouver: Oolichan Books/Institute for Research on Public Policy 1992
–, ed. *Reaching Just Settlements: Land Claims in British Columbia.* Vancouver: Oolichan Books/Institute for Research on Public Policy 1991
Cater, Douglas. *Power in Washington.* New York: Random House 1964
Cavanagh, Michael. "Offshore Health and Safety Policy in the North Sea: Policy Networks and Policy Outcomes in Britain and Norway." In *Comparing Policy Networks,* ed. D. Marsh, 90-109. Buckingham: Open University Press 1998
Cavanagh, Michael, David Marsh, and Martin Smith. "The Relationship between Policy Networks at the Sectoral and Sub-Sectoral Levels: A Response to Jordan, Maloney and McLaughlin," *Public Administration* 73 (Winter 1995): 627-33
Clogg, Jessica. "Tenure Reform for Ecologically and Socially Responsible Forest use in British Columbia." MES thesis, York University 1997
Coates, K., ed. *Aboriginal Land Claims in Canada: A Regional Perspective.* Toronto: Copp Clark Pitman 1992
Cobb, Roger W., and Marc Howard Ross. "Denying Agenda Access: Strategic Considerations." In *Cultural Strategies of Agenda Denial: Avoidance, Attack and Redefinition,* ed. Cobb and Ross. Lawrence: University Press of Kansas 1997

Cohen, David S. "A Report on the Social and Economic Objectives of the Crown that the Minister of Forests Should Communicate to the Chief Forester in Relation to the Setting of Allowable Annual Cuts under Section 7 of the Forest Act." 1994

Cohen, David S., and Brian Radnoff. "Regulation, Takings, Compensation, and the Environment: An Economic Perspective." In *The Wealth of Forests: Markets, Regulation, and Sustainable Forestry*, ed. C. Tollefson. Vancouver: UBC Press 1998

Coleman, William D., and Grace Skogstad. "Neo-Liberalism, Policy Networks, and Policy Change: Agricultural Policy Reform in Australia and Canada," *Australian Journal of Political Science* 30 (1995): 242-63

Coleman, William D., and Grace Skogstad, eds. *Policy Communities and Public Policy in Canada: A Structural Approach*. Mississauga, ON: Copp Clark Pitman 1990

Laurence Copithorne, "Natural Resources and Regional Disparities: A Skeptical View," *Canadian Public Policy* 5, 2 (Spring 1979): 181-94

CORE (Commission on Resources and the Environment). "1992-93 Annual Report to the Legislative Assembly." Victoria: June 1993

Cottam, Barry. "The Twentieth Century Legacy of the St. Catherine's Case: Thoughts on Aboriginal Title in Common Law." In *Co-Existence? Studies in Ontario-First Nations Relations*, ed. B.W. Hodgins, S. Heard, and J.S. Milloy. Peterborough: Frost Centre for Canadian Heritage and Development Studies 1992

Council of Forest Industries. "A Blueprint for Competitiveness." Vancouver: Council of Forest Industries 1999

–. "British Columbia Forest Industry Fact Book: 1998." Council of Forest Industries 1999

–. "COFI Submits 30-Day List." Council of Forest Industries 1998

Crenson, Matthew A. *The Un-Politics of Air Pollution: A Study of Non-Decisionmaking in the Cities*. Baltimore: Johns Hopkins University Press 1971

Cubbage, Frederick W., Jay O'Laughlin, and Charles S. Bullock. *Forest Resource Policy*. New York: John Wiley and Sons 1993

Daugherty, W.E. *Maritime Indian Treaties in Historical Perspective*. Ottawa: Indian and Northern Affairs Canada 1983

Daugherty, W.E., and D. Madill. *Indian Government Under Indian Act Legislation, 1868-1951*. Ottawa: Indian Affairs and Northern Development 1980

David, Paul A. "Clio and the Economics of QWERTY," *American Economic Review* 75,2 (1985): 332-7

David Suzuki Foundation. "Fundamental Change Needed for Forest Industry Survival." Vancouver: David Suzuki Foundation 1998

Davis, Charles. *Western Public Lands and Environmental Politics*. Boulder, CO: Westview Press 1997

Debnam, Geoffrey. "Nondecisions and Power: The Two Faces of Bachrach and Baratz," *American Political Science Review* 69,3 (1975): 889-900

deLeon, Peter. "The Stages Approach to the Policy Process: What Has It Done? Where Is It Going?" In *Theories of the Policy Process*, ed. P. Sabatier. Boulder, CO: Westview Press 1999

Dellert, Lois H. "Sustained Yield: Why Has It Failed to Achieve Sustainability?" In *The Wealth of Forests: Markets, Regulation, and Sustainable Forestry*, ed. C. Tollefson. Vancouver: UBC Press 1998

Dellert, Lois H. "Sustained Yield Forestry in British Columbia: The Making and Breaking of a Policy (1900-1993)." MES thesis, York University 1994

Dery, David. "Policy by the Way: When Policy Is Incidental to Making Other Policies," *Journal of Public Policy* 18,2 (1999): 163-76

Douglas, Mary, and Aaron Wildavsky. *Risk and Culture*. Berkeley: University of California Press 1982

Doyle, Aaron, Brian Elliott, and David Tindall. "Framing the Forests: Corporations, the BC Forest Alliance, and the Media." In *Organizing Dissent: Contemporary Social Movements in Theory and Practice*, ed. W. Carroll. Toronto: Garamond Press 1997

Drushka, Ken. "Forest Tenure: Forest Ownership and the Case for Diversification." In *Touch Wood: BC Forests at the Crossroads*, ed. K. Drushka, B. Nixon, and R. Travers, 1-22. Madeira Park, BC: Harbour Publishing 1993

–. *H.R.: A Biography of H.R. MacMillan*. Madeira Park, BC: Harbour Publishing 1995
–. *In the Bight: The BC Forest Industry Today*. Madeira Park, BC: Harbour Publishing 1999
Dryzeck, John S. *The Politics of the Earth: Environmental Discourses*. Oxford: Oxford University Press 1997
Duerden, Frank, Sean Black, and Richard G. Kuhn. "An Evaluation of the Effectiveness of First Nations Participation in the Development of Land-Use Plans in the Yukon," *Canadian Journal of Native Studies* 16,1 (1990): 105-24
Dunlap, Riley E., George H. Gallup Jr., and Alec M. Gallup. *Health of the Planet: Results of a 1992 International Environmental Opinion Survey of Citizens in 24 Nations*. Princeton, NJ: The George H. Gallup International Institute 1993
Dunning, R.W. "Some Aspects of Governmental Indian Policy and Administration," *Anthropologica* 4,2 (1962): 209-31
Eccles, W.J. "Sovereignty-Association, 1500-1783," *Canadian Historical Review* 65,4 (1984): 475-510
Egan Ecological Services. "Environmental Impacts of Forest Management Practices in Tree Farm Licence 44 and Tree Farm Licence 46." Victoria: Sierra Club of Western Canada 1991
Eidsvik, Harold. "Canada in a Global Context." In *Endangered Spaces: The Future for Canadian Wilderness*, ed. M. Hummel. Toronto: Key Porter 1989
Ellefson, Paul V. *Forest Resource Policy: Process, Participants and Programs*. New York: McGraw-Hill 1992
Ervin, A.M. "Contrasts between the Resolution of Native Land Claims in the United States and Canada Based on Observations of the Alaska Native Land Claims Movement," *Canadian Journal of Native Studies* 1 (1981): 123-40
Evans, Peter B., Dietrich Rueschemeyer, and Theda Skocpol. *Bringing the State Back In*. Cambridge: Cambridge University Press 1985
F.L.C. Reed and Associates. *Forest Management in Canada*. Vol. 1. Ottawa: Canadian Forestry Service 1978
Feit, Harvey A. "Negotiating Recognition of Aboriginal Rights: History, Strategies and Reactions to the James Bay and Northern Quebec Agreement," *Canadian Journal of Anthropology* 1 (1980): 159-70
Ferreira, Darlene Abreu. "Oil and Lubicons Don't Mix: A Land Claim in Northern Alberta in Historical Perspective," *Canadian Journal of Native Studies* 12,1 (1992)
Flumerfelt, A.C. "Forest Resources." In *Canada and Its Provinces: A History of the Canadian People and Their Institutions by One Hundred Associates*. Vol. 22, *The Pacific Province Part II*, ed. A. Shortt and A.G. Doughty. Glasgow: Brook and Company 1914
Forest Alliance of British Columbia. *Analysis of Recent British Columbia Government Forest Policy and Land Use Initiatives*. Vancouver: Price Waterhouse 1995
Forest Renewal BC. "Business Plan 1998-99." Victoria: Forest Renewal BC 1988
–. "Business Plan 1999-2000." Victoria: Forest Renewal BC 1999
–. "Business Plan, 2000-1." Victoria: Forest Renewal BC 2000
–. "Five Year Report 1994-99." Victoria: Forest Renewal BC 1999
Forest Resources Commission. "The Future of Our Forests." Victoria: Forest Resources Commission 1991
–. *Providing the Framework: A Forest Practices Code*. Victoria: Forest Resources Commission 1992
Forester, John. "Bounded Rationality and the Politics of Muddling Through," *Public Administration Review* 44,1 (1984): 23-31
Frank, Darrell, and Kim Allan. "Community Forests in British Columbia: Models That Work," *The Forestry Chronicle* 70 (1994): 720-4
"Fraser River Headwaters at Risk," *BC Environmental Report* 9,1 (Spring 1998): 12-13
Frey, Frederick W. "Comment: On Issues and Nonissues in the Study of Power," *American Political Science Review* 65 (1971): 1081-101
Frideres, James S. "Government Policy and Indian Natural Resource Development," *Canadian Journal of Native Studies* 4,1 (1984): 51-66
Garrett, Geoffrey, and Barry Weingast. "Ideas, Interests, and Institutions: Constructing the European Community's Internal Market." In *Ideas and Foreign Policy*, ed. J. Goldstein and R.O. Keohane. Ithaca: Cornell University Press 1993

Gillis, R. Peter, and Thomas R. Roach. *Lost Initiatives: Canada's Forest Industries, Forest Policy and Forest Conservation*. New York: Greenwood 1986
Gosnell, R.E. "The Year Book of British Columbia and Manual of Provincial Information." Victoria: Legislative Assembly 1903
Grant, Wyn, and Anne MacNamara. "When Policy Communities Intersect: The Cases of Agriculture and Banking," *Political Studies* 43 (1995): 509-15
Grant, Wyn, William Paterson, and Colin Whitson. *Government and the Chemical Industry: A Comparative Study of Britain and West Germany*. Oxford: Clarendon Press 1988
Gray, J.A. "Royal Commission and Forest Policy in British Columbia: A Review of the Pearse Report," *Canadian Public Policy* 3,2 (1977): 219-23
Gray, Kevin R. "The Nunavut Land Claims Agreement and the Future of the Eastern Arctic: The Uncharted Path to Effective Self-Government," *University of Toronto Law Review* 52,2 (1994): 300-45
Great Britain. "Report on the Indians of Upper Canada: 1839." Toronto: Canadiana House 1968
Greenpeace Canada, Greenpeace International, and Greenpeace San Francisco. "Broken Promises: The Truth about What's Happening to British Columbia's Forests." Vancouver: Greenpeace Canada, Greenpeace International, Greenpeace San Francisco 1997
Groen, James P. "British Columbia's International Relations: Consolidating a Coalition-Building Strategy," *BC Studies* 102 (1994): 25-59
Hajer, Maarten A. *The Politics of Environmental Discourse: Ecological Modernization and the Policy Process*. Oxford: Clarendon Press 1995
Haley, David. "The Forest Tenure System as a Constraint on Efficient Timber Management: Problems and Solutions," *Canadian Public Policy* 11 (1985): 315-20
–. "A Regional Comparison of Stumpage Values in British Columba and the United States Pacific Northwest," *Forestry Chronicle* 56 (October 1980): 225-30
Haley, David, and Martin K. Luckert. "Tenures as Economic Instruments for Achieving Objectives of Public Forest Policy in British Columbia." In *The Wealth of Forests: Markets, Regulation, and Sustainable Forestry*, ed. C. Tollefson. Vancouver: UBC Press 1998
Haley, David, and Martin K. Luckert. *Forest Tenures in Canada, Canadian Public Policy*. Forestry Canada 1990
Hall, Peter. "Policy Paradigms, Social Learning, and the State: The Case of Economic Policymaking in Britain," *Comparative Politics* 25,3 (April 1993): 275
Hall, Peter, and Rosemary Taylor. "Political Science and the Three New Institutionalisms," *Political Studies* 44,5 (1996): 936-57
Hall, T. "Self-Government or Self-Delusion? Brian Mulroney and Aboriginal Rights," *Canadian Journal of Native Studies* 6 (1986): 77-89
Hammond, Herb. "Community Control of Forests," *Forest Planning Canada* 6,6 (1990): 43-6
–. *Public Forests or Private Timber Supplies? The Need for Community Control of British Columbia's Forests*. Winlaw, BC: Silva Ecosystem Consultants 1989
–. *Seeing the Forest among the Trees*. Winlaw, BC: Polestar Press 1991
–. *Wholistic Forest Use*. Winlaw, BC: Silva Ecosystem Consultants 1991
Hammond, Herb, and Susan Hammond. "Sustainable Forest Planning and Use," *Forest Planning Canada* 1,4 (1985): 8-10
Hansen, A.J., T.A. Spies, F.J. Swanson, and J.L. Ohmann. "Conserving Biodiversity in Managed Forests: Lessons from Natural Forests," *BioScience* 41 (1991): 382-91
Harrison, Kathryn. *Passing the Buck: Federalism and Canadian Environmental Policy*. Vancouver: UBC Press 1996
–. "Talking with the Donkey: Cooperative Approaches to Environmental Protection," *Journal of Industrial Ecology* 2,3 (1999): 51-72
Heclo, Hugh. *Modern Social Policies in Britain and Sweden*. New Haven, CT: Yale University Press 1974
Hessing, Melody, and Michael Howlett. *Canadian Natural Resource and Environmental Policy: Political Economy and Public Policy*. Vancouver: UBC Press 1996

Hoberg, George. "The British Columbia Forest Practices Code: Formalization and Its Effects." In *Canadian Forest Policy: Regimes, Policy Dynamics, and Institutional Adaptations*, ed. M. Howlett. Toronto: University of Toronto Press, forthcoming 2001
–. "The Coming Revolution in Regulating Our Forests," *Policy Options* 20 (December 1999): 53-6
–. "Distinguishing Learning from Other Sources of Policy Change: The Case of Forestry in the Pacific Northwest." Paper presented at the Annual Meeting of the American Political Science Association, Boston, 3-6 September 1998
–. "Environmental Policy: Alternative Styles." In *Governing Canada: Institutions and Public Policy*, ed. M.M. Atkinson. Toronto: Harcourt Brace Jovanovich Canada 1993
–. "From Localism to Legalism: The Transformation of Federal Forest Policy." In *Western Public Lands and Environmental Politics*, ed. C. Davis. Boulder, CO: Westview Press 1997
–. "How the Way We Make Policy Governs the Policy We Make." In *Sustaining the Forests of the Pacific Coast: Forging Truces in the War in the Woods*, ed. D. Salazar and D. Alper. Vancouver: UBC Press 2000
–. "The Politics of Sustainability: Forest Policy in British Columbia." In *Politics, Policy, and Government in British Columbia*, ed. R.K. Carty. Vancouver: UBC Press 1996
–. "Putting Ideas in Their Place: A Response to 'Learning and Change in the British Columbia Forest Policy Sector,'" *Canadian Journal of Political Science* 29,1 (March 1996): 135-44
Hoberg, George, and Edward Morawski. "Policy Change through Sector Intersection: Forest and Aboriginal Policy in Clayoquot Sound," *Canadian Public Administration* 40,3 (1997): 387-414
Hodgetts, J.E. *The Canadian Public Service: A Physiology of Government 1867-1970*. Toronto: University of Toronto Press 1973
–. *Pioneer Public Service: An Administrative History of the United Canadas, 1841-1867*. Toronto: University of Toronto Press 1955
Hodgins, Bruce W., Shawn Heard, and John Sheridan Milloy. *Co-existence? Studies in Ontario-First Nations Relations*. Peterborough, ON: Frost Centre for Canadian Heritage and Development Studies 1992
Hogwood, Brian W., and B. Guy Peters. *Policy Dynamics*. Brighton, Sussex: Wheatsheaf 1983
Howlett, Michael. "Acts of Commission and Acts of Omission: Legal-Historical Research and the Intentions of Government in a Federal State," *Canadian Journal of Political Science* 19,2 (1986): 363-70
–. "Policy Paradigms and Policy Change: Lessons from the Old and New Canadian Policies towards Aboriginal Peoples," *Policy Studies Journal* 22,4 (1994): 631-52
–. "Understanding Policy Change and Policy Stability: Elements of a Vector Theory of Policy Dynamics." Paper presented to the Annual Meeting of the British Columbia Political Studies Association. Vancouver 1999
–, ed. *Canadian Forest Policy: Regimes, Policy Dynamics and Institutional Adaptations*. Toronto: University of Toronto Press, forthcoming 2001
Howlett, Michael, and M. Ramesh, *Studying Public Policy: Policy Cycles and Policy Subsystems*. Toronto: Oxford University Press 1995
Howlett, Michael, and Jeremy Rayner. "Do Ideas Matter? Policy Network Configurations and Resistance to Policy Change in the Canadian Forest Sector," *Canadian Public Administration* 38 (1995): 382-410
HST Consortium, Forest Resources Commission. "A Summary of Technical Reviews of Forest Inventories and Allowable Annual Cut Determinations in BC." Victoria: Forest Resources Commission 1991
Humphreys, David. "The Global Politics of Forest Conservation since the UNCED," *Environmental Politics* 5,2 (1996): 231-56
Hunt, Constance. "Approaches to Native Land Settlements and Implications for Northern Land Use and Resource Management Policies." In *Northern Transitions*, ed. R.F. Keith and J.B. Wright. Ottawa: Canadian Arctic Resources Committee 1978
Innis, Harold A. *Problems of Staple Production in Canada*. Toronto: Ryerson Press 1933

Isaac, Thomas. "Balancing Rights: The Supreme Court of Canada, *R.* v. *Sparrow*, and the Future of Aboriginal Rights," *Canadian Journal of Native Studies* 13,2 (1993): 199-219
IWA Canada. "Toward a Value-Added Lumber Industry: Tenure and Stumpage Reform in British Columbia" Vancouver: IWA Canada 1999
Jenkins-Smith, Hank, and Paul Sabatier. "Evaluating the Advocacy Coalition Framework," *Journal of Public Policy* 14,2 (1994): 175-203
Johnson, I. *Helping Indians to Help Themselves: The 1951 Indian Act Consultation Process.* Ottawa: Indian and Northern Affairs Canada 1984
Jones, Charles O. *An Introduction to the Study of Public Policy,* 3rd ed. Monterey, CA: Brooks-Cole 1997
Jones, Bryan D. *Reconceiving Decision-Making in Democratic Politics: Attention, Choice and Public Policy.* Chicago: University of Chicago Press 1994
Jordan, Andrew, and John Greenaway. "Shifting Agendas, Changing Regulatory Structures and the 'New' Politics of Environmental Pollution: British Coastal Water Policy 1955-1995," *Public Administration* 76 (1998): 669-94
Jordan, Grant, William A. Maloney, and Andrew M. McLaughlin. "Characterizing Agricultural Policy-Making," *Public Administration* 72 (1994): 505-26
Kamieniecki, Sheldon. "Testing Alternative Theories of Agenda Setting: Forest Policy Change in British Columbia, Canada." Paper presented at the Annual Meeting of the American Political Science Association, Boston, 3-6 September 1998
Keeping, J. *The Inuvialuit Final Agreement.* Calgary: Canadian Institute of Resources Law 1989
Kingdon, John. *Agendas, Alternatives, and Public Policies.* Boston: Little Brown 1984
–. *Agendas, Alternatives, and Public Policies.* 2nd ed. New York: HarperCollins College Publishers 1995
Knafla, Louis A. *Law and Justice in a New Land: Essays in Western Canadian Legal History.* Calgary: Carswell 1986
Knoepfel, , Peter, and Ingrid Kissling-Naef. "Social Learning in Policy Networks," *Policy and Politics* 26, 3 (1998): 343-67
KPMG, Thorau and Associates, Perrin, and H.A. Simons and Associates. "Financial State of the Forest Industry and Delivered Wood Cost Drivers." Vancouver: KPMG 1997
Krasner, Stephen D. "Approaches to the State: Alternative Conceptions and Historical Dynamics," *Comparative Politics* 16,2 (1984): 223-46
Krehbeil, Keith. *Pivotal Politics: A Theory of US Lawmaking.* Chicago: Chicago University Press 1998
Lawrence, Joseph Collins. "Markets and Capital: A History of the Lumber Industry of British Columbia (1778-1952)." MA thesis, University of British Columbia 1957
Légaré, André. "The Process Leading to a Land Claims Agreement and Its Implementation: The Case of the Nunavut Land Claims Settlement," *Canadian Journal of Native Studies* 16,1 (1996): 139-63
Lee, Christopher A., and Phil Symington. "Land Claims Process and Its Potential Impact on Wood Supply," *Forestry Chronicle* 73,3 (1997): 349-52
Leigh, Duane E. *Does Training Work for Displaced Workers?* Kalamazoo, MI: W.E. Upjohn Institute for Employment Research 1990
Lertzman, Ken, Jeremy Rayner, and Jeremy Wilson. "Learning and Change in the BC Forest Policy Sector: A Consideration of Sabatier's Advocacy Coalition Framework," *Canadian Journal of Political Science* 29,1 (March 1996): 111-33
Leslie, J.F. *Commissions of Inquiry into Indian Affairs in the Canadas: 1828-1858.* Ottawa: Indian and Northern Affairs Canada 1985
–. "Vision Versus Revision: Native People, Government Officials and the Joint Senate/House of Commons Committee on Indian Affairs 1946-1948 and 1959-1961." In *Papers Presented to the Canadian Historical Association Annual Meeting.* Victoria: Canadian Historical Association 1984
Lewis, Karen, and Susan Westmacott. "Provincial Overview and Status Report." Victoria: Land Use Coordination Office 1996
Lindblom, Charles. *Politics and Markets.* New York: Basic Books 1977

–. "The Science of Muddling Through," *Public Administration Review* 19 (1959): 79-88
Long, J. Anthony. "Political Revitalization in Canadian Native Indian Societies," *Canadian Journal of Political Science* 23 (1990): 751-74
Long, J. Anthony, and Menno Boldt. "Introduction." In *Governments in Conflict? Provinces and Indian Nations in Canada,* ed. Long, Boldt, and Little Bear. Toronto: University of Toronto Press 1988
–. "Leadership Selection in Canadian Native Communities: Reforming the Present and Incorporating the Past," *Great Plains Quarterly* 7 (1987): 103-15
Lowi, Theodore. "American Business, Public Policy, Case Studies, and Political Theory," *World Politics* 16 (1964): 677-93
–. *The End of Liberalism.* New York: W.W. Norton 1969
Lysyk, K. "Approaches to Settlement of Indian Title Claims: The Alaska Model," *UBC Law Review* 8 (1973): 321-42
–. "The Indian Title Question in Canada: An Appraisal in the Light of Calder," *Canadian Bar Review* 51 (1973): 450-80
Mackintosh, W.A. "Economic Factors in Canadian History." In *Approaches to Canadian Economic History,* ed. W.T. Easterbrook and M.H. Watkins. Toronto: McClelland and Stewart 1967
Macklem, Patrick, and Michael Asch. "Aboriginal Rights and Canadian Sovereignty: An Essay on *R.* v. *Sparrow,*" *Alberta Law Review* 29 (1991): 498-517
MacLachlan, L. "The Gwich'in Final Agreement," *Resources* 36 (1991): 6-11
MacMillan Bloedel. "MacMillan Bloedel to Phase Out Clearcutting." Press release, Vancouver, 10 June 1998
–. "A White Paper for Discussion: Stumpage and Tenure Reform in BC." Vancouver: MacMillan Bloedel 1998
Magnusson, Warren, W.K. Carroll, C. Doyle, et al., eds. *After Bennett: A New Politics for British Columbia.* Vancouver: New Star Books 1986
Maitland, Alice. "Forest Planning Canada." In *Forest Industry Charter of Rights* 6,2 (1990): 5-9
Manning, E.C. "The Administration of Crown Lands in British Columbia," *Journal of Forestry* 36, 10 (1938): 314-25
Marchak, M. Patricia. *Green Gold: The Forestry Industry in British Columbia.* Vancouver: UBC Press 1983
Marchak, M. Patricia, Scott L. Aycock, and Deborah M. Herbert. *Falldown: Forest Policy in British Columbia.* Vancouver: David Suzuki Foundation and Ecotrust Canada 1999
Marris, Robert Howard. "Pretty Sleek and Fat: The Genesis of Forest Policy in British Columbia, 1903-1914." MA thesis, University of British Columbia 1979
Marsh, David. *Comparing Policy Networks.* Buckingham: Open University Press 1998
Mazmanian, Daniel, and Paul Sabatier. *Implementation and Public Policy.* Glenview, IL: Scott, Foresman 1983
McAskill, Ian. "Public Charges and Private Values: A Study of British Columbia Timber Pricing." MA thesis, Simon Fraser University 1984
McConnell, Grant. *Private Power and American Democracy.* New York: Alfred Knopf 1969
McConnell, W.H. "The Calder Case in Historical Perspective," *Saskatchewan Law Review* 38 (1974): 88-122
McCubbins, Matthew, Roger Noll, and Barry Weingast. "Structure and Process, Politics and Policy: Administrative Arrangements and the Political Control of Agencies," *Virginia Law Review* 75 (1989): 431-82
McGee, H.F. *The Native Peoples of Atlantic Canada.* Toronto: McClelland and Stewart 1974
McKee, Chris. *Treaty Talks in British Columbia: Negotiating a Mutually Beneficial Future.* Vancouver: UBC Press 1997
McNeil, Kent. "The Meaning of Aboriginal Title." In *Aboriginal and Treaty Rights in Canada: Essays on Law, Equity, and Respect for Difference,* ed. M. Asch. Vancouver: UBC Press 1997
McRoberts, Kenneth, and Patrick Monahan. *The Charlottetown Accord, the Referendum, and the Future of Canada.* Toronto: University of Toronto Press 1993

Meidinger, Errol. "Look Who's Making the Rules: International Environmental Standard Setting by Non-Governmental Organizations," *Human Ecology Review* 4,1 (1997): 52-4
M'Gonigle, Michael. "Developing Sustainability: A Native/Environmentalist Prescription for Third-Level Government," *BC Studies* 84 (1989/90): 65-99
–. "From the Ground Up: Lessons from the Stein River Valley." In *After Bennett: A New Politics for British Columbia*, ed. W. Magnusson et al. 169-91. Vancouver: New Star Books 1986
–. "Local Economies Solve Global Problems," *The New Catalyst* (Spring 1987): 17-21
M'Gonigle, Michael, and Ben Parfitt. *Forestopia: A Practical Guide to the New Forest Economy*. Madeira Park, BC: Harbour Publishing 1994
Mickenberg, Neil, and Peter Cumming, eds. *Native Rights in Canada*. Toronto: General Publishing 1972
Miller, K.T., and G. Lerchs. *The Historical Development of the Indian Act*. Ottawa: Indian and Northern Affairs Canada, Treaties and Historical Research Centre 1978
Mirbach, Martin von. "Demanding Good Wood," *Alternatives* 23 (Summer 1997): 8-25
Moe, Terry. "The Politics of Bureaucratic Structure." In *Can the Government Govern?* ed. J.E. Chubbs and P.E. Peterson, Washington, DC: Brookings Institution 1989
Moore, Keith. *Coastal Watersheds: An Inventory of Watersheds in the Coastal Temperate Forests of British Columbia*. Vancouver: Earthlife Canada Foundation, Ecotrust/Conservation International 1991
–. "Presentation to House of Commons Standing Committee on Natural Resources and Government Operations." 1999. www.fpd.gov.bc.ca/background/MPs.htm
Moore, Patrick. "Review of 'Forestopia: A Practical Guide to the New Forest Economy' by Michael M'Gonigle and Ben Parfitt." Vancouver: Forest Alliance of BC 1994
Morris, Alexander. *The Treaties of Canada with the Indians of Manitoba, the North-West Territories, and Kee-wa-tin*. Toronto: Belfords, Clarke and Co., 1880
Muir, Magdalena A.K. *Comprehensive Land Claims Agreements of the Northwest Territories: Implications for Land and Water Management*. Calgary: Canadian Institute of Resources Law 1994
Mulholland, F.D. *The Forest Resources of British Columbia*. Victoria: King's Printer 1937
Nagel, G.S. "Economic and Public Policy in the Forestry Sector of British Columbia." PhD diss., Yale University, 1970
Nathan, Holly. "Aboriginal Forestry: The Role of the First Nations." In *Touch Wood: BC Forests at the Crossroads*, ed. K. Drushka, B. Nixon, and R. Travers, 137-70. Madeira Park, BC: Harbour Publishing 1993
National Aboriginal Forestry Association. "Aboriginal and Treaty Rights to Renewable Resources." In *Forest Land and Resources for Aboriginal Peoples: An Intervention Submitted to the Royal Commission on Aboriginal Peoples by the National Aboriginal Forestry Association*. Ottawa, ON: National Aboriginal Forestry Association July 1993
–. "Co-Management and Other Forms of Agreement in the Forest Sector [5 Part Strategy: Tools for Aboriginal Communities to Promote Good Forest Land Management Practices]." http://sae.ca/nafa/comanage.htm
–. Introduction to *Forest Land and Resources for Aboriginal Peoples: An Intervention Submitted to the Royal Commission on Aboriginal Peoples by the National Aboriginal Forestry Association*. Ottawa, ON: National Aboriginal Forestry Association 1993
–. "Provincial Forest Resource Access Policies, a Submission to the Royal Commission on Aboriginal Peoples by the National Aboriginal Forestry Association." Ottawa, ON: National Aboriginal Forestry Association 1993
New Democratic Party. "A Better Way for British Columbia." Vancouver: New Democratic Election Platform 1991
–. "Sustainable Development: 1989 Legislative Program for Sustainable Development." Victoria: New Democratic Party 1989
Niskanen, William. *Bureaucracy and Representative Government*. Chicago: Aldine-Atherton 1971
Nixon, R. "Forestry and the End of Innocence," *Forest Planning Canada* 7,6 (1991): 18-22
Notzke, Claudia. *Aboriginal Peoples and Natural Resources in Canada*. Toronto: Centre for Aboriginal Management, Education and Training 1994

O'Reilly, J. "The Courts and Community Values: Litigation Involving Native Peoples and Resource Development," *Alternatives* 15 (1988): 40-8

Paget, Gregg, and Barbara Morton. "Forest Certification and 'Radical' Forest Stewardship in British Columbia, Canada: The Influence of Corporate Environmental Procurement." Paper presented at Greening of Industry Network Conference, 1999, Kenan-Flagler Business School, University of North Carolina, Chapel Hill, November 1999

Pal, Leslie. *Beyond Policy Analysis: Public Issue Management in Turbulent Times.* Scarborough, ON: ITP Nelson 1997

Parfitt, Ben. "Province Can Give MacBlo Squat," *Georgia Straight*, 6-13 May 1999

Patterson, E.P. "A Decade of Change: Origins of the Nishga and Tsimshian Land Protests in the 1880's," *Journal of Canadian Studies* 18 (1983): 40-54

Pearse, Peter H. "Conflicting Objectives in Forest Policy: The Case of British Columbia," *Forestry Chronicle* 46 (1970): 281-7

–. *Crown Charges for Early Timber Rights: Royalties and Other Levies for Harvesting Rights on Timber Leases, Licences and Berths in British Columbia – First Report of the Task Force on Crown Timber Disposal.* Victoria: Ministry of Forests 1974

–. *Forest Policy in Canada.* Victoria: UBC Forest Economics and Policy Analysis Project 1985

Pedersen, Larry. "The Big Picture Policy Framework." Address to Southern Interior Silviculture Committee Winter Workshop. Penticton, BC, 10 March 1999

–. "The Truth Is Out There." Address to the Annual Convention of the Northern Forest Products Association. Prince George, BC, 3 April 1997

Percy, Michael, and Christian Yoder. *The Softwood Lumber Dispute and Canada-US Trade in Natural Resources.* Halifax: Institute for Research on Public Policy 1987

Peters, E.J. *Existing Aboriginal Self-Government Arrangements in Canada: An Overview.* Kingston: Queen's University Institute of Intergovernmental Relations 1987

Phillips, Susan. "Discourse, Identity, and Voice: Feminist Contributions to Policy Studies." In *Policy Studies in Canada: The State of the Art*, ed. L. Dobuzinskis, M.P. Howlett, and D.H. Laycock. Toronto: University of Toronto Press 1998

Piché, Gilbert. "Forestry Notes in Connection with Stumpage Dues in Quebec." In *Report of the Minister of Lands and Forests of the Province of Quebec for the Twelve Months Ending 30th June 1910.* Quebec: King's Printer 1911

Pierson, Paul. "Increasing Returns, Path Dependence, and the Study of Politics," *American Political Science Review* 94,2 (2000): 251-68

–. "When Effect Becomes Cause: Policy Feedback and Political Change," *World Politics* 45, 4(July 1993): 595-628

Pinkerton, Evelyn. "Taking the Minister to Court: Changes in Public Opinion about Forest Management and Their Political Expression in Haida Land Claims," *BC Studies* 57 (1983): 68-85

–, ed. *Co-operative management of Local Fisheries: New Directions for Improved Management and Community Development.* Vancouver: UBC Press 1989

Pollock, Philip H., Stuart A. Lilie, and M. Elliot Vittes. "Hard Issues, Core Values and Vertical Constraint: The Case of Nuclear Power," *British Journal of Political Science* 23,1 (1993): 29-50

Ponting, J. Rick, ed. *Arduous Journey*. Toronto: McClelland and Stewart 1986

Ponting, J. Rick, and Roger Gibbins. *Out of Irrelevance*. Toronto: Butterworths 1980

PricewaterhouseCoopers. *The BC Forest Industry: Unrealized Potential.* Vancouver: PricewaterhouseCoopers, 2000

Pross, A. Paul. *Group Politics and Public Policy*. Toronto: Oxford University Press 1992

Quebec. *The James Bay and Northern Quebec Agreement: Agreement between the Government of Quebec, the Société d'Énergie de la Baie James, the Société de Développement de la Baie James, the Commission Hydroélectrique de Quebec (Hydro-Quebec) and the Grand Council of the Crees (of Quebec), the Northern Quebec Inuit Association and the Government of Canada.* Quebec: Éditeur Officiel du Québec 1976

Rajala, Richard Allan. *Clearcutting the Pacific Rain Forest: Production, Science, and Regulation.* Vancouver: UBC Press 1998

Raunet, D. *Without Surrender, Without Consent: A History of the Nishga Land Claims*. Vancouver: Douglas and McIntyre 1984

Rayner, Jeremy. "Priority-Use Zoning: Sustainable Solution or Symbolic Politics?" In *The Wealth of Forests: Markets, Regulation, and Sustainable Forestry*, ed. C. Tollefson. Vancouver: UBC Press 1998

RBC Dominion Securities. "BC Forest Products Sector." Vancouver: 1998

Reid, J.A.K. *Significant Events and Developments in the Evolution of Timberland and Forestry Legislation in British Columbia*. Victoria: Ministry of Forests 1985

Reid, Keith, and Don Weaver. "Aspects of the Political Economy of the BC Forest Industry." In *Essays in BC Political Economy*, ed. P. Knox and P. Resnick. Vancouver: New Star Books 1974

Rhodes, R.A.W. "Policy Networks: A British Perspective," *Journal of Theoretical Politics* 2,3 (1990): 293-317

Riddell, Darcy. "Sierra Report." Victoria: Sierra Club of British Columbia 1999

Robin, Martin, *The Rush for Spoils: The Company Province, 1871-1953*. Toronto: McClelland and Stewart 1972

Rose, Richard. "Inheritance before Choice in Public Policy," *Journal of Theoretical Politics* 2,3 (1990): 263-91

Royal Commission on Aboriginal Peoples. *Treaty Making in the Spirit of Co-Existence: An Alternative to Extinguishment*. Ottawa: Minister of Supply and Services 1994

Royal Commission on Aboriginal Peoples. *Restructuring the Relationship*. Vol. 2, chap. 4, pt. I. Ottawa: Canada. Minister of Supply and Services 1996

Rugman, Alan M., John J. Kirton, and Julie A. Soloway. *Environmental Regulations and Corporate Strategy: A NAFTA Perspective*. New York: Oxford University Press 1999

Sabatier, Paul. "Knowledge, Policy-Oriented Learning, and Policy Change: An Advocacy Coalition Framework," *Knowledge: Creation, Diffusion, Utilization* 8,4 (1987): 648-92

–. "Policy Change over a Decade or More." In *Policy Change and Learning: An Advocacy Coalition Approach*, ed. Sabatier and H.C. Jenkins-Smith. Boulder, CO: Westview Press 1993

–. "Toward Better Theories of the Policy Process," *PS: Political Science and Politics* 24 (1991): 449-76

–, ed. *Theories of the Policy Process*. Boulder, CO: Westview Press 1999

Sabatier, Paul, and Hank C. Jenkins-Smith. "The Advocacy Coalition Framework: Assessment, Revisions, and Implications for Scholars and Practitioners." In *Policy Change and Learning: An Advocacy Coalition Aproach*, ed. Sabatier and Jenkins-Smith. Boulder, CO: Westview Press 1993

Sahajananthan, Sivaguru, David Haley, and John Nelson. "Planning for Sustainable Forests in British Columbia through Land Use Zoning," *Canadian Public Policy* 24, Supplement 2 (1998): S73-S81

Salasan Associates Ltd. *Summary of Public Response to the Proposed Forest Licence Conversion Policy*. Victoria: Ministry of Forests 1990

Sanders, Douglas. "The Indian Lobby." In *And No One Cheered: Federalism, Democracy and the Constitution Act*, ed. R. Simeon and K. Banting. Toronto: Methuen 1983

–. "The Nishga Case," *BC Studies* 19 (1973): 3-20

–. "The Queen's Promises." In *Law and Justice in a New Land: Essays in Western Canadian Legal History*, ed. L.A. Knafla. Toronto: Carswell 1986

–. "The Supreme Court of Canada and the 'Legal and Political Struggle' over Indigenous Rights," *Canadian Ethnic Studies* 22 (1990): 122-9

–. "An Uncertain Path: The Aboriginal Constitutional Conferences." In *Litigating the Values of a Nation*, ed. J.M. Weiler and R.M. Elliot. Toronto: Carswell 1986

Sanjayan, M.E., and M.A. Soule. *Moving beyond Brundtland: The Conservation Value of British Columbia's 12 Percent Protected Area Strategy*. Vancouver: Greenpeace 1997

Sawchuk, Wayne. "Northern Rockies Victory," *BC Environmental Report* 8,4 (1997): 21

Scarfe, Brian L. "Timber Pricing Policies and Sustainable Forestry." In *The Wealth of Forests: Markets, Regulation, and Sustainable Forestry*, ed. C. Tollefson. Vancouver: UBC Press 1998

Schneider, Anne, and Helen Ingram. "Social Construction of Target Populations: Implications for Politics and Policy," *American Journal of Political Science* 87,2 (1993): 334-47

Schwindt, Richard. "Report of the Commission of Inquiry into Compensation for the Taking of Resource Interests." Victoria: Queen's Printer 1992

–. "The British Columbia Forest Sector: Pros and Cons of the Stumpage System." In *Resource Rents and Public Policy in Western Canada,* ed. Thomas Gunton and John Richards. Halifax: Institute for Research on Public Policy 1987

Schwindt, Richard, and Steven Globerman. "Takings of Private Rights to Public Natural Resources: A Policy Analysis," *Canadian Public Policy* 22, 3 (1996): 205-24

Schwindt, Richard, and Terry Heaps. "Chopping Up the Money Tree: Distributing the Wealth From British Columbia's Forests." Vancouver: David Suzuki Foundation 1996

Schwindt, Richard, with Adrienne Wanstall. "The Pearse Commission and the Industrial Organization of the British Columbia Forest Industry," *BC Studies* 41 (Fall 1979): 3-25

Scientific Panel for Sustainable Forest Practices in Clayoquot Sound. "Sustainable Ecosystem Management in Clayoquot Sound: Planning and Practices." British Columbia: Scientific Panel for Sustainable Forest Practices in Clayoquot Sound 1995

Scott, D.C. "Indian Affairs 1840-1867." In *Canada and Its Provinces,* ed. A. Shortt and A.G. Doughty. Toronto: Publishers Association of Canada 1914

Scott, Ian G., and J.T.S. McCabe. "The Role of the Provinces in the Elucidation of Aboriginal Rights in Canada." In *Governments in Conflict? Provinces and Indian Nations in Canada,* ed. J.A. Long and M. Boldt. Toronto: University of Toronto Press 1988

Senez, Paul. "Timber Grab on Vancouver Island," *Watershed Sentinel* 10,3 (2000): 2-5

Sherman, Paddy. *Bennett.* Toronto: McClelland and Stewart 1966

Shortt, Adam, and Arthur G. Doughty. *Canada and Its Provinces: A History of the Canadian People and Their Institutions by One Hundred Associates.* Vol. 22: *The Pacific Province Part II.* Toronto: Glascow, Brook and Company 1914

Sierra Club of British Columbia. "Beyond Timber Targets: A Balanced Vision for Vancouver Island." Victoria: Sierra Club of BC 1997

–. "Government Withholds Key Report from Public." Victoria: Sierra Club of BC 2000

–. "More Forest Privatization in the Works." Victoria: Sierra Club of BC 1999

Sierra Legal Defence Fund. "British Columbia's Clear Cut Code: Changing the Way We Manage Our Forests? Tough Enforcement." Vancouver: Sierra Legal Defence Fund 1996

–. "British Columbia's Forestry Report Card." Vancouver: Sierra Legal Defence Fund 1998

–. "Forest Practices Code Rollbacks Confirmed." Media release, Vancouver, 2 April 1998

–. "Going Downhill Fast: Landslides and the Forest Practices Code." Vancouver: Sierra Legal Defence Fund 1997

–. "Profits or Plunder: Mismanagement of BC's Forests." Vancouver: Sierra Legal Defence Fund 1998

–. "Stream Protection under the Code: The Destruction Continues." Vancouver: Sierra Legal Defence Fund 1997

–. "Wildlife at Risk." Vancouver: Sierra Legal Defence Fund 1999

Skocpol, Theda. "Bringing the State Back In: Strategies of Analysis in Current Research." In *Bringing the State Back In,* ed. P.B. Evans, D. Rueschemeyer, and T. Skocpol. Cambridge: Cambridge University Press 1986

Slattery, Brian. "The Constitutional Guarantee of Aboriginal and Treaty Rights," *Queen's Law Journal* 8 (1983): 232-73

–. "The Hidden Constitution: Aboriginal Rights in Canada," *American Journal of Comparative Law* 32 (1984): 361-76

Sloan, Gordon. *Report of the Commissioner on the Forest and Resource of British Columbia.* Victoria: Charles F. Banfield 1945

Smith, Peggy, and Geoff Quaile "An Aboriginal Perspective on Canada's Progress toward Meeting Its National Commitments to Improve Aboriginal Participation in Sustainable Forest Management." Paper Presented to the Eleventh World Forestry Congress, Antalya, Turkey, 13-22 October 1993

Smith, Richard A. "Decision Making and Non-Decision Making in Cities: Some Implications for Community Structural Research," *American Sociological Review* 44,1 (1979): 147-61
Spranca, Mark, Elisa Minsk, and Jonathan Baron. "Omission and Commission in Judgement and Choice," *Journal of Experimental Social Psychology* 27 (1991): 76-105
Stanbury, W.T., and Ilan Vertinsky. "Boycotts in Conflicts over Forestry Issues: The Case of Clayoquot Sound," *Commonwealth Forestry Review* 76,1 (1997): 18-24
–. "Governing Instruments for Forest Policy in British Columbia: A Positive and Normative Analysis." In *The Wealth of Forests: Markets, Regulation, and Sustainable Forestry,* ed. C. Tollefson. Vancouver: UBC Press 1998
Steinmo, Sven, and Jon Watts. "It's the Institutions, Stupid! Why Comprehensive National Health Insurance Always Fails in America," *Journal of Health Politics, Policy and Law* 20 (1995): 329-72
Stephens, Tom. "White Paper for Discussion: Stumpage and Tenure Reform in BC." In *Forest Resources Commission, Background Papers.* Vol. 3. Vancouver: Macmillan Bloedel 1998
Sterling Wood Group. "Review of Forest Tenures in British Columbia." Victoria: Forest Resources Commission 1991
Sterritt, Neil. "The Nisga'a Treaty: Competing Claims Ignored," *BC Studies* 120 (1998/9): 73-98
Stone, Deborah. "Causal Stories and the Formation of Policy Agendas," *Political Science Quarterly* 104 (1989): 281-300
Taylor, Duncan, and Jeremy Wilson. "Ending the Watershed Battles: BC Forest Communities Seek Peace through Local Control," *Environments* 22,3 (1994): 93-102
Taylor, G.W. *Timber: History of the Forest Industry in British Columbia.* Vancouver: J.J. Douglas 1975
Taylor, J. *Canadian Indian Policy during the Interwar Years, 1918-1939.* Ottawa: Indian Affairs and Northern Development 1984
Tennant, Paul. "Aboriginal Peoples and Aboriginal Title in British Columbia Politics." In *Politics, Policy, and Government in British Columbia,* ed. R.K. Carty. Vancouver: UBC Press 1996
–. *Aboriginal Peoples and Politics: The Indian Land Question in British Columbia, 1849-1989.* Vancouver: UBC Press 1990
–. "Delgamuuk'w and Diplomacy: First Nations and Municipalities in British Columbia." Paper presented at the Fraser Institute Conference co-hosted by the Canadian Property Rights Research Institute, Ottawa, 26-7 May 1999
–. "Native Indian Political Activity in British Columbia 1969-1983," *BC Studies* 57 (1983): 112-36
–. "Native Indian Political Organization in British Columbia 1900-1969: A Response to Internal Colonialism," *BC Studies* 55 (1982): 3-49
Tin Wis Coalition. "Community Control, Developing Sustainability, Social Solidarity." Vancouver: Tin Wis Coalition 1991
Titley, E. Brian. *A Narrow Vision: Duncan Campbell Scott and the Administration of Indian Affairs in Canada.* Vancouver: UBC Press 1986
Toovey, Jack. "Forest Renewal: British Columbia." In *Tomorrow's Forests ... Today's Challenge? Proceedings of the National Forest Regeneration Conference.* 117-25. Quebec City: Canadian Forestry Association 1977
Torgerson, Douglas. "Power and Insight in Policy Discourse: Post-Positivism and Policy Discourse." In *Policy Studies in Canada: The State of the Art,* ed. L. Dobuzinskis, M.P. Howlett, and D.H. Laycock. Toronto: University of Toronto Press 1998
Tough, Frank. *As Their Natural Resources Fail: Native Peoples and the Economic History of Northern Manitoba 1870-1930.* Vancouver: UBC Press 1996
Townsend, Roger "Specific Claims Policy: Too Little Too Late." In *Nation to Nation: Aboriginal Sovereignty and the Future of Canada,* ed. D. Engelstad and J. Bird, 67-77. Toronto: Anansi 1992
Travers, O. Ray. "Stewardship of Tree Farm Licenses 44 and 46 in the Proposed Management and Working Plans: An Evaluation." Victoria: Prepared for the Sierra Club of Western Canada 1991

Tripp, Derek. *The Use and Effectiveness of the Coastal Fisheries Forestry Guidelines in Selected Forest Districts of Coastal British Columbia*. Nanaimo, BC: Tripp Biological Consultants 1994

Tripp, D., A. Nixon, and R. Dunlop. *The Application and Effectiveness of the Coastal Fisheries Forestry Guidelines in Selected Cut Blocks on Vancouver Island*. Nanaimo, BC: Tripp Biological Consultants Ltd. Prepared for the Ministry of Environment, Lands and Parks 1992

Truck Loggers Association. "BC Forests: A Vision for Tomorrow, Working Papers." Vancouver: TLA 1990

–. "Options for the Forest Resources Commission: Review, Reconsideration, Recommendations." Vancouver: TLA 1990

Turpel, Mary Ellen. "The Charlottetown Discord and Aboriginal Peoples' Struggle for Fundamental Political Change." In *The Charlottetown Accord, the Referendum, and the Future of Canada*, ed. K. McRoberts and P. Monahan. Toronto: University of Toronto Press 1993

United States. Department of Agriculture, Committee of Scientists, *Sustaining the People's Lands: Recommendations for Stewardship of the National Forests and Grasslands into the Next Century*. Washington, DC: US Department of Agriculture 1999

–. Department of Agriculture, Forest Service. "Ecosystem Management: 1993 Annual Report of the Forest Service." Washington: US Department of Agriculture Forest Service 1994

–. General Accounting Office. "Forest Service Priorities: Evolving Mission Favors Resource Protection over Production." Washington, DC: General Accounting Office 1999

United States and Forest Ecosystem Management Assessment Team. "Forest Ecosystem Management: An Ecological, Economic, and Social Assessment." Washington, DC: US Government Printing Office 1993

Upton, L. "The Origins of Canadian Indian Policy," *Journal of Canadian Studies* 8 (1973): 51-61

Usher, Peter. "Some Implications of the Sparrow Judgement for Resource Conservation and Management," *Alternatives* 18 (1991): 20-2

Usher, P.J., F.J. Tough, and R.M. Galois. "Reclaiming the Land: Aboriginal Title, Treaty Rights and Land Claims in Canada," *Applied Geography* 12 (1992): 109-32

Utzig, Greg, and Donna Macdonald. *Citizens' Guide to AAC Determinations: How to Make a Difference*. Vancouver: BCEN 2000

Kordes-de Vaal, Johanna H. "Intention and the Omission Bias: Omissions Perceived as Nondecisions," *Acta Psychologica* 93 (1996): 161-72

van Kooten, G. Cornelis, and Ilan Vertinsky. "Introduction: Framework for Forest Policy Comparisons." In *Forest Policy: International Case Studies*, ed. B. Wilson, van Kooten, Vertinsky, and L. Arthur. Oxon, UK: CABI 1999

Vance, Joan E. *Tree Planning: A Guide to Public Involvement in Forest Stewardship*. Vancouver: BC Public Interest Advocacy Centre 1990

Village of Hazelton. "Framework for Watershed Management (formerly the Forest Industry Charter of Rights)." Hazelton, BC: The Corporation of the Village of Hazelton 1991

Vold, Terje, and Andy MacKinnon. "Old-Growth Forests Inventory for British Columbia," *Natural Areas Journal* 18,4 (1998): 308-18

Wagner, M.W. "Footsteps along the Road: Indian Land Claims and Access to Natural Resources," *Alternatives* 18 (1991): 22-8

Weatherbe, Steve. "Tenure Reform: Thinking the Unthinkable," *Business Examiner*, 18 May 1999

Weaver, S. "The Joint Cabinet/National Indian Brotherhood Committee: A Unique Experiment in Pressure Group Relations," *Canadian Public Administration* 25 (1982): 211-39

–. "A New Paradigm in Canadian Indian Policy for the 1990's," *Canadian Ethnic Studies* 22 (1990): 8-18

Weiler, Joseph M., and Robin M. Elliot. *Litigating the Values of a Nation: The Canadian Charter of Rights and Freedoms*. Toronto: Carswell 1986

Weir, Margaret. "Ideas and the Politics of Bounded Innovation." In *Structuring Politics: Historical Institutionalism in Comparative Analysis*, ed. S. Steinmo, K. Thelen, and F. Longstreth. Cambridge: Cambridge University Press 1992
Westland Resource Group. "A Review of the Forest Practices Code of British Columbia and Fourteen Other Jurisdictions." 1995. www.for.gov.bc.ca/pab/publctns/westland/1-10.htm
Whitford, H.N., and R.D. Craig. "Forests of British Columbia." Ottawa, ON: Canada Commission of Conservation, Committee on Forests 1918
Wildsmith, B.H. *Aboriginal People and Section 25 of the Canadian Charter of Rights and Freedoms*. Saskatoon: University of Saskatchewan Native Law Centre 1988
Wilkinson, Alan K. "British Columbia." In *Report of Proceedings of the Twenty Eighth Tax Conference*, ed. D.A. Wilson. Toronto: Canadian Tax Foundation 1976
Wilsford, David. "The Conjoncture of Ideas and Interests," *Comparative Political Studies* 18,3 (1985): 357-72
–. "Path Dependency, or Why History Makes It Difficult but Not Impossible to Reform Health Care Systems in a Big Way," *Journal of Public Policy* 14,3 (1994): 251-84
Wilson, David A. "Forestry Taxes and Tenures in Canada." In *Report of Proceedings of the Twenty-Eighth Tax Conference*. Toronto: Canadian Tax Foundation 1976
Wilson, James Q. *Bureaucracy*. New York: Basic Books 1989
–, ed. *The Politics of Regulation*. New York: Basic Books 1980
Wilson, Jeremy. "Forest Conservation in British Columbia 1935-85: Reflections on a Barren Political Debate," *BC Studies* 76 (1987/8): 3-32
–. "Implementing Forest Policy Change in British Columbia: Comparing the Experiences of the NDP Governments of 1972-75 and 1991-?." In *Troubles in the Rainforest: British Columbia's Forest Economy in Transition*, ed. T.J. Barnes and R. Hayter. Victoria: Western Geographical Press 1997
–. *Talk and Log: Wilderness Politics in British Columbia*. Vancouver: UBC Press 1998
–. "Wilderness Politics in BC: The Business Dominated State and the Containment of Environmentalism." In *Policy Communities in Canada: A Structural Approach*, ed. W.D. Coleman and G. Skogstad. Mississauga, ON: Copp Clark Pitman 1990
Wolfinger, Raymond E. "Nondecisions and the Study of Local Politics," *American Political Science Review* 65 (1971): 1063-80
Wouters, Gary. *Shaping Our Future: BC Forest Policy Review*. Victoria: Office of the Jobs and Timber Accord Advocate, 2000
Wright, Debra. "Self-government, Land Claims and Silviculture: An Aboriginal Forest Strategy," *Forestry Chronicle* 70,3 (1994): 238-41
Yaffee, Steven Lewis. *The Wisdom of the Spotted Owl: Policy Lessons for a New Century*. Washington, DC: Island Press 1994
Young, W. *Timber Supply Management in British Columbia: Past, Present and Future*. Victoria: Ministry of Forests 1981
Zakreski, Sheldon, and Cory Waters. "Forestry Jobs and Timber." Victoria: Prepared for the Sierra Club of BC 1997
Zelditch, Morris Jr., William Harris, George M. Thomas, and Henry A. Walker. "Decisions, Nondecisions and Metadecisions," *Research in Social Movements, Conflict and Change* 5 (1983): 1-32
Zelditch, Morris Jr., and Joan Butler Ford. "Uncertainty, Potential Power and Nondecisions," *Social Psychology Quarterly* 57,1 (1994): 64-74
Zhang, Daowei. "Welfare Loss of Log Export Restrictions in British Columbia." Auburn, AL: Forest Policy Center, School of Forestry and Wildlife Sciences Auburn University 2000
Zlotkin, Norman K. "Post-Confederation Treaties." In *Aboriginal Peoples and the Law*, ed. B. Morse. Ottawa: Carleton University Press 1985

Index

Note: Page numbers in semi-bold indicate that the information cited appears in a figure or table.

Set in Stone by Bamboo and Silk Design

Printed and bound in Canada

Copy editor: Maureen Nicholson

Proofreader: Joanne Richardson

Indexer: Heather Ebbs